The Last Tasmanian Tiger
THE HISTORY AND EXTINCTION OF THE THYLACINE

This book is the most complete and up-to-date examination of the history and extinction of one of Australia's most enduring folkloric beasts – the thylacine, otherwise affectionately known as the Tasmanian tiger. Bob Paddle challenges conventional theories explaining the behaviour and eventual extinction of the thylacine, arguing that rural politicians used the Tasmanian tiger as a scapegoat to protect local agricultural enterprise from the consequences of mismanagement. After the population of thylacines was decimated through a government bounty scheme, ineffective political action by scientists finally resulted in the extinction of a once proud species. Paddle also uncovers a deeper intellectual snobbery that set the scene for the thylacine's eventual extinction. *The Last Tasmanian Tiger* offers new perspectives on the subjective nature of scientific investigation and the politics of preservation.

Bob Paddle's background is in the study of animal behaviour and the history and philosophy of science. He currently lectures in Psychology at Australian Catholic University. This is his first book.

In affectionate memory of two musical primates:

Ian Stewart (1938–85)

and

Lewis Brian Hopkin Jones (1942–69)

I am a brother to dragons, and a companion to owls.

Job 30: 29

the world is full of light and life,
and the true crime is not to be interested in it.

A. S. Byatt

The Last Tasmanian Tiger

THE HISTORY AND EXTINCTION
OF THE THYLACINE

Robert Paddle

Australian Catholic University

CAMBRIDGE UNIVERSITY PRESS
Cambridge, New York, Melbourne, Madrid, Cape Town, Singapore, São Paulo, Delhi

Cambridge University Press
The Edinburgh Building, Cambridge CB2 8RU, UK

Published in the United States of America by Cambridge University Press, New York

www.cambridge.org
Information on this title: www.cambridge.org/9780521531542

First published 2000
First paperback edition 2002

A catalogue record for this publication is available from the British Library

National Library of Australia Cataloguing in Publication data
Paddle, Robert N. (Robert Norman), 1949–
The last Tasmanian tiger: the history and extinction
of the thylacine.
Bibliography.
Includes index.
ISBN 0 521 53154 3 (pbk.).
1. Thylacine. 2. Rare animals – Tasmania. I. Title.
599.27

ISBN 978-0-521-78219-7 hardback
ISBN 978-0-521-53154-2 paperback

Transferred to digital printing 2009

Contents

Illustrations

I owe a debt of gratitude to many more people than can personally be listed   vii
here. First, I need to thank all those individuals I interviewed and cajoled into
writing to me in the course of my research. The basis for this book is to be found
in my PhD thesis at the University of Melbourne (1997), whose completion
represented a productive team effort between my three supervisors – Professor
Rod Home, Professor Graham Mitchell, Joan Dixon – and myself. I would also
like to acknowledge the encouragement and interest shown in my work by
Professor Homer Le Grand and Drs Anni Dugdale, Linden Gillbank and Neil
Thomason. For their generous, supportive and constructive comments made on
my text thanks to Professor Harriet Ritvo, of the Massachusetts Institute of Tech-
nology, and Professor Michael Archer, now of the Australian Museum. Similar
acknowledgment goes to the editorial staff at Cambridge University Press, Paul
Watt, Sharon Mullins and Lee White, for their patience and guidance.

My thanks go to all the many people who have aided my research in their
professional capacities, without whom this book could not have been written.

For providing access to museum collections, data records, files and correspondence, thanks: to Tim Kingston, Mary Cameron, Kaye Dowling, Judy
Rainbird, Julie McGowan, John Leeming, Riccardo Bartoli and Allison Pickford
of the Queen Victoria Museum, Launceston; to Joan Dixon, Lina Frigo, Tom
Rich, Frank Job and Val Hogan of the Museum of Victoria; to Linda Gibson of
the Australian Museum, Sydney; to Catherine Kemper, Tim Flannery, L. Queale
and Phillip Jones at the South Australian Museum, Adelaide; to David Ride of
the Australian National Museum, Canberra; to John Long, of the Western
Australian Museum, Perth; to Bill Cox and Allan Bain at the Smithsonian
Institution, Washington; to Anton Van Helden and Ross O'Rourke of the
National Museum, Wellington; to Amanda Freeman and Geoff Tunnecliffe of
the Canterbury Museum, Christchurch; to Brian Gill, Brian Henderson and

John Darby of the Auckland, Wanganui and Dunedin Museums respectively; and to the late Phil (A. P.) Andrews of the Tasmanian Museum and Art Gallery, Hobart, who died unexpectedly in 1997. Phil was always open and generous to a fault, sharing his time and his knowledge with those undertaking vertebrate research – he is sorely missed. Thanks also to the Director of the Museum, Pat Sabine, and to the Curator of Vertebrate Zoology, Kathryn Medlock.

For providing access and information about the archival records of the zoological gardens, thanks: to Christina MacDonald of the Adelaide Zoo; to Graham Mitchell, Peter Myroniuk, Chris Stephens, Angas Martin and Chris Larcombe of the Royal Melbourne Zoo, and also to the zoo's historian, Catherine de Courcy; to Rhonda Hamilton, Ross Smith and Lorraine MacKnight of the Community History Museum, Launceston; to Margaret Miller and Carol Bach of Taronga Park Zoo, Sydney; to Steve Johnson of the New York Zoological Park; to Ron Goudswaard and Kerry Muller of Wellington Zoo; and to Gavan McCarthy and Tim Sherratt of the Australian Science Archives Project in Melbourne and Canberra.

For the knowledge of their own collections and suggestions of alternative directions in which to search when other attempts had failed, thanks: to Judith Hollingsworth, Lou Foley, Marian Sergeant and Leonie Provost of the State Library of Tasmania, Launceston; to Fran Voss and Virginia German at the *Launceston Examiner* (and also to the editor, Lloyd Wish-Wilson, for initial permission to use the *Examiner*'s archival collection, and in a special category of his own, for indexing some of the nineteenth-century *Examiners*, thanks to Dennis Hodgkinson); to Tony Marshall, Dora Heard, Barbara Valentine and Gillian Winter of the Tasmaniana Library, Hobart; to Ian Pearce, Margaret Bryant, Brian Diamond, David Benjamin, Patricia Quarry, Alexander Knight, Robin Eastley and Jill Waters of the Archives Office of Tasmania, Hobart; to Sue MacDonald of the State Library of Tasmania, Burnie, and to J. M. Bruce whose collected and privately printed papers on the records of the Van Diemen's Land Company (when I eventually discovered them) saved me months of work on the original and microfilmed documents in Hobart; to Peter and Kathleen Burns of the St Helens' Local History Room; to Fiona Preston of the library of the Department of Parks, Wildlife and Heritage, Hobart; to Jan Brazier of the Australian Museum Library, Sydney; to Frank Job and Val Hogan of the library of the Museum of Victoria; to John Lowe, in particular, at the State Library of Victoria (who also, kindly, helped with German translations) and to all the staff at the Public Records Office, Laverton, Victoria, who were universally helpful; to Tania Milanko of the Mortlake Library, State Library of South Australia; to Judith Jeffery of the State Records, Adelaide; to Jacki Kahui at the Wellington City Council Record Office; to Naomi Highfield of the National Archives Records Centre, Wellington; to Martin Beckert and his fellow staff at the Mitchell Library, State Library of New South Wales, Sydney; and to all the staff at the Archives Authority of New South Wales.

A number of significant individuals have served as sounding boards for discussion and debate upon the thylacine. Particular thanks to the late Alison Reid, the last curator of Hobart Zoo; Rex Hesline, of Launceston, with his infectious enthusiasm for anything thylacinic; and Nick Mooney of the Department of Parks, Wildlife and Heritage, Hobart. Thanks also to Steven Smith, for access to some of the audiotape recordings that he made courtesy of the World Wildlife Trust (Australia). Special mention needs to be made of Neil Murray and Peter Tilly for providing me with their translations of the writings of visiting European scientists, before they published this material themselves.

In attempting to understand thylacine behaviour I have stood unashamedly on Eric Guiler's shoulders. Necessarily, this book often concentrates on the differences in interpretation between Eric (1985) and myself. Such criticism should not blind the reader to the reality, expressed more in silences of this text, that I agree with most of Eric's interpretations of thylacine anatomy, physiology and behaviour.

The Australian Catholic University provided me with four different research grants to assist in regular travel to Tasmania. Amongst my colleagues I would like to single out the following for their general support and commitment towards research: Kevin Bourke, Peter Carpenter, Gabrielle McMullen, Bernie Daffey, Wolfgang Grichting, Marie Joyce, Michael McKay, Stuart Sharlow and Peter Sheehan. The campus librarians also deserve special thanks for their help in locating and retrieving information: Daryl Bailey, Diane Brebner, Lyn Coles, Jenny Dewar, Willow King and Vera Pohl.

I also wish to thank the following people for their suggestions, criticisms, interest and support: Eric Badcock, Col Bailey, Bruce and Barbara Baird, Jack Branagan, Jo Calaby, Peter Carter, Cäsar Claude, Glenis Davey, Maureen Dietz, John Edwards, Dorothy Gould, Mary Hammond, Lloyd Harris, Phillipa Hodder, Deborah Jordan, Simon Kennedy, Hazel Laird, Edwin Lambert, Rex Lane, Margaret Leane, Virginia Lowe, A. M. Lucas, Carmel and Paul Macpherson, Bob Mesibov, Heinz Moeller, Neil Murray, Greg Parry, Amy Pulford, Mary Ramsay, Peter Rendell, Will Rolland, Liz Rowell, Irene Semens, Liz Simpson, Dean Southwell, June and Derry Stewart, David Tiley, Monty Turner jnr, Hendrik Van den bergh, Kim Van Haaster, Pat Vickers-Rich, Michael Westerman, Cecilia Winkelman and Steve Wroe.

Finally, there is my long-suffering family, Sarah, Seth and Hermione, who with good will have put up with my absences interstate and overseas and, at times, my rather restricted topics of conversation and interest. Seth and Hermione have spent days of their lives at photocopying machines in Tasmanian libraries, having been bribed with the promise of an interstate trip if they would take on the task of research assistants. Sarah, as a professional historian, has done her best to tolerate her partner's blundering intrusion into her own academic discipline, encouraging and supporting the production of this text. As an historian she has not felt threatened by this, merely remarking, in a typically back-handed compliment passing between long-term partners that, if I restrict myself to science in the future, it is likely to be of advantage to both disciplines! Thanks to all of you for your support.
</p>

x To help distinguish between published and unpublished source material in the Bibliography, in-text citations for unpublished source material are presented in day/month/year, for example, an interview conducted on 25 January 1981 is rendered 25/1/1981.

It is not uncommon for the age of young marsupials to be recorded, not from their date of birth, but from the date of their independence from the pouch. However, ages given in this text (quotations aside) are always reckoned from time of birth.

Monetary amounts are presented in the original sterling measures of pounds.shillings.pence, such that £4.12.6 refers to four pounds, twelve shillings and sixpence. There were 12 pennies in one shilling, and 20 shillings in one pound. The relative value of the £1 government bounty payment that ran from 1888–1909 may be determined by considering that the salary for the 'tiger man' employed at Woolnorth in 1903 was only £20 per annum.

Lengths and weights for historical data are reported in their original imperial units (with metric equivalents in brackets). For comparison: 1 inch = 2.54 cm; 1 foot = 30.48 cm; 1 yard = 0.91 m; 1 mile = 1.61 km; and 1 acre = 0.41 ha. One pound weight (1 lb) = 0.45 kg.

Chapter 1

Introduction

SCIENCE AND THE SPECIES FROM A EUROPEAN PERSPECTIVE

Species finally depart the biota, not with a bang but a whimper. The thylacine,   1
Tasmanian tiger or marsupial wolf, *Thylacinus cynocephalus*, is one of a handful of
species where that whimper has a precise date. The thylacine became extinct on
7 September 1936 when the last known specimen died in captivity in the
Beaumaris Zoo, Hobart.

This book summarises and interposes the known biology and behaviour of
the species with its recent history and the contingent events that led to its
extinction. Necessarily, it concentrates upon the recent historical records of the
thylacine in Tasmania, but prior to the repeated human invasions of Australia
the marsupial wolf was widespread in distribution. It was found throughout the
one continental landmass, from New Guinea in the north to Tasmania in the
south, hence the factors involved in the mainland extermination of the species
need to be considered in order to understand how the final extinction process
came about.

In constructing a narrative of the thylacine I have focussed primarily upon
changing scientific perceptions of the thylacine's predatory behaviour. Within
Tasmania I consider how human–thylacine interactions were constructed in a
social, political and cultural context; and how these imperialistic constructions
became incorporated into the changing face of Australian colonial science,
coming to represent the accepted body of knowledge on the species. These
constructions and their results (for example, a government bounty scheme)
were more than adequate to drive the species to extinction, unaffected by the
widespread growth of environmental concerns evident in Tasmania in the
nineteenth century and throughout Australia in the early twentieth century.

As a dynamic and creative intellectual exercise, scientific constructions never
stand still. Nevertheless, it might be expected that the extinction of a species
would be associated with some stability of scientific knowledge construction. This

expectation, however, does not appear to hold for the thylacine. With human–thylacine constructions and interactions so patently involved in the deliberate extinction of the species – 'bangs' very much preceded the whimper – it is, perhaps, not surprising to find a shifting of emphasis, interpretation and responsibility apparent in both popular and scientific constructions of the animal after its extinction. While these changes are explicable in terms of the typical behaviour of the human species, their 'naturalness' nevertheless strikes at the heart of scientific endeavour, which ideally attempts to construct an objective developmental history and analysis of the observable world.

This chapter briefly introduces the species from a European perspective, covering its recent discovery and classification, and reflecting on the methodology used in data construction and the operation of science in colonial cultures.

A nineteenth-century classification

Tasmania (formerly known as Van Diemen's Land) is an island state of some 68 300 square kilometres – a little smaller than Ireland – that represents the southernmost part of the Australian continent, from which it is now separated by the relatively shallow Bass Strait, established by rising sea levels about 12 000 years ago. It has been affectionately called 'The Apple Isle', reflecting both the rough outline of its shape, and one of its best-known agricultural exports. Lying between 40° and 43$^{1}/_{2}$° south of the equator, it has much in common with similarly placed – and mountained – latitudes in the northern hemisphere, such as northern California and north-west Spain.

The mountains of western Tasmania are weathered ranges of pre-Cambrian quartz metamorphics and conglomerates. Although rarely exceeding 1500 metres they are rugged and covered with cool temperate rainforest, descending to high rainfall sedgeland near the coast. The centre, east and south-east of Tasmania consists of the central plateau (over 700 metres above sea level) based on horizontal Permian and Triassic sediments. Prior to European colonisation, the central plateau was dominated by sclerophyll woodland. Both the western mountains and the central plateau contain thousands of lakes, the majority of which are glacial in origin. To the east of the central plateau, essentially running between Launceston and Hobart, lie the Midland Plains, based on Triassic sediments, and consisting of open grassland with sclerophyll forest remnants. The peaks of the Devonian mountains in the north-east are covered by rainforest, descending to sclerophyll forest at the coast. Coastal plains around the island are limited, usually restricted to only a few hundred metres in width (Davies, 1965; Guiler, 1985). In the highlands snow may fall at any time of the year, but peak falls occur in late winter and spring. Maximum temperatures in the island are experienced in the low-lying areas of the Midland Plains in the east and south-east and on the coastal plains, where daily summer temperatures may rise to over 100°F (37.8°C).

European contact with thylacines possibly dates from the seventeenth century. Abel Tasman sighted Tasmania on 24 November 1642 and named it Anthony Van Diemens Land. An on-shore investigation was carried out by Pilot-Major Francoys Jacobsz on 2 December: 'They saw the footing of wild beasts having claws like a *Tyger*' (Rembrantse, 1682, p. 180). Given the dubious nature of post-1936 claims of thylacine footprints, the possibility of footprint misidentification needs to be entertained. Apart from its date, however, this record is not unique. Analysis of the archives of the Dutch East India Company in Australian waters has identified additional records of 'tyger' footprints and also 'tyger' sightings in the seventeenth and eighteenth centuries: Heuvelmans (1958); Lang (*ca* 1910); Moeller (1990, 1970); Whitley (1970).

Eighteenth-century French expeditions to Tasmania also apparently met with thylacines. du Fresne, in the *Mascarin*, arrived in Tasmania in March 1772, and met with a 'tiger cat' (Roth, 1891). La Billardière was naturalist aboard *La Récherche* on the expedition captained by D'Entrecasteaux, that visited Tasmania in April and May 1792. While collecting ashore, he 'found . . . the upper jaw-bone of a large animal of the carnivorous tribe' and 'heard the cry of a beast of prey' (Labillardière, 1800, p. 114), indicating contact with either a thylacine, Tasmanian devil or spotted-tailed quoll. Specific contact with a thylacine is now known to have taken place on 13 May 1792. The English translation has been confusing, referring to an encounter with 'a beast of prey . . . a quadruped the size of a large dog . . . of a white colour spotted with black' (1800, pp. 118–19). As the basic background colour of many thylacines was a light pale brown or grey, the 'white colour' presents no problem, but 'spotted with black' is a decidedly unhelpful description of a thylacine's stripes. However, Neil Murray (personal communication) has returned to the original French publication of La Billardière's journal, and suggests that the translation of the phrase *tacheté de noir* (La Billardière, 1799, p. 163) in this context, should more appropriately be rendered as 'marked' or 'streaked' with black.[1]

It was not until the early nineteenth century that the thylacine was first described scientifically. Tasmania's Lieutenant-Governor Paterson sent a detailed description of the marsupial wolf to Sydney for publication in the *Sydney Gazette and New South Wales Advertiser* (21 April 1805). The species was officially described and named (*Didelphis cynocephala*) by Tasmania's Deputy Surveyor-General George Harris. He sent 'drawings & descriptions from the life, of two Animals of the genus Didelphis' (31 August 1806) to Sir Joseph Banks, who read Harris' description of the thylacine (and Tasmanian devil) before the Linnean Society on 21/4/1807, prior to publication in 1808:

> The length of this animal from the tip of the nose to the end of the tail is 5 feet 10 inches, of which the tail is about 2 feet. . . . Head very large, bearing a near resemblance to the wolf or hyæna. Eyes large and full, black, with a nictant membrane, which gives the animal a savage and malicious appearance. . . .
> Tail much compressed, and tapering to a point . . . Scrotum pendulous, but partly concealed in a small cavity or pouch in the abdomen.

3

The whole animal is covered with short smooth hair of a dusky yellowish brown . . . On the hind part of the back and rump are 16 jet-black transverse stripes, broadest on the back, and gradually tapering downwards . . .

Only two specimens (both males) have yet been taken. It inhabits amongst caverns and rocks in the deep and almost impenetrable glens in the neighbourhood of the highest mountainous parts of Van Diemen's Land, where it probably preys on the brush Kangaroo, and various small animals that abound in those places. (Harris, 1808, pp. 174–5)

The basic background dorsal colour varied between specimens from an intense deep brown (Gould, 1851) to a sandy colour (L. R. Green, *ca* 1975), with softer, more washed-out coats described as varying from pale brown (Waterhouse, 1841), to greyish brown (Lydekker, 1894) to grey (Gunn, 1838x). Variation was also shown in the contrasting stripes, described in selected specimens as black (Gould, 1851), brown-black (Thomas, 1888), dark brown (Dransfield, 25/8/1981) or darker grey (Gunn, 1838x) or darker sandy colour (L. R. Green, *ca* 1975). Colbron-Pearse (1968) recalled a captive specimen in which the banding itself was not uniform, but changed from dark to light brown.

Selected vital dimensions from over forty different live, or recently killed, specimens of marsupial wolf are known, considered here in the British Imperial units in which they were originally expressed. Specimens with nose-to-tail lengths of less than 4 feet (121.9 cm) are defined as juvenile specimens, following Cunningham (1882). However, it could take some considerable time to achieve even this length in captivity. The male thylacine that died at London Zoo on 25 September 1853 had been caught in Tasmania in November 1849 (Gunn, 1850w). After four years in captivity, on its death it still measured only 4 ft $^{1}/_{2}$ in (Crisp, 1855). While this may be a straight-line measure, rather than an along-the-body measure, the difference in the two measures for a specimen of this size would only amount to 6 inches (Waterhouse, 1841, 1846). Either way, it was still only a moderately sized specimen. A slight, but noticeable sexual dimorphism was present in the species, with adult males having a slightly longer body length than females (Gunn, 1852w; Ride, 1964). Removing the juvenile specimens from the sample suggests that the average direct length, nose-to-tail, of adult male thylacines, was around 5 ft 4 in (162.6 cm), and for females, 5 ft $^{1}/_{2}$ in (153.7 cm), with along-the-body measures for specimens this size likely to give readings $7^{1}/_{2}$ to 8 inches (19.1–20.3 cm) longer. Adult specimens up to 6 ft 6 in, probably measured along the curve of the body, were not uncommon (*Hobart Town Gazette*, 5 April 1817; Lord and Scott, 1924; Meredith, 1852; Paterson, 1805; *Tasmanian*, 16 September 1871) and no doubt the occasional specimen was even larger. However, Oscar's (1882) description of a freshly killed 8-foot specimen was merely an estimate of body size. It was obviously a large and impressive specimen, but it was not accurately measured.

Specifically measured weights of adult specimens have been provided by Paterson (1805), Gunn (1838x) and *Launceston Examiner* (14 March 1868). In addition, after the arrival of three adult thylacines at Melbourne Zoo in 1874 and 1875, the Secretary of the Gardens, A. A. C. Le Souëf provided an adult weight range of 60 to 70 lb (Commissioners of the Victorian Intercolonial Exhibition, 1875). The above measures suggest the average weight of an adult thylacine was about 65 lb (29.5 kg).

The 'tyger' was well known from the earliest days of colonial settlement (Knopwood: 20 August 1803, 18 June 1805; Paterson, 1805) before its official scientific description. It was acknowledged as rare and uncommon, and to retreat from the areas of settlement and land alienation (Oxley, 1810). In terms of distribution it was only ever considered to be locally common in the north-west (Gray, 1841; Gunn, 1838w) – principally in response to the claims of significant sheep predation emanating from the Van Diemen's Land Company. Gunn (1852x, p. 80) considered: 'This animal is found all over the island, from the sea coast to the summits of the mountains', but the suggested distribution of the thylacine into the dense rainforest and sedgeland of south-western and southern Tasmania is now known to have been incorrect (Guiler, 1991). The thylacine was described as common along the west coast and inland, down to Macquarie Harbour (Robinson, 29/9/1830). For south-western Tasmania, there are only occasional nineteenth-century and early twentieth-century records: from Cape Sorell lighthouse keepers in 1932 (Williams, 10/9/1981); ferry and hut builders at the mouth of the Hibbs River in 1914 (Slebin, 22/4/1937); and survivors of the shipwrecks of the *Acacia* in 1905 at Mainwaring Inlet (Guiler, 1985), and the *Moyne* at Low Rocky Point in 1867 (Mollison, 25/11/1951). These records reflect the species' attraction for the coastal environment, not the denser sedgeland and rainforest of the south-western and southern mountains (see Map 1.1).

In Harris' original description (1808), reflecting the practice in British scientific circles, the thylacine was placed in the genus *Didelphis*, originally created by Linnaeus for American opossums. Harris' summarised description of the species was presented in Latin – 'Didelphis fusco-flavescens supra postice nigro-fasciata, caudâ compressâ subtus lateribusque nudâ' (p. 174) – and with the Latin term *cynocephala* translated as dog-headed, the original nomination of the marsupial wolf described it as the 'dog-headed opossum'. Amongst conti-nental scientists it was early recognised that Australian marsupials were significantly different from their American cousins, and from the 1790s onwards Geoffroy Saint-Hilaire, Lacépède, Temminck and de Blainville between them-selves – Shaw and *Macropus* excepted – established the basic genera and classifi-cation scheme against which all Australian marsupial species are referred to this day. It was some considerable time before British scientists reluctantly accepted the necessity of using this intrusive continental classification into what they saw as their own colonial scientific domain. Geoffroy Saint-Hilaire established the genus *Dasyurus* for Australian marsupi-carnivores in 1796, into which he placed the thylacine (1810). The Tasmanian tiger was finally separated into its own genus *Thylacinus* by Temminck (1824). As the mixing of Latin and Greek roots in scientific nomenclature was (and still is) considered extremely poor taxo-nomic form, once the marsupial wolf was placed in the Greek-rooted genus of *Dasyurus* there was a necessity to alter its specific name from the Latin-based *cynocephala* to the Greek-based *cynocephalus* (Geoffroy Saint-Hilaire, 1810).

5

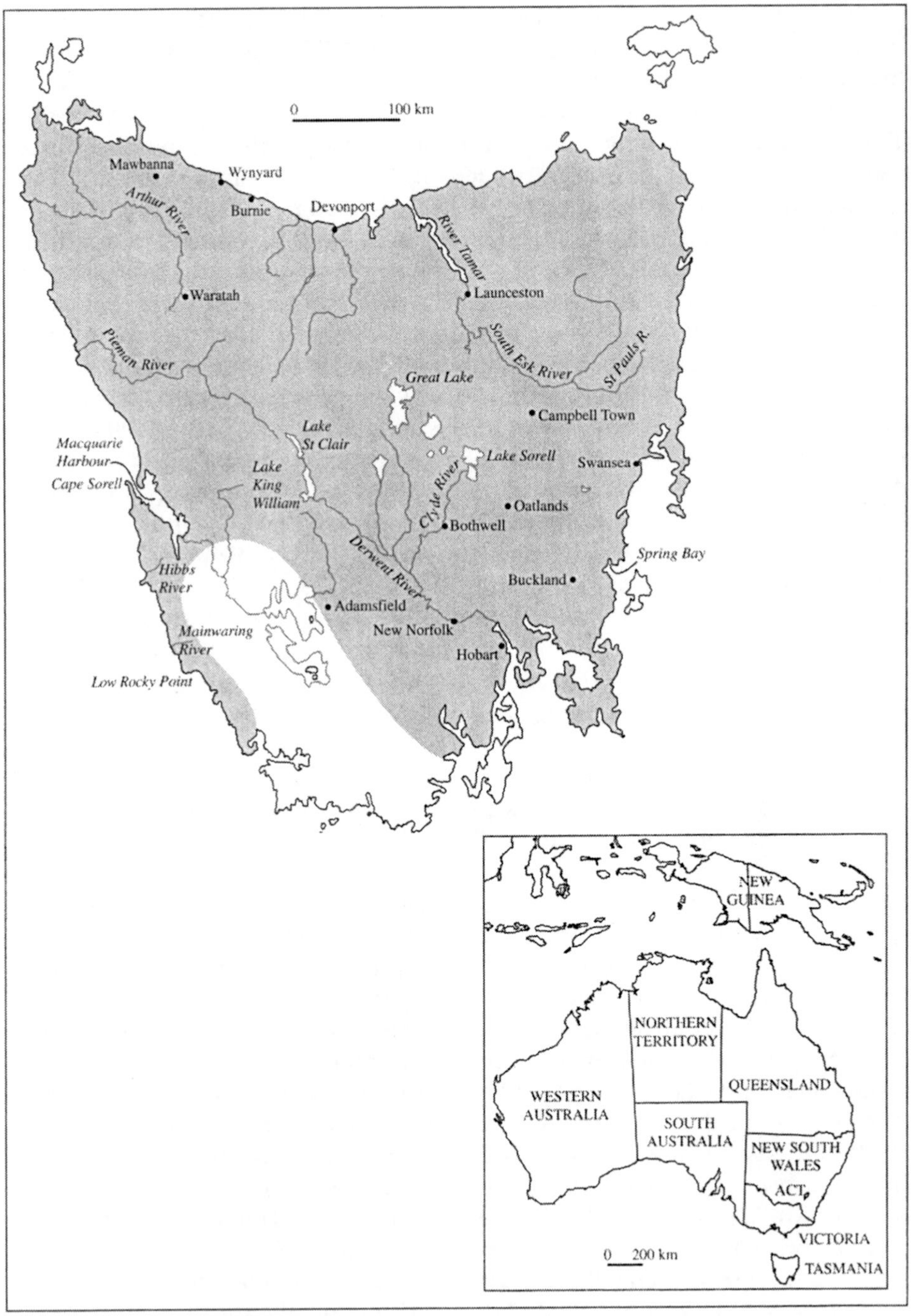

Map 1.1 The location of Tasmania and probable recent range of the thylacine.

Twentieth-century designations

The most common translation of meaning for the genus *Thylacinus* is 'pouched-dog' (Strahan, 1981). This has given rise to a certain redundancy in translating the marsupial wolf's scientific name as 'the dog-headed pouched-dog' (Aflalo, 1896; Moeller, 1990; W, 1855). Its very obviousness is inelegant, and borders upon the stupid and crass. In common with the rest of humanity, scientists as a whole do not appreciate impositions of crassness and stupidity, particularly when they emanate from within their own boundaries. It is possibly for this reason alone that it became increasingly fashionable, throughout the twentieth century, to strangely render the translation of *cynocephalus* as 'wolf-headed', despite there being no lupine reference within the specific name. While blatantly incorrect, the suggested translation of the marsupial wolf's name as 'the pouched dog with a wolf-like head' (Lydekker, 1915, p. 216) is free from redundancy, inelegance and stupidity, and thus became the favoured translation in the scientific literature of the twentieth century. The problem of redundancy in translation, however, actually lies with the incorrect translation of *Thylacinus* rather than *cynocephalus*. As Malcolm Smith (16/9/1972) and Gotch (1979) have pointed out, Temminck constructed the genus *Thylacinus* from the single Greek word 'thylakos' (θυλακος), meaning a leather pouch. Hence the literal translation of the marsupial wolf's name reads as 'the pouched thing with a dog head'. Only a few twentieth-century scientists have swum against the majority 'wolf's head' tide: R. Brown (1973); Guiler (1958x); Hickman (1955); Sayles (1980); and M. Smith (1982).

7

The marsupial wolf is currently taxonomically placed in the Family Thylacinidae, which has a 24 000 000-year history behind it (Muirhead and Wroe, 1998). Queensland and Northern Territory deposits have produced five additional genera (*Badjcinus*, *Muribacinus*, *Ngamalacinus*, *Nimbacinus* and *Wabulacinus*), and within the genus *Thylacinus* there are currently recognised five extinct sister species to *T. cynocephalus*.

Having briefly considered the description and initial classification of the marsupial wolf, an additional word is required on the nomination and nomenclature used in this text. Considerable pressure is being exerted, certainly in an Australian context, to no longer colloquially refer to the two largest Australian mammalian groups as marsupials and placentals. Admittedly, the distinction between these two different reproductive types cannot be made on the simple possession or absence of a placenta. A placenta is an organ of easy operational definition that can be demonstrated, through observation, analysis and measurement, to occur in all marsupials as well as in all placentals (and in the numerous live-bearing reptiles as well). It is, therefore, inaccurate and imprecise to differentiate one mammal type with a label referring to an easily defined and measured organ that is possessed by both mammal types. Hence, there is a move to change the common nomenclature of marsupial and placental away from the incorrectly implied placental versus non-placental possession, in favour of a return to the old Huxleyan labels of eutheria, for placentals; and metatheria, for marsupials. Unfortunately, these words carry significant cultural connotations: eutherian relating to the true, well-formed beasts, the most highly developed mammals; while being metatherian infers only an attempted approximation to such a eutherian condition. Metatherians are supposedly located in a mid-way position, their live birth representing an advanced stage beyond that of the first egg-laying mammals (the prototheria), but the small birth size, and development within the marsupium, purportedly represents an earlier, serial stage in the evolution of mammals. Despite this cultural baggage of second-class citizenship, the advocacy of change to eutherian and metatherian has been trumpeted (Marshall Graves, Hope and Cooper, 1990, pp. 2–3).

The willingness of contemporary Australian scientists to accept a cultural labelling of the secondary and inferior status of marsupials, in order to preserve a purity of measurement in the operational definition of a reproductive organ, demonstrates an amazing lack of scientific sensitivity to the cultural values involved in such labelling and their reflection in the European treatment of Australian biodiversity. Historically, the perceptions of marsupial inferiority posited by progressive evolutionists were powerful and significant arguments, allowing scientists to justify the inevitability of extinction for the Tasmanian tiger and other indigenous species, and, arguably, these false perceptions still limit scientific research on Australian marsupials, and influence public perceptions of their status, to the present day. Richard Owen was largely responsible for defining the terms Marsupialia and Placentalia – as representatives of equal archetypes – and I purposefully use the words marsupial and placental throughout this text, hoping to see the anachronistic labels of eutherian and metatherian fall rapidly into disfavour.

Methodology

Constructing a narrative of the thylacine involves comparing and contrasting different thylacine sources. There are ambiguities, silences, real omissions, pretended omissions and contradictions in the literature. Oral history, visual images and photographs exist beside the written word. In any consideration of the thylacine's behaviour these realities need to be acknowledged, argued and explained.

Recent literature on the thylacine tends to construct an hierarchy of sources, with an emphasis favouring the use of the most recent, twentieth-century accounts and publications on the species. This is fairly traditional scientific behaviour: assuming, as knowledge increases with time and theories become refined, the more recent observations and publications will reflect this growth and change, and thus possess greater accuracy and validity over older-published observation and comment. For many objects of scientific analysis this appears a quite reasonable assumption. In so far as any object of scientific curiosity is capable of further observation and experimentation these orientations may well be justified.

With a recently extinct animal, however, such orientation, assumptions and assignment of value towards source material are less easily justified. In the years prior to extinction, any species, but particularly a social species such as the thylacine, is likely to be operating under severe psychological stress. With the decimation of the species' population comes a disruption of the entire repertoire of social behaviour in the species. As a result of such stress, observations of behaviour on individual specimens in the last years of a species' existence are far more likely to report aspects of abnormal behaviour in inappropriate contexts and environments. Despite this, the operation of a selective categorisation of sources with an hierarchical preference for twentieth-century records is very much evident in the current scientific literature on the thylacine. Three examples are chosen to illustrate problems associated with this hierarchical categorisation.

Only one major scientific book has been published in English on the thylacine, Eric Guiler's (1985) *Thylacine: The tragedy of the Tasmanian tiger.* Its 207 pages represent the most detailed description of the species' anatomy and behaviour previously published, but it is unashamedly twentieth century in its orientation and perspective and contains less than sixty nineteenth-century source references on the species. At first I thought this was a genuine reflection of the scientific resources available. It is, after all, a commonly expressed opinion that nineteenth-century scientists showed little interest in the species. As I set out to consider the narrative of the thylacine in its scientific context I found this not to be the case (as a cursory examination of the reference material associated with this volume shows).

9

The second example is more disturbing, for it shows a readiness amongst some recent authors to accept post-extinction 'sightings' and their associated accounts of the species' behaviour at face value as valid scientific data. I do not consider it to be a myopic retreat into naïve inductionism to demand a body before accepting any post-1936 account of the species. This principle has been an operative, methodological guideline throughout the writing of this text, which is based entirely on pre-1937 observations as sources for constructing the species' behaviour. This principle is not always apparent in the twentieth-century hierarchical organisation of the literature. The different types of lairs and dens in which thylacines lived are well described in the pre-extinction literature. No photograph of a genuine thylacine lair was known or published in the literature until my discovery of a 1902 photograph of a lair from which a thylacine was caught for Melbourne Zoo (Paddle, 1992). Nevertheless, prior to my publication at least three different photographs of supposed thylacine 'lairs' were available in the scientific literature, all associated with unsubstantiated 'sightings' of the species post-1936 (Guiler, 1991; Park, 1986; *Science News*, 1966).

The third example of hierarchical source categorisation relates to modern scientific constructions of the social behaviour – or rather the absence of it – in the species. Unsurprisingly, most observations of adult thylacines in the twentieth century (hunting behaviours excepted) were restricted to that of lone, isolated individuals. By the time the twentieth century arrived, the thylacine had been persecuted and slaughtered in the state bounty scheme, commencing in 1888, as well as being the victim of an epidemic disease. The cessation of the bounty in 1909 left a handful of relict individuals in Tasmania, in a population under significant psychological stress that was heading rapidly towards extinction, and with the characteristic social behaviour of the species dramatically disrupted. The acceptance of an hierarchical categorisation of sources has resulted in the fashionable, twentieth-century scientific construction of the thylacine as a solitary animal and essentially asocial species (except for an adult female with young). It is even possible, with this social organisation in mind, to suggest that nineteenth-century accounts of the species are conformable with this viewpoint. From the suggestion that it 'flies at the approach of man' (Oxley, 1810) and 'is scarcely heard of in the located districts' (Widowson, 1829, p. 180) the thylacine in the nineteenth century was constructed as 'solitary', in the sense that it lived a retiring existence, remote from human society: 'it is a solitary animal, and does not approach the thickly settled parts of the colony' (Mudie, 1829, p. 176); see also Bischoff (1832); Ireland (1865); Lloyd (1862); Lycett (1824); Parker (1833) and Wentworth (1819). Modern constructions of thylacine social behaviour have, however, distorted the meaning of the word 'solitary' from its original nineteenth-century construction as referring to a 'retiring species', to an alternative reading, as if the word 'solitary' was indicative of an asocial species. This change in translated meaning appears to be almost entirely a twentieth-century phenomenon, one that runs counter to the commonly recorded nineteenth-century observations and descriptions of the species as primarily socially orientated, living (and hunting) in small family groups.

10

In terms of an hierarchy of sources, I have consistently favoured nineteenth-century over twentieth-century records of the species as likely to be more representative of the species' behaviour.

This is not to suggest that all nineteenth-century accounts of the species are conformable, far from it. The politics of the production of knowledge on the species needs to be considered in the nineteenth century, particularly with respect to the species' predatory behaviour. Most of those involved in agricultural enterprise in the growing colony had little affection for native Tasmanian flora or fauna, desiring to get rid of them as soon as possible, and replace them with imported, domesticated plants and animals. In dealing with the thylacine as Tasmania's dominant, indigenous predator, there were significant and powerful interest groups in the community, outside the scientific sorority, with vested interests towards (or rather against) the species. While in the early European responses to the initial contact with the species there are no differences between scientific and popular constructions of the thylacine, over time a growing polarisation becomes evident in the nineteenth century between scientific constructions of the thylacine and the popular mythology of the species, fuelled by economically motivated considerations. While admitting a preference for scientific over economically motivated constructions, nevertheless, whenever possible, the polarisation of published sources on the species needs to be investigated by a consideration of the available unpublished sources.

Importantly, if dissimulation is hypothesised against one or more competing sources on the behaviour of the thylacine, can such dissimulation be demonstrated and exposed by considering unpublished or previously ignored records of the thylacine's behaviour?

There is a wealth of unrecognised published information on the species in the nineteenth century, both in the traditional scientific sense of published papers and comments in scientific journals, as well as in the more popular sense of newspaper and magazine articles. This lack of recognition relates to both the twentieth-century hierarchical classification of sources on the species, as well as ingrained assumptions that judge the worth of different sources of information, on the name of the author (the 'great man' hypothesis), the place, the publisher or mode of publication. In dealing with colonial constructions of the thylacine it needs to be acknowledged that publications by British and European scientists in British or European journals are frequently assumed to be of greater value than publications by Australian scientists in Australian journals; which, in turn, are held to be of greater value than Australian newspaper reports. While aspects of the latter comparison are certainly true, it also needs to be recognised that newspaper reports, for all their scientific limitations, as well as frequently expressing popular mythology, may nevertheless contain important, first-hand, primary source observations on a species' behaviour.

There is also a wealth of unpublished nineteenth-century data on the species as well. These unpublished records cover the letters, diaries and unpublished manuscripts of (a) those naturalists and scientists who had already published on the species; (b) those in professional positions in contact with the species (museum and zoological garden curators and collectors); (c) explorers and surveyors investigating the Tasmanian environment; (d) those who kept thylacines as pets; and (e) those with an economic interest in the destruction of the species. Individuals encompassed in the last category may even display differences between publicly and privately held positions.

A further source of unpublished records exists. After the extinction of the species, numerous oral history interviews and recordings were made, essentially from 1950 onwards, of either trappers and old-timers who knew the thylacine in the past, or people who claimed to have seen the thylacine in the present. I have conducted and recorded interviews with knowledgeable and generous old-timers myself. Unfortunately, most recent oral history records relate to, or were initiated by, supposed 'sightings' of the species post-1936. While I have read and listened to much of this material, little is presented here, as I have applied strict criteria to all oral history records and found most of them, for my purposes, wanting. I have only accepted recent oral history records when the individuals concerned have made it quite clear that they were (a) capable of actually identifying the species under discussion, either through reference to captive thylacines known to exist at the time, or, alternatively, through records of thylacines killed by family or friends that prove traceable through the surviving bounty records; and (b) able to provide pre-1937 dates for their observations and comments.

11

Finally, in terms of source material used, in considering the social behaviour of the species, wherever possible, I have made the effort to integrate information from Aboriginal sources into discussion, recognising the limitation that, on most occasions, this information has been filtered through white observers and obtained through white publication.

Theorising colonial science

Various models have been suggested for the historical development of science. A broad view of science that encompasses the historical sciences with their beyond-the-laboratory, real world, ecological orientation, prompts models that reflect ecological orientations and encompass the broad spectrum of the effects of the social, cultural and political environments upon scientific development. As a result of the first attempts to provide comparative studies of the development of scientific disciplines in Australia, Canada and America, Fleming (1964) suggested that there were common parallels in the behaviour of colonial science investigators.

Basalla (1967) used Fleming's (1964) identification of a colonial stage in the development of science as the middle level of a three-phase model he designed to describe the development of western science traditions in European colonies and non-European cultures. It is a global, broad-based model. Stage 1, the Non-scientific Stage, essentially covers the period in which European explorers and observers themselves obtain the specimens or samples and take this scientific information back with them to Europe. Scientific data at this stage consist largely of biological, geographical and geological investigations. This situation is superseded by Stage 2, the Colonial Science Stage, where colonial investigators now gather and send the raw data back to Europe, to be pored over, sifted through and interpreted by European scientists. A strong tendency is evident in these European-based scientists tending to view their colonial colleagues as mere collectors. The natural, historical sciences still dominate the early years of the colonial stage. Stage 3 occurs with the establishment of Independent Scientific Traditions, covering the replacement of the externally, European-orientated scientist by the nationally orientated scientist. This is only capable of being achieved with the teaching of science at all levels of education; with the ability of individuals to be employed as scientists, with a supportive (or at least neutral) government; and with the establishment of national scientific societies and national scientific journals (Home, 1988).

When attempts have been made to place the timing of Basalla's stages in an Australian context, there has been an orientation that considers the development of the Australasian Association for the Advancement of Science in the Australian colonies of the 1880s as the start of stage 3. It is then argued that significant changes in the independence of Australian science took place in the early twentieth century, firmly establishing stage 3 science (MacLeod, 1988). I confess to being rather unimpressed at the idea that Australian science had successfully entered a post-colonial, independent stage from 1916 (Inkster, 1985). On the suggestion that stage 2 'has passed its peak when its practitioners begin a deliberate campaign to strengthen institutions at home and end their reliance upon the external scientific culture' (Basalla, 1967, p. 611), then within the scientific disciplines of animal behaviour, the designation of the successful attainment of stage 3 appears a little premature. At best, I believe that with the formation of viable, independent, national societies and journals, stage 2 had passed its peak by the 1950s and 1960s, but it still has a long way to go. When Australian zoologists and comparative psychologists, as a whole, cease making, or believing in, statements suggesting an *a priori* state of marsupial inferiority, and this viewpoint ceases to be taught in schools and universities, then possibly stage 3 may have arrived for the behavioural sciences within Australia.

Considering the operation of science within a colonial culture provides valuable insights into the construction of perspectives in science. Considering the behavioural sciences in an Australian context, I suggest that, because of its reasonably easy identification, no species provides a better data base for observing the behaviour of scientists than the extensive historical records associated with the marsupial wolf.

In Basalla's model the establishment of colonial settlement signals the ending of the non-scientific stage, and the commencement of the colonial stage where scientific material is obtained by resident colonials and sent back to Europe. In the nineteenth century few European scientists were able to observe thylacines in their natural environment: knowledge about the animal was dependent on anatomical dissection of specimens sent to Europe, and the reports of colonial witnesses, most of whom, from a European perspective, lacked adequate scientific training. In other words, there was a situation where field research and laboratory investigations were being undertaken by different people on different sides of the globe. Unsurprisingly, colonial field observers developed theories which differed significantly from those developed by anatomists and zoologists in Europe, because they were confronted with different experiences and different 'facts'. European perspectives on marsupials after all, were initially dominated by perspectives based upon the previously known South American marsupial fauna. The European scientist, unable to personally verify Australian data that violated accepted (European) scientific theories, had to reject such field observation data as unproven, or accept data originating from non-traditional sources outside the scientists' intellectual and cultural metropolis. Needless to say, instances of the latter were rare.

But field observations and laboratory studies have to be reconciled before a consistent explanation and scientific perspective are placed upon an animal and its behaviour. This task falls essentially on the colonial and post-colonial scientist, as the colonised country, along with its potential political independence also attempts to establish its own independent scientific traditions as well.

The thylacine holds such a signal place in the history of Australian science because it became extinct shortly after these very first attempts were made to deal with contradictions and anomalies from different data sources. The history

13

of some of the scientific constructions of the thylacine prior to 1936 may be used to demonstrate the interactions and relationships between European and colonial scientists and observers. The history of scientific constructions of the thylacine, after its extinction in 1936, reworks these and other contradictions and anomalies, in the absence of any further 'objective' data, delightfully exposing the cultural myth behind the naïve inductionist's belief in scientific objectivity and a progression towards the 'truth'.

Scientific cultures are not independent from general academic culture or contemporary popular culture. The real operation of scientific culture reflects these other contemporary realities. 'Ideal' scientific processes are regularly compromised through the intellectual culture in ascendancy at specific points in time, including, in an Australian context, the acceptance of imperialistic assumptions of our scientific inferiority. There is the 'great man' hypothesis, involving assumptions about the status of previous significant authors and their data, leading to a common replication of the same without validation. There is the assumption that, because information on a subject is added to over time, that more recently published information is likely to better approximate reality than older observations and publications. Finally, there is the assumption that some information is better than others, based upon the mode or place of publication.

14

Note

1   While Murray has provided confirmation of a thylacine encounter in May 1792, his re-examination of the original journal has destroyed another. While the English translation suggests earlier thylacine sightings in April, 'a few wild dogs were observed' (1800, p. 94); the 'few wild dogs' is in fact a mistranslation of *quelques canards sauvages* (1799, p. 127), a 'few wild ducks'.

Chapter 2

Constructing Objectivity

This chapter places the species in the context of the physical environment of the Australian continent, experiencing repeated human invasions from Asia, South-East Asia and finally Europe. It considers the marsupial wolf's former distribution on the Australian mainland and examines both the intellectual construction of the dominant, contemporary scientific views on the cause of mainland extinction, as well as the likelihood of such attribution, when measured against the known behaviour of the different species concerned. This serves as a parallel with what was to happen in Tasmania. The thylacine is then viewed in the context of the European intellectual environment associated with the rapid and deliberate 'species-cleansing' that took part in Tasmania in the late nineteenth and early twentieth centuries, thanks to the government bounty. This destruction of the animal was accompanied by changing colonial perspectives on the species, both popular and scientific. Two examples of changing perspectives on the behaviour of the thylacine are used to illustrate the changing orientations of Australian science.

15

On the behaviour of dominant predators

The exact arrival date of Australia's first human invaders is unknown. While radio-carbon techniques, operating at the limit of their sensitivity, suggest a date of about only 40 000 years, thermoluminescence techniques suggest dates about 60 000 years, and palaeoecological evidence has been used to suggest human occupation around 120 000 to 125 000 years ago (Fullagar, Price and Head, 1996). During these times, because of a slightly lower sea level, the Australian continent was a single joined landmass that incorporated the present-day islands of Tasmania to the south, and New Guinea to the north. Whenever the time of their first arrival, as an unspecialised predator the whole continent stood open before them. Accepting, for argument's sake, the thermoluminescence estimate of 60 000 years, for a considerable period of time the biota absorbed a human population increasing in size and distribution, but the ecological fabric of the continent was finally torn apart 40 000 years ago when, pushed to their limits of existence, the extinction of the megafauna began.[1] In terms of estimated body weight, all Australian terrestrial mammal species over 60 kg in weight became extinct – the maximum weight suggested has been up to 2 tonnes for the largest diprotodontids – and 75 per cent of those species above 10 kg in body weight also disappeared (Murray, 1991; White and Flannery, 1992). The avian mega-fauna was equally decimated, with the destruction of all the large flightless species of *Genyornis*, with estimated body weights up to 300 kg, along with other species, such as the giant megapode *Progura* (Flannery, 1994).

The demise of the major indigenous mammalian predators on the main-land, the marsupial lion and marsupial wolf, was paralleled in the destruction of the major predators in the reptilian megafauna: *Megalania prisca*, the giant monitor lizard, up to 6 metres in length; and the two large crocodile genera *Pallimnarchus* and *Baru*, similar in size to the present *Crocodylus porosus*, but potentially more terrestrial and wide-ranging, being less bound to aquatic habitats and tolerant of a temperate existence. Other representatives of the reptilian megafauna, such as the giant terrestrial turtle *Miolania*, weighing up to 200 kg, were also exterminated (Molnar, 1991; Murray, 1991).

One of the themes taken up in the penultimate chapter of this book relates to post-extinction politics, examining the reluctance of scientists to significantly credit to themselves, or their species, any responsibility for failing to prevent the extinction of the thylacine. At a broader level, however, there appears to be a ready cultural and intellectual acceptance that extinction may be a necessary part of the establishment and spread of economically 'advanced' human cul-tures, strangely contrasted with a reluctance to acknowledge that such destruc-tive behaviour may be characteristic of our species as a whole when removed from its natural (African) environment. Progressive assumptions about the nature of humankind, with a supposedly unique, religiously justified dominance of the world, and racist assumptions about different capacities of intellectual and physical development posited between different human cultures, have legitimised the destruction of the biota through time, in a furtherance of the population growth of 'advanced' civilisations, in contrast to a perceived stability of population and harmony with the environment supposedly possessed by the 'noble savage'. If the suggestion that, outside of Africa, there is no evidence that any human society or culture has ever lived in balance or harmony with its environment, is seen as a radical statement, it can only be because of the implication that within Africa things have been a little different (Martin, 1967). Undeniably, the only continent still possessing a significant proportion of its Pleistocene mammalian megafauna is Africa. Equally undeniably, the spread of humans out of Africa has been followed by environmental destruction and the extinction of continental as well as insular megafauna in Eurasia and North and

South America, as well as Australia (Eldredge, 1991; Flannery, 1994; P. S. Martin, 1973; Murray, 1991; Wilson, 1992).

Important political ends are at work in the attempted preservation of the concept of the 'noble savage'. Racist apologists for contemporary materialistic human cultures cite the primitiveness of the 'noble savage' as evidence for the superiority of contemporary civilisations; while many indigenous peoples have used the concept of the 'noble savage' to provide another intellectual argument for the return of land to the care and control of a country's original human inhabitants. Such political ends have so powerfully constrained scientific thinking on the global extinction of the Pleistocene megafauna as to lead to a suspension in the reasonable operation of Occam's razor (*entia non sunt multiplicanda praeter necessitatem*), seeking the simplest, single, causal interpretation to account for established phenomena. Commonly, attempts have been made to find explanations for the extinction of the world's Pleistocene megafauna in terms of local climatic change, epidemic diseases and resulting species psychological stress; suggesting that the incidence of human invasion with the disruption of different continental ecosystems and mass extinctions represent unfortunate, accidental correlations. While climatic change is a conveniently neutral causal factor that could exterminate the megafauna without any emotive connotations about human behaviour, the supporting scientific evidence for this perspective in Australia is slight:

> Episodic droughts severe enough to result in ecological tethering . . . have been occurring in Australia since the late Miocene . . . These droughts did not result in the extirpation of the megafauna then, and there seem to be no compelling reasons why drought alone should have become any more effective eight million years later. (Murray, 1991, p. 1142)

Given the wide climatic tolerance exhibited in the fossil distribution of the thylacine – from the tropical rainforests of northern Australia and New Guinea and the cool-temperate rainforests of southern Australia, to the dry sclerophyll forests and deserts of inland Victoria, South Australia and Western Australia – explanations for the continental extermination of this species, based on arguments of climatic change, prove difficult to entertain.

While it may be tempting to find some comfort in the depiction of the Australian Aboriginal hunter as a conservation-minded 'noble savage' who 'made very little impact on the Australian scene . . . living in a state of ecological balance with the natural world' (McMichael, 1968, p. 6), this is a viewpoint that has its solid foundation in the most offensive and racist attitudes held about Australia's original human inhabitants, suggesting that their ecological harmony was a result of their primitiveness, lack of intelligence and lack of cultural specialisation: 'Australia is the present home and refuge of creatures often crude and quaint, that have elsewhere passed away and given place to higher forms. This applies equally to the aboriginal as to the platypus and kangaroo' (Spencer, 1927, p. 126).

17

While the commonest fossils associated with human activity are, as would be expected, herbivorous members of the megafauna – simply because for any ecosystem to work towards equilibrium there must be a larger proportion of herbivores producing the food for a smaller population of carnivores – nevertheless, the effects of the invasion of human predators into Australia were felt equally by the major carnivores with whom the Aborigines were in direct competition. The large reptilian predators were the first to disappear. The demise of the crocodile species *Baru*, *Pallimnarchus* and *Quinkana* was associated with the major extinctions of the herbivorous megafauna (Willis, 1997) and probably attained by about 28 000 years ago. *Megalania*, the giant monitor survived until 19 000 years ago. *Thylacoleo*, the marsupial lion survived, at least in southern Australia, until close to 10 000 years ago (Finch and Freedman, 1982; Tyndale-Biscoe, 1973). The most recent mainland fossil remains of the thylacine are dated to about 3090 years ago (Archer, 1974), and sub-fossil mainland remains of Tasmanian devils have been dated as recently as about 430 years ago (Archer and Baynes, 1972). The indigenous carnivores were destroyed through direct competition with humans for a finite food resource.

The demands of human predation, accompanying an increasing human population, caused the complete disappearance of the previous ecological balance and the decimation of whole prey species. Aboriginal artefacts, stone tools and hand-axes dated at slightly more than 30 000 years have recently been identified as possessing traces of blood and been found in association with the bones of megafaunal *Diprotodon*, *Genyornis* and *Sthenurus* species (Dodson, Fullagar, Furby, Jones and Prosser, 1993). These demands alone would result in a drastic drop in the numbers of indigenous predatory species in Australia. In all likelihood, however, paralleling this was the deliberate destruction of competing predatory species through direct killing, predation and consumption.

Thylacines no doubt suffered deliberate predation and consumption from Aborigines, both on the mainland as well as in Tasmania. There is a well-preserved Aboriginal rock painting from the Northern Territory of a thylacine with a three-pronged spear in its back (Plate 2.1a). That the thylacine had some importance as a food item to the Tasmanian Aborigines is reflected in the writing of George Augustus Robinson on his travels throughout Tasmania, establishing a treaty and the supposed temporary removal of Aborigines to the Bass Strait islands (H. Reynolds, 1995). Robinson records different attitudes to the thylacine from different Aboriginal groups with whom he came into contact, mentioning in his diary five thylacines being killed by Aborigines (21/11/1831, 17/6/1834, 18/6/1834), another potentially killed by Aborigines (29/8/1832)

18

Plate 2.1 Aboriginal illustrations of thylacine behaviour: (a) speared thylacine; (b) raised-tail thylacine; (c) solid feeding of semi-independent young; (d) suckling of semi-independent young; (e) detail of d. (Illustrations (a) and (b) drawn from Guiler and Godard, 1998; (c) drawn from Anon/*Northern Territory Newsletter*, April 1974; (d) and (e) from Lewis, 1977)

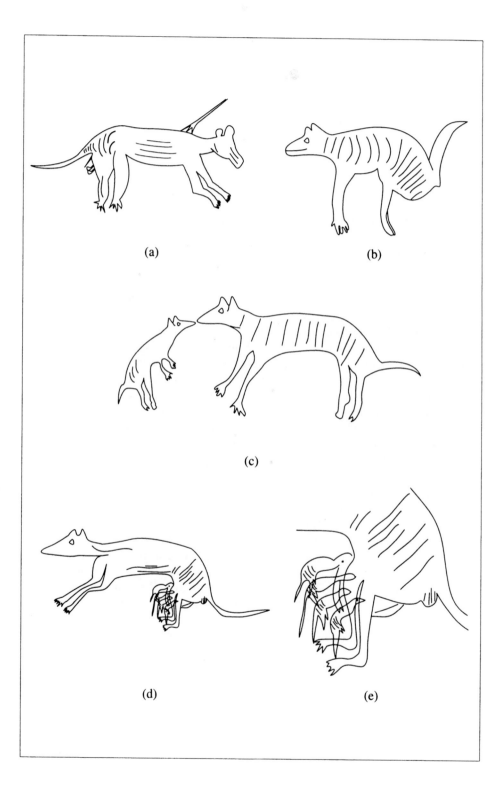

(a)

(b)

(c)

(d)

(e)

19

and quotes from a conversation with Aborigines who 'said they had speared plenty' (15/7/1832). Three thylacine cubs were killed and their carcasses taken away and eaten by the Aborigines (21/11/1831). However, while some Tasmanian Aboriginal groups ate thylacines, other groups and individuals revered the species and refused to do so. On 18 June 1834 a female thylacine was killed, skinned and her skull saved, but the body was not eaten by Aborigines. On the 18th the weather was 'showery', on the 19th it was 'raining', the 20th 'pleasant', but the 21st was 'hazy . . . with heavy and incessant rain and hail'. Robinson recorded the explanation given for this misfortune by his Aboriginal companions: 'The cause of this bad weather is attributed to the circumstance of the carcase of the hyaena being left exposed on the ground and the natives wondered I had not told the white men to have made a hut to cover the bones, which they do themselves, [they] make a little house' (21/6/1834). In similar manner to European attitudes to the thylacine, different Aboriginal groups had different perspectives upon the value, importance, use and nature of the animal.

Returning to the mainland extinction of the thylacine, a possible additional cause was the importation of another predator, with new colonists, arriving around 3500 years ago. Many recent authors, desiring to keep the 'noble savage' model of our species, with its implied suggestion of 'innocence' in the face of extinction, have attempted to lay the blame for much of the decimation of the mainland fauna on the dingo (*Canis familiaris*), a domestic dog – and supposedly superior placental carnivore – gone wild: 'The dingo . . . apparently drove both the thylacine and the Tasmanian devil to extinction on mainland Australia' (Flannery, 1994, p. 115). While the arrival of a second new predator into Australia did not help the local faunal populations and biota in any way, it is rather unrealistic to blame the dingo for the continuing extinguishable outcomes of an ecological disintegration that had been taking place for at least the previous 55 000 years.

The dingo, with its potential for pack hunting and ability to exist on an omnivorous diet (Calaby, 1971), has often been presented as a superior hunter to the strictly carnivorous thylacine, frequently described as a solitary hunter; with the dingo both out-competing the thylacine for prey, and potentially directly attacking thylacines as well. This suggestion was first made in the nineteenth century by Ogilby in a paper read before the Linnean Society in 1837 and published in 1841. Initially, however, the idea of dingo competition received little support. It was even inferred, contrariwise, that the thylacine was responsible for the extermination of the dingo in Tasmania (Krefft, 1868x). But, as an example of how scientific perceptions can change, and change rapidly, the superiority of the dingo suddenly became a popular point of view in the 1950s amongst scientists seriously forced to entertain the likely loss of the species and apparently desperate to attribute the thylacine's extinction to non-human causes.[2]

That such significant changes can be seen to occur in the thylacine literature demonstrates clearly that extinction, by itself, need not put a brake

20

upon changing scientific constructions of an animal's behaviour. In so far as these changing constructions relate to improved knowledge of departed Pleistocene environments and departed species not historically known, this is fair enough, and may be used to illustrate the positive side of changing scientific constructions in an attempt to improve a supposedly objective understanding of the world. However, when these constructions are at variance with the historical record of the species, it is appropriate to question the objectivity of such change. After all, those individuals who identify the dog as the primary cause for the extermination of the thylacine from the mainland need to confront the dissonance obvious in the historical records of thylacine–dog relationships in Tasmania. There is little evidence that the European introduction of dogs into Tasmania was a direct factor in the thylacine's extinction. While thylacine–dog relationships are considered further in chapters 4 and 5, in summary, nineteenth-century records suggest one adult thylacine was more than a match for one or two dogs. Most dogs evinced fear of thylacines, and almost without exception, historical records of the successful killing of thylacines by dogs relate to multiple dogs attacking a single thylacine accompanied by one or more active humans with clubs or guns. A single dog, forced to confront a single thylacine, was in considerable trouble. As suggested by Cooper (14/10/1970), 99 per cent of all dogs were afraid of thylacines, and the ones that were not, were stupid.

21

Ecological contingencies exterminated the thylacine in New Guinea and the Australian mainland. The temptation to label the thylacine as an inferior, less intelligent or efficient predator, predestined to be out-competed by a placental carnivore, ignores many of the historical contingencies involved in that extermination. In no sense can one reasonably argue that the dingo directly caused the extermination of the thylacine on the mainland. The burden of responsibility cannot be shifted so easily from human shoulders. Human destruction of the environment; competitive extinctions of the other megafaunal carnivores; and extinction of the herbivorous mammalian and avian megafauna predicted the extermination of the thylacine. With all the major terrestrial reptilian predators and the marsupial lion already extinct, being in direct competition with human predators, the population numbers of the thylacine would have been under similar pressures. It needs to be recognised that the thylacine, as the last remaining large-sized marsupi-carnivore, was already significantly under stress well before the importation of the dingo. In a small way, the arrival of the dingo may have speeded up the extinction process. But the recent elevation of the dingo in status as the primary cause of the thylacine's mainland extinction represents an unstated, even unconscious, scientific desire to absolve humans – in this case the 'noble savage' – of responsibility for the consequences of their behaviour. As developed in chapter 9 a similar reluctance is shown to attributing responsibility to Europeans for the rapid extinction of the species in Tasmania.

On one other level, the simple contrast of dingo versus thylacine predatory success fails to deal with historical contingency. The dingo was brought to Australia with a new wave of human invaders for a purpose. Aborigines today still keep dingos as companion animals and use them for hunting game (Breckwoldt, 1988). Four thousand years ago the major competitive stress factor on thylacines was predatory competition with humans. With the arrival of the dog, this competitive stress increased, not because of any simple direct competition between the two quadrupedal predators, but because the thylacines' major stress factor, competing with humans, was significantly increased, with humans now hunting co-operatively with their dogs (Breckwoldt, 1988; Kolig, 1973). While some dingos no doubt escaped captivity and bred and spread before the arrival of Europeans, in terms of wild population numbers, this process was dramatically accelerated after the European invasion, as a side-effect of the wholesale slaughter of Aborigines, the attempted destruction of their culture, and the alienation of the surviving individuals from their land and possessions. The present situation of wild dingos existing by themselves at high population density levels throughout the Australian mainland is probably the product of only the last 200 years, and the effects of this population explosion and spread have yet to be ascertained on the Australian biota.

At the time of the last major human invasion into the Australasian region, the thylacine was common only in Tasmania. Currently there is no mainland fossil or sub-fossil thylacine material dated much after the arrival of the dingo (Archer, letter 17/5/1998), however, radio-carbon dates are not the only source of evidence for the thylacine's recent mainland existence. This history of the thylacine has been constructed upon an interrogation of the texts of early scientists and naturalists writing on the species. Such critical analysis, combined with the records of indigenous peoples, leads me to suggest the possibility that at least two relict populations of thylacines existed on the mainland at the time of early European colonisation.

The Victorian naturalist Cambrian wrote a series of articles on the natural history of mainland Australian marsupials (1855w,x,y). While he noted that he 'never saw a living specimen' he records having personally examined the remains of two mainland thylacines, one of them being the head, feet and skin of a thylacine killed in the Blue Mountains, near Sydney, New South Wales. Cambrian also refers to his examination of part of the skin of a female thylacine which, by implication from the additional sources he quotes as recording mainland thylacines – the South Australian explorer Charles Sturt and South Australian Governor Sir George Grey – together with the mainland distribution limits he suggested for the species 'from the Sydney district . . . to Lake Torrens' (1855y, p. 360), probably came from the Lake Torrens/Flinders Ranges region. I have been unable to locate either Sturt's or Grey's comments on observations of South Australian thylacines, but I have rediscovered detailed descriptions of South Australian thylacines made by another contemporary Australian scientist.

Dr John Palmer Litchfield, medical practitioner, surgeon and scientist, was the widely travelled Chairman of the Infirmary Board and Inspector of Hospitals for South Australia. On 13 September 1839 he presented a lecture 'On the Natural History of South Australia' before members of the Adelaide Literary Association and Mechanics' Institute:

> The dog-faced Dasyurus, or native dog, is a marsupial animal . . . covered with a dirty yellowish brown fur, with transverse stripes of a brownish black colour on its back. These animals occasion much annoyance to the first settlers of a country. In Van Diemen's Land it was found necessary to offer a reward . . . In the province [of South Australia] it was also found necessary to offer a reward for destroying them, but their ravages are now pretty much confined to the thinly settled districts. (*South Australian Record and Australasian Chronicle*, 21 March 1840)

Significant corroborative evidence for European observations of South Australian thylacines at this time is to be found in the oral history records of the Adnyamathanha people of the Flinders Ranges, which suggest that thylacines survived in the Flinders Ranges region, to the east of Lake Torrens, into the 1830s:

> Today **marrukurli** [thylacines] are known only as mammals of the Dreaming. That is, there is no one living who claims to have seen one. Both the Dreaming and other oral tradition, however, suggest the possibility of their bodily existence in the region in the not-too-distant past. (Tunbridge, 1991, p. 48)

The centenarian, Mount Serle Bob, who died in 1919, was the last Adnyamathanhan who claimed to have seen a live thylacine, having done so in his childhood (Tunbridge, 1991, p. 93).

There are many well-meaning individuals, not just in Tasmania, but in all mainland Australian states as well, who, to the present day, return from driving or walking in wilderness areas with stories of thylacine sightings. But the criterion for establishing the existence of the thylacine beyond 1936 can only be met through the production of a body, either dead, or, preferably, alive. However, the assessment of evidence for the existence of the thylacine on the Australian mainland in the 1830s cannot reasonably be so constrained by such a twenty-first-century, post-extinction criterion. First, and most importantly, the species was certainly in existence elsewhere in the 1830s and 1840s. Secondly, although none of the physical remains described have survived as museum specimens, they were examined by a significant, Victorian naturalist, whose competence may be ascertained by a frank appraisal of the accuracy of the non-marsupi-carnivore material presented in the published articles, in the light of European marsupial knowledge at the time. Thirdly, claimed sightings of the species cross cultural boundaries. The published observations and records of thylacine sightings by South Australia's leading scientist and naturalist are corroborated by the indigenous people's knowledge at the time. Fourthly, and just as significantly, the absence of physical evidence for the existence of South Australian thylacines beyond the 1840s or 1850s is paralleled, not just by its absence from generated local species lists by contemporary South Australian naturalists and scientists, but at the same time, it also disappears as a living species from the oral history records of the indigenous peoples.

23

Notwithstanding the absence of museum specimens, the continued existence of the thylacine on mainland Australia to the 1830s and 1840s needs to be seriously entertained.

There is some cause for concern, however, that even given the presence of nineteenth-century museum specimens of mainland thylacines, scientists would still refuse to acknowledge their existence!

Some scientific beliefs, held unquestioningly as tenets of faith, are strong enough to allow scientists to ignore the most basic and obvious objective data. Placental chauvinism, holding to the superiority of our own group, the placental mammals over other mammal types (indeed, a dominance over all other living creatures), is one such powerful belief system. No better example of the blind operation of this belief system may be found than in its obvious constraints imposed on scientific thinking reflected in the amazing attitude of contemporary Australian zoologists to the records of mainland populations of Tasmanian devils, supposedly exterminated on the mainland, along with the thylacine, in competition with the superior, placental dingo.

The only zoologists to credit the continued existence of Tasmanian devils in mainland Australia were Wood Jones (1923) and Troughton, who considered its likelihood from the first edition of his classic *Furred Animals of Australia* (1941). Other zoologists who have published catalogues or monographs on Australian, or specifically south-eastern Australian mammals in the twentieth century, found the idea an anathema: Brazenor, 1950; Le Souëf and Burrell, 1926; Lucas and Le Souëf, 1909; Menkhorst, 1995; Ride, 1970; Strahan, 1983; and Triggs, 1984. All these publications appear in conformity, and comfort, with the most recently dated sub-fossil devil material at 430 years before the present, and the idea that these remains necessarily represent the very last mainland representatives of the species, going down before the superiority of the dingo.

Nineteenth-century records exist for mainland Tasmanian devils, *Sarcophilus harrisii* (also at times referred to as *S. ursinus*, referring to its bear-like nature and original classification by Harris [1808] as *Didelphis ursina*). These records parallel those of mainland thylacines. Cambrian (1855) records the presence of the Tasmanian devil in New South Wales, Victoria and South Australia, and was personally familiar with both live and dead specimens. While widespread in distribution, Cambrian recognised it as uncommon: 'during the whole of our stay in South Australia we only saw two'. Unfortunately, the preserved fragments of Litchfield's public lecture in Adelaide in 1839 include the mention of only one marsupi-carnivore, the thylacine. His descriptions of other South Australian marsupi-carnivores are lost. Also missing from mainland Tasmanian devil reports is that, to date, no oral indigenous records of the species are known to exist. However, occasional references to the continued existence of the species are to be found in contemporary newspaper reports. In a nature column in 1896 the *Adelaide Observer* noted that 'a colony of Tasmanian bears' was still living around Lake Albert (28 November 1896). In Victoria, the presence of a significant black-coloured predator upon stock in the Grampians and Dandenong Ranges in the 1880s, soon gave rise to the first newspaper reports of the existence of a supposed black panther in the bush (*Age*, 14 June 1885), records of which continue to the present day (Chapple, 3/6/1999; Healy and Cropper, 1994). Additionally, skeletal material of Tasmanian devils, collected on the surface of the land and unassociated with Aboriginal kitchen middens, were deposited in the National Museum of Victoria from three different Victorian localities; these bones were 'so recent that they might easily have belonged to an animal which lived but a few years ago' (Kershaw, 1912, p. 76). At face value, records for the continuing existence of mainland Tasmanian devils in the nineteenth century appear weaker than the records for nineteenth-century mainland thylacines: no museum specimens or indigenous records, a handful of newspaper reports and the published observations of but one credible European naturalist. All this was soon to change.

In 1912 a live Tasmanian devil was captured at Tooborac in Victoria, and described in scientific publication by the curator of zoological collections at the National Museum of Victoria (Kershaw, 1912). The skin and skeleton were preserved in the museum, and while its conformity with the nineteenth-century skeletal material was emphasised, the less likely explanation of it being an escaped captive was also entertained. But since then the bodies of four more Victorian specimens of devil have been submitted to, preserved and catalogued by, the Mammalogy Department of the Museum of Melbourne – the most recent specimens being two road kills in 1991, from localities 150 km apart. There would have to be a phenomenally high population of incompetent Victorian naturalists illegally keeping Tasmanian devils as pets, in order for as many as five of their lost captives to become museum specimens.

Records of the Tasmanian devil on the mainland in Victoria date back to that colony's first establishment, and the five museum specimen localities match the distributions suggested from nineteenth-century observations. Yet the stranglehold that placental chauvinism has upon the minds of Australian scientists has been so strong that, until now, no recent professional zoologist has been prepared to face up to the evidence, reject the prevailing paradigm of the superiority of the dingo (and hence all museum specimens as escaped captives), and proudly claim in publication an extension of the Tasmanian devil's distribution to Victoria to this day. Scientific objectivity? What scientific objectivity? How can such data be ignored? The blind belief in placental superiority is so monumentally powerful as to prevent even the easy recognition and acceptance of the hardest of all biological data – preserved and frozen specimens in museum collections.

In examining the history of scientific constructions on the behaviour of the Tasmanian tiger, one needs to recognise and be wary of just how easily and uncritically records of thylacine behaviour that conform to a belief in placental superiority may have been incorporated into the accepted canon on the species' behaviour.

25

Myth construction and deconstruction in colonial science

The problems and assumptions identified in the Introduction as indicative of the operation of scientific culture have all played their part in two extreme examples I have chosen to illustrate changing scientific perceptions of the thylacine.

The first European encounters with the thylacine were all in coastal localities. Obviously, it could hardly have been otherwise.

In his original description of the thylacine Harris (1808, p. 175) mentioned that the tail was 'much compressed'. This characteristic was soon seen to be of vast significance by European scientists, and anatomical analogies were made with other mammals possessing laterally flattened tails, such as the South American water opossum (*Chironectes minimus*) and placental otters. The compressed tail was mentioned by Geoffroy Saint-Hilaire (1810) and Temminck (1824). Temminck also noted the thylacine's common occurrence around the coast and suggested that it fed upon crabs. Cuvier expanded on the idea in full and provided the first detailed argument and description of the thylacine as a marine predator:

> Its compressed tail seems to indicate that it is a swimmer, and it is known to be an inhabitant of the rocks on the sea shore of Van Diemen's Land, and to feed on flesh . . . fish and insects . . . They also seek, with avidity, the half-corrupted bodies of Seals and cetaceous animals on the sea-shore . . .
> (Cuvier, 1827, pp. 36, 37)

Mudie continued the argument of marine scavenging and active marine predation, suggesting that the thylacine 'is found on the sea shore, where it preys upon the remains of animals, and probably swims after fish' (1829, p. 175). The position became entrenched in the literature, and the popular mind, with the continued reprinting of Swainson's (1834) entry in Murray et al's *Encyclopædia of Geography*: 'the tail is compressed, which suggests the supposition that it is used in swimming, particularly as this animal inhabits the rocks on the sea shore, and is known to feed upon fish' (see Swainson, 1834, p. 1485).

Not all Tasmanian naturalists were prepared to swallow this European construction. Convinced of the error of European perspectives on the basis of his own observations, Gunn wrote a forthright comment on the concept of marine predation in a private letter accompanying a collection of specimens sent to W. J. Hooker. Hooker was duly impressed with the collection and the information provided and, without consulting the author or editing the text, published Gunn's letter with its thylacine comments in detail:

> In Murray's Ency. of Geography it is stated . . . that its *tail is compressed*, which suggests the supposition that it is used in swimming. The tail is not compressed, neither is it at all aquatic in its habits. They are most numerous inland . . . As to their feeding on fish, I hardly know how it could have been ascertained, unless

the fish had been previously caught and given to one, when, like many carnivorous quadrupeds, it is probable it would eat them. Deductions are frequently too hastily drawn by naturalists (or persons professing to be such) from isolated facts. That the *Thylacinus* may often be seen on the sea-coast, as also every other species of our quadrupeds, is quite probable, and may once or twice have been seen eating a dead fish thrown up by the sea; but as to its *fishing*, it is out of the question. (Gunn, 1838w, pp. 101–2)

A denial of marine predation written by Edward Blyth was immediately introduced into the next posthumous edition of *Cuvier's Animal Kingdom* (1840, p. 103), but Gunn's criticism apparently made no difference to either Swainson or Murray and the description of the thylacine as a marine predator continued to be reprinted in Murray's *Encyclopædia of Geography* until 1846 at least.

While Swainson was fair game for critical comment by his fellow European professionals – for example, Waterhouse (1846, pp. 456–7) took Swainson to task for his incomplete knowledge of thylacine dentition – it was another matter when an unknown colonial attempted the same thing. This was particularly so, as Gunn's comments were not restricted to the minor figure of Swainson alone. Gunn clearly believed that the primary source in knowledge construction about Australian marsupials should be found in the observations and analyses made by colonial Australians, over the isolated anatomical facts and analogies obtained by sedentary Europeans. While Gunn may have assumed Swainson to be the originator of the marine predator construction, nevertheless, by accident or design, amongst the 'naturalists (or persons professing to be such)' coming under his critical fire were two of the most significant 'great men' of biological science: Geoffroy Saint-Hilaire and Cuvier.

27

This upstart colonial needed to be put in his place, and thanks to the editors of the major British scientific publications, this was achieved. Gunn continued to travel widely, observe, collect and publish articles in Australasian journals about Tasmanian, Victorian and New Zealand flora and fauna, both recent and extinct. He sent extensive collections of botanical specimens accompanied by observational notes to the Hookers in the Royal Botanical Gardens, Kew, which were influential in affecting Darwin's developing ideas on evolution (Baulch, 1961; Gulson, 1991); and collections of zoological specimens, accompanied by notes on wild and captive observations and the anatomical dissection of fresh specimens, to Owen and Gray at the British Museum – but these observations and comments did not receive the acknowledgment of any further British publication. Gunn was effectively 'black-balled' from the most influential zoological journals of his age, and remained so for the next eleven years.

When his existence was again acknowledged in a British zoological publication (Gunn, 1849), it amounted to a mere 150 words extracted from a letter to Gray on the locality and probable affinities of the impressively coloured and sized cowrie *Cypraea umbilicata*. Realistically, the publication of Gunn's comments probably had a lot to do with the fact that, of the two cowrie specimens he had sent to London, one was sold to a private collector for the then extraordinary price of £30 – as mentioned in the lengthier article on the cowrie by Gray (1849) that followed Gunn's brief remarks. Finally, Gunn sought atonement and, after the successful arrival of two live thylacines he sent to the Zoological Society of London – for which he received acknowledgment in *The Times* (21 May 1850) – he was eventually granted absolution by the zoological establishment. Significantly, when his letter on the behaviour of thylacines was published in the *Proceedings of the Zoological Society of London,* over a quarter of the published text was a grovelling apology for his earlier 1838 remarks:

> An observation of mine, contained in a letter to Sir W. Hooker, and which was not meant for publication, has been misunderstood, and has led to the propagation of error – for which I am very sorry. In it I said the Thylacine's tail was not compressed – in reference to an observation of Mr. Swainson's in the 'Encyclopædia of Geography' . . . that the tail of the Thylacine was compressed, *which suggested the supposition that it was used in swimming,* &c. It was to the latter part of this observation that my remarks were particularly applied . . . I meant that the tail was not compressed to such an extent as to have justified the inference that it was useful in swimming; and thus that the animal obtained its food principally from the sea, which the paragraph in the 'Encyclopædia of Geography' implied. The tail is obviously slightly compressed, but not, I think, more so than the tails of the Dasyures, to which aquatic habits are not attributed. In writing hurriedly – and not for publication – I did not express myself with the precision I ought to have done. I mainly wished to point out that the tail would not justify the inference of Mr. Swainson (which I thought very far strained), that the animal was aquatic in its habits and piscivorous. Pray set me right whenever you have an opportunity. (Gunn, 1850w, p. 91)[3]

No matter how benign one's interpretation may be of the controlling power of British scientists and British scientific institutions over the work of their colonial colleagues, this picture of a colonial scientist being punished, embarrassed by and needing to apologise for publishing 'the truth' is decidedly unattractive.

From 1850 onwards all but a handful of authors writing on the thylacine accepted Gunn's position – his denial of marine predation and swimming becoming a part of the entry on the thylacine in the *Encyclopædia Britannica* (W, 1857). There are, however, some authors whose contact with the thylacine literature appears limited to the writings only of the 'great men' of science, who failed to find either of Gunn's publications, or chose to ignore his colonial correction.

Figuier (1870) persisted with the idea of marine scavenging, as well as active predation upon crabs. Aflalo (1896, p. 66) denied marine predation – although accepting the compressed tail and swimming ability argument, he thought it unlikely that thylacines would actively swim after marine prey because of the likelihood that they would be eaten by sharks! – but he supported marine scavenging, suggesting thylacines possessed 'a taste for ambergris'. In the twentieth century Beatty (1952), Brogden (1948), Hall (1910) and Haswell

(1926) have persisted with the construction of the thylacine as a significant marine scavenger; while Cuppy (1950) has continued to support general marine predation and Rœdelberger and Groschoff (1967) specific predation upon seals.

There are two important realities about scientific constructions obtainable from this history of the construction of the thylacine as a marine predator. First, despite the naïve inductionist's faith in the objectivity of science, prevailing constructions, as outlined by the 'great men' of science, are powerful – no matter how poor the original data source may be (in this case anatomical analogy in the laboratory) – and not easily overturned by persons of supposedly less scientific consequence, no matter how accurate their observations in the field may be. Secondly, once the 'great men' have decided upon a construction and it enters the literature, it may persist for a very long time.

If the successful denial of thylacine marine predation in the nineteenth century represents Australian science operating at its best, contrariwise, the elevation and acceptance of the popular mythology of the thylacine, as a blood-sucking vampire and fearsome werewolf of Tasmania, throughout the twentieth century, represents Australian science operating at its worst.

The factors involved in such a remarkable hijacking of a supposed commitment to objective reality in the formation of scientific constructs provide insights into the subjective reality behind much scientific concept formation. Nor is this a unique case. A similar adoption of Gothic constructs (and the difficulty of their removal) may be found in the history of scientific constructions on the behaviour of the gorilla (Wendt, 1956).

By 1855 W (p. 247) had suggested that the thylacine had become 'a great object of dread' amongst the settlers. While the origin for this 'dread' may be found in popular perceptions of a threat (real or imagined) to colonial husbandry, such suggested fear and loathing of the species, once acknowledged, soon took on a life of its own. Popular perceptions of the thylacine became based on a 'penny-dreadful' mentality, and the desire of the 'civilised' city dweller to be shocked, horrified or frightened by the wild and primitive, accompanied by colonial and European attitudes – both rural and urban – of the Australian bush as 'the enemy' (Horton, 1996; A. J. Marshall, 1966).

Even in the 1920s, with the species rapidly heading for extinction, A. S. Le Souëf and Burrell noted that: 'The inhabitants seem to have a superstitious dread of the "hyena," . . . and will kill the wolf whenever the opportunity offers. Indeed, some will even smash the wolf to pulp afterward, thus depriving science of the skeleton and skin' (1926, p. 319). Joines (1983), Serventy (1966) and Troughton (1941) echo this point of view. Various old-timers have expressed this in their own words. In an unpublished letter Kath Doherty directly touches upon the mythology and fantasy associated with the thylacine: 'One of the queer things is that I cant remember ever having any remorse about its destruction. Almost like we were heroes and it a ferocious beast' (Doherty, 16/12/1972). As a way of demonstrating prowess in killing the 'ferocious beast', skulls from killed specimens would be nailed to barn doors or to the outside of bush huts (Fleay, 1956; Guiler, 1985).

Distrust of the species also extended to captive specimens: '50 years ago Mrs. Harrison's brother had a tiger in a cage, at Forest. It . . . was quite healthy, but the neighbours were scared of it, and poisoned it after several weeks' (Harrison and Harrison, 9/10/1952). In addition, the 'hysterical loathing' of the thylacine (Joines, 1983, p. 5) persisted long after the species became extinct. Mrs E. Holmes recalled:

> Every time I hear the Tas. Tiger mentioned I am filled with remorse when
> I think of a thing that my mother & I did. As early as I can remember my
> collector Uncle lived with us. Occupying the largest bedroom for his displays,
> of mineral, & marine, and the 'Tiger skin'. It held pride of place in front of the
> fireplace. He loved to take visitors to his room to see his various collections.
> How my mother hated that skin, she was really scared of it . . . At my Uncle's
> passing in 1956, that treasure was first on the fire, as my mother got rid of his
> rubbish, as she put it . . . I can see it lying there on the floor as tho it were
> yesterday . . . and hating it because my mother did. (26/8/1981)

However manifested, the marsupial wolf obviously aroused in many lay people a strong, negative, emotional response. Considering that the history of popular constructions of the placental wolf (*Canis lupus*) has involved wide-spread imputations of vampirism, witchcraft and lycanthropy (Day, 1981; Jenkinson, 1980), this should, perhaps, come as no surprise.

This emotional response to the thylacine was certainly in place by the mid-nineteenth century. How important a role it played in the construction of the thylacine as a blood-sucking vampire is unclear. By itself, there was nothing unusual or significantly 'vampire-like' about a thylacine's attraction for the neck of its prey. Most large carnivores preferentially direct their attack at the neck of their victim. This has the potential to kill the prey rapidly, as Pitti-Sing mythologises in *Mikado*, to 'Cut cleanly through his cervical vertebræ', and/or markedly reduce the blood flow to the head. If neither of these two effects has been achieved, merely hanging on to the neck will eventually suffocate the prey anyway. However, a blood-feeding reputation has been given to the thylacine by growing numbers of scientists and naturalists throughout the twentieth century, suggesting that thylacines suck or lap the blood from the throat of freshly killed prey and then may, or may not, also eat some of the vascular tissues (Guiler, 1985, pp. 82, 140); and that the source for this information is to be found in records and trappers' tales from the nineteenth century.

Taking the last point first, I have been unable to find any mention of blood-drinking thylacines in nineteenth-century sources.[4] Apparently, all records of vampirism only come from the twentieth century, and while the oral history records of trappers' and old-timers' reports, both published and unpublished, are inconsistent on this issue, the vast majority of old-timers do not accept the blood-feeding proposition.[5] As this proposition exists in some early twentieth-century scientific literature, recent Australian scientists were confronted with a situation in which contradictions and anomalies existed between different reported field observations of data. Consequently, some decision, for or against vampirism, was required. At one level, the majority decision to accept vampirism may be viewed proudly as an example of scientific thinking that works against popular stereotypes of science as a dry and conservative profession, reflecting a narrow and constrained way of reasoning. At another level, this spectacular inclusion of twentieth-century popular mythology into critical twentieth-century scientific constructions deserves investigation for what it may tell us about how scientists choose between different and competing sets of data. This is not a unique problem. Historical scientists, dealing with contingent events occurring in time, are often faced with incomplete data obtained from non-replicatable situations.

Illustrating the minority viewpoint amongst trappers and old-timers is the position of Kathleen Griffiths. During an audio-taped interview she mentioned that her brothers brought home a snared thylacine which was kept as a pet and later sold to Mary Roberts of Beaumaris Zoo in Hobart, and commented:

> They were very mischievous, the way they got among sheep and killed them. They were not satisfied to kill one and eat it, but they'd kill four or five and get the blood, that's all they wanted . . . They'll come quite close to houses where there are sheep in the paddocks, sneak down out of the country, get a sheep and kill it and drink the blood. They were a blood feeder, wherever it came from they'd suck the blood. Blood was their main food . . . (Griffiths, 21/5/1980)

31

By way of contrast, representing the majority viewpoint of trappers and old-timers opposed to the concept of blood-feeding, are the comments of Adye Jordan, trapper, who with his brother snared three thylacines, one of which was alive and became the last thylacine purchased by Melbourne Zoo:

> Its eating and drinking habits are easily observed and I strongly contradict these people who have so often described the tiger as a blood-sucking animal. Many people . . . have claimed that a tiger had killed a sheep, kangaroo, rabbit and other animals and sucked the blood from their neck, some people thrive on myths, they prefer myths because nobody has anything to prove. (Jordan, 1987, p. 95)

Jordan makes the obvious comment with respect to all recent changes in popular mythology (or scientific constructions) of the animal: the inability to prove – or rather disprove – such construct changes, given the present population status of the species.

Before leaving these contrasting positions held by old-timers with demonstrable knowledge of the species for a consideration of the scientific adoption of vampirism, it is worth searching for possible indications as to which of these two sources may possess the greater reliability. One inconsistency is apparent between Griffiths' claim that 'blood was their main food' and the diet they offered their captive specimen. It was not fed on buckets of blood, or even the blood-filled vascular organs, but rather 'kangaroo carcasses . . . mutton [and] meat of any description' (Griffiths, 21/5/1980).

I have been able to locate only five references to blood-feeding in the scientific literature during the first half of the twentieth century. In chronological order they are: G. Smith (1909w), C. E. Lord (1927), C. Barrett (1943), Harper (1945) and Harman (1949). The 1950s, however, saw three publications, one from each of the professionally acknowledged experts on Tasmania's animals, that were to have significant effects upon scientific constructions of the thylacine for the rest of the century.

First, V. V. Hickman, the Professor of Biology at the University of Tasmania, wrote: 'A single thylacine often killed several sheep in a night, merely sucking their blood and eating a little of the kidney fat' (Hickman, 1955, p. 10). Then Tasmania's pre-eminent naturalist, Michael Sharland, published an article about a captive thylacine that refused to eat either dead wallaby flesh, or to kill and eat a live wallaby offered to it, but 'ultimately it was persuaded to eat by having the smell of blood from a freshly killed wallaby put before its nose' (Sharland, 1957w, p. 26). Finally, Eric Guiler, of the Zoology Department at the University of Tasmania, wrote that the thylacine 'is primarily a blood feeder, sucking blood from the severed jugular vein of its kill' (Guiler, 1958x, p. 354).

Guiler repeated the idea in 1960 (p. 21), the Australian Museum picked it up and included it in its published leaflet on the species (1964), then the professional naturalist Serventy wrote that 'The animals showed a preference for sucking the blood of their victims' (1966, p. 37) and after that, it was on for young and old. Allen Keast, at that time Professor at Queen's University, Kingston, Ontario, referred to blood-feeding in his book on the natural history of Australasia (1966), and Richard Sadlier, Professor in the Department of Biological Sciences at Simon Fraser University, British Columbia, included blood-feeding as a characteristic in his book on Australasian animals (1970, p. 115). After an interview with Bill Cotton, Sharland published blood-sucking comments in newspaper (*Hobart Mercury*, 6 March 1971) and book formats (1971x). The zoologist, Jeremy Griffith of the Thylacine Expeditionary Research Team, accepted 'They feed primarily on blood, vascular tissue, and occasionally muscle' (1972, p. 73), and Larry Collins, author of the Smithsonian's monograph on *Monotremes and Marsupials* also recognised blood-feeding (1973, p. 138). Serventy referred to it again (Serventy and Raymond, 1973) and Phil Andrews, zoologist at the Tasmanian Museum noted it was 'a blood feeder' (*ca* 1975, p. 2). Hier accepted blood-feeding (1976, p. 29) and this point of view was also expressed by the zoologist Rod Pearse in a paper presented at the Australian Mammal Society Conference (1976, p. 2). Arthur Woods, lecturer in zoology at the University of New South Wales, wrote that the thylacine 'prefers blood to flesh, and it takes its fill by severing the jugular vein of its victim' (1977). With the inclusion of blood-feeding in the Australian National Parks and Wildlife Service's definitive statement on the thylacine (1978) in its *Rare and Endangered Species Leaflet* series – copies of which were sent to many second-ary school science departments, and were probably held in most municipal and school libraries – blood-feeding could be seen not only to have obtained main-stream scientific support, but to have become a widely known and acknowl-edged characteristic. Unsurprisingly, the blood-feeding construction became included in further significant overseas reviews of the thylacine: Joines (1983) and Marshall (1984).

Not all recent scientific authors have been equally enthusiastic about blood-feeding, although usually their dissatisfaction is expressed via omission of the construct, rather than frontal attack. After all, Guiler, as the recognised senior authority on the species, still holds firmly to the position: 'they used their huge gape to bite out the throat [of sheep and kangaroos] and then they drank the blood' (1991, p. 14). In his review of the species M. Smith (1982, p. 249) downplayed the importance of blood-feeding and did not consider it 'a normal practice'. However, Czechura (1984), Grzimek (1976) and Salvadori (1978) have come out strongly against the concept, considering blood-feeding to be a fabricated story by sheep farmers wishing to magnify the extent of stock losses caused by thylacines, suggesting such ideas had been introduced into the literature through enthusiastic, if gullible, scientists. As this work develops it will be obvious that I am perfectly prepared to criticise sheep farmers for their attitudes towards the thylacine. Vampirism, however, is one area in which I consider such criticism to be patently unfair.

Considering the five references to blood-feeding published in the first half of the twentieth century, Barrett (1943), Harper (1945) and Harman (1949) refer to the sucking of blood from the jugular vein and eating the kidney fat, and are just essentially repeating the position expressed in earlier publications.[6] C. E. Lord, Director of the Tasmanian Museum, overviewed the characteristics of the thylacine in three scientific publications: two journal articles and one book. Only with his last publication, a paper on 'Existing Tasmanian marsupials' (1927) is there any mention of blood-feeding. His first scientific paper 'Notes on the mammals of Tasmania' (1919) and the book on Tasmanian animals written in conjunction with the curator of the Queen Victoria Museum in 1924, make no mention of blood-feeding. However, three years later, for the first time in an Australian publication, Lord decided that: 'If a thylacine kills a sheep it will usually only suck the blood, and may also take a little of the kidney fat' (1927, p. 21).

Lord provides no explanation in his 1927 text for the inclusion of blood-feeding, and the paper does not include a reference list. The reason why Lord adopted blood-feeding is therefore unknown. It may have been just that he came into contact with a farmer or trapper whose statements and arguments for blood-feeding Lord found totally convincing. But it is more likely that, given the similar manner in which he described blood-feeding, Lord, as Director of the museum, had become aware of the importance of Geoffrey Watkins Smith's sojourn at the museum in 1907–08 and his opinions and resulting publications.

33

For it was Geoffrey Smith in 1909 who first published the idea that the thylacine was a blood-feeder: 'a Tiger will only make one meal of a sheep, merely sucking the blood from the jugular vein or perhaps devouring the fat round the kidneys' (1909w, p. 96). Why, after being ignored for nearly twenty years, Smith's ideas were treated seriously, first by Lord, and then later by a growing population of professional zoologists, possibly relates more to the perceived status of Smith as an individual, rather than to his competence of observation. As a Fellow of New College, Oxford; with the length of his visit to Tasmania, and his resulting publications, Smith arguably provided an important building block upon which Australian zoologists were able to stand in their attempts to establish their own independent scientific traditions in the newly formed nation. The work of Smith, as a former Oxford academic and war hero (he was killed on the Somme on 10 July 1916) became possessed of considerable status for Tasmanian zoologists, and with his book published by no less than Clarendon Press, Oxford, this presented more than enough for Smith to be locally granted a 'great man' status, and for his ideas to be taken seriously and unquestionably by the fledgling Australian academy, when they resurfaced, courtesy of Lord in 1927, leading to their adoption by a new generation of 'great men' in the 1950s.

34   It is possible to go further than just suggest that blood-drinking became accepted into the scientific literature through the operation of the 'great man' hypothesis. The reason why blood-drinking appears in Smith's book can, I think, be inferred from his scientific publications and personal letters (privately published by White, 1917) sent to his relatives back home in England.

Geoffrey Smith was principally a crustacean expert (1909x). He arrived in Tasmania on 1 November 1907, full of enthusiasm: 'There are all sorts of legends of strange animals about, especially in the lake district' (2 November 1907), and, sure enough, it was Lake District mythology that he swallowed in the end.

In his time in Tasmania Smith failed to see a thylacine at all (thanks to the government bounty scheme). All of his details regarding the species are second-hand. Somewhat lost in the bush, discovering that 'the traveller in Tasmania should not trust implicitly to the Government maps' and that, unexpectedly and disturbingly, 'we could not reach Lake St Clair that evening' (Smith, 1909w, pp. 93, 95) on 14 January 1908 Smith was fortunate to stumble across and find shelter in a shepherd's hut belonging to David Temple. Temple certainly had some practical knowledge of the thylacine: between 1891 and 1907 he received payment for nineteen adult and four juvenile thylacines submitted to the government bounty scheme. Smith also had some practical knowledge of the Australian fashion for 'spinning yarns over the campfire' (31 December 1907), but for some reason, on the night of 14 January 1908 he chose to suspend disbelief.

In his private letters home Smith describes how 'These colonial people want rather delicate handling and they are very independent' (20 February 1908). He was entertained at the University and by the Royal Society, and provided with a room, laboratory and equipment for his use in the Tasmanian

Museum. His fellow colonial professionals arranged introductions, itineraries and activities for Smith, with significant crustacean emphasis. They were patronisingly judged by Smith as a 'very nice simple sort of people' (2 November 1907). The day before chance led him to seek shelter with Temple, from his own lofty, Oxford position he had written: 'The shepherds up here are rather primitive sort of people' (13 January 1908). I think these comments in private letters home provide revealing insights into Smith's personal attitude towards Australia and Australians.

I venture to suggest that the unexpected arrival of a misdirected British academic, full of his own self-assured superiority and arrogance, would be too much for any self-respecting Australian shepherd to pass up the opportunity to see just how far he could stretch the limits of credibility in his guest, with stories of blood-sucking thylacines, Tasmanian werewolves and the like. Smith started his journey expecting 'legends of strange animals . . . in the lake district' and, sure enough, he found them. It was obviously a great night. Temple's unexpected visitor believed everything he was told! Not only that, but, he published it!

The effects from this single incident of a single academic in the wrong place at the right time, aided by a popular mythology prepared to accept fantastic behaviours attributed to the animal, have crazily entered into late twentieth-century scientific constructions of the animal. What is fascinating about this is not just what it says about the formation of contemporary scientific constructions of the thylacine, or the way in which Australian scientists tend to behave when presented with Oxbridge generated data; but how it serves as a further illustration of the questionable assumption made by naïve inductionists, of the automatic progress and improved objectivity in the operation of science over time.

Having considered the narrow scientific constructions of the thylacine as a marine predator and vampire, the broad question arises, upon what did the species actually feed? This is the matter for discussion in the next three chapters.

Notes

1 Richard Owen was the first scientist to propose that the extinction of the Australian megafauna was due to the 'hostile agency of man' (1877, p. ix.)

2 In my reading of the literature, for the 108 years following the publication of Ogilby (1841), from 1842 to 1950, I have found only nineteen additional references, suggesting the responsibility of the dingo for the mainland extinction of the thylacine. In contrast, in the fifty years since then over eighty references to the competitive superiority of the dingo have been found (Paddle, November 1997).

3 The effect of this apologetic paragraph was immediate. Gunn was elected a fellow of the Linnean Society in 1850 and a fellow of the Royal Society of London in 1854 (Burns and Skemp, 1961).

4 For the first half of the nineteenth century Tasmanians did not have to turn to
their indigenous fauna to support the vampire myth. A proportion of escaped
convicts turned to their companions for their energy requirements. The earliest
known perpetrator of convict cannibalism, William Pearce, was presented to the
public as a 'vampire' in the Tasmanian press (*Hobart Town Gazette*, 25 June 1824).
It was some considerable time after convict transportation ceased before the
vampire label was awarded to a different Tasmanian resident.

5 I have identified published and unpublished source material comments from
more than 100 snarers and old-timers who had contact with the species. Out of
this sample I found only five individuals who subscribe to the blood-feeding
proposition: Cotton, Griffiths, Sawford, Skelly and George Stevenson.

6 Ian Harman introduced into the scientific literature the use of the word 'vampire'
to describe the thylacine's feeding behaviour (1949, p. 87). Additional resonance
with Bram Stoker's 1897 novel *Dracula* may be found in the original name of
'Transylvania' for the mountainous area between Lake St Clair and the west coast
of Tasmania.

Chapter 3

Of Signal Importance

This chapter considers the behaviour of the marsupial wolf in the context of its social relationships across the life-span of the species. It is based on nineteenth-century perspectives, and concentrates on hunting behaviours and vocal communication.

Social and hunting behaviours in thylacine development

Initial European contact with a new, retiring species was always far more likely to have been an unexpected observation of a single individual thylacine, rather than a family group. However, it was soon recognised that multiple thylacines were usually involved in predatory activities (Grant, 1831; T. Scott, 1829, 1830).

George Augustus Robinson, based on information from his Aboriginal companion 'Black Dick' and his own observations in pursuing and killing thylacines in the wild, was the first European to record the social structure of the species as consisting of an adult male and female plus young, and their combined hunting of prey and possession of an home base or lair to which they retired (6 April 1834). By the mid-nineteenth century, it was widely accepted that co-operative hunting took place between family group members, expressed as either the adult male and female together, or the adult female and up to four young (Gunn, 1850x, 1852w; Meredith, 1852). Observations of adult thylacine pairs (J. Smith, 1862) and their co-operative hunting of kangaroos (Meredith, 1880; see also Loone, 1928, for an account from the 1870s) gave rise to a popular perception that, as a species, thylacines hunted together 'with the pertinacity of a pack of wolves on the steppes of frozen Russia' (Oscar, 1882).

The most detailed descriptions of thylacine family group structure and behaviour may be found in the reminiscences of Archie Wilson (*ca* 1922w,x,y,z)

– who came from a family of successful bounty claimants – and A. R. ('Dick') Rowe (25/11/1951, 26/9/1980) – who, with his brothers, caught three young thylacines that were sold to Melbourne Zoo in 1914. Rowe noted the existence of a family and its location within a specific home range or territory – 'absolutely forced to imagine that they had an area to themselves' (26/9/1980) – and that when a particular family group had been destroyed, no more thylacines would be heard of for some time (until a new female or pair appeared and commenced breeding). Similar conclusions come from the analysis of the Van Diemen's Land Company diaries from 1874–1914:

> the evidence quite strongly suggests that the animal made little attempt to move away from an area even in the face of persecution. This implies thylacine has a home range, if not a defined territory and is a view supported by A. Youd of Deloraine, who trapped in the Lake Adelaide – Golden Valley region, and said that *'once you found where they lived then all you had to do was to stick at it until you caught them'.* (Guiler and Godard, 1998, p. 22)

Thylacine family dynamics and structure were described in handwritten notes by Wilson:

38

> Re the two old tigers & young ones keeping together: When they leave the pouch till they are about three parts grown they keep together. [In] my own experience I have found them that way & I have many times tracked them in the snow, the two old ones & two young ones & sometimes three young ones, where they have been hunting Wallaby Rat [*Bettongia gaimardi*] & have killed and eaten these rats, & I have seen the same where they have been killing sheep. But after they pass that stage in their life, they live a solitary life . . . in [the] remote & roughest parts of the mountains & are a very shy animal, even when there were plenty of them about, it was very rare to see one in the bush. (Wilson, *ca* 1922x)

The first generation of young left their parents at variable times, related to the onset of further adult reproduction. The first generation of young commenced leaving the family in the months before the next breeding cycle began, and departure was completed (and probably parentally encouraged) before the next generation of young became fully independent of the pouch. I have, as yet, found no definitive observational record of a family group consisting of independent young of more than one generation. Occasional references to the occurrence of more than four young thylacines together, where no size differences are recorded (Batchelor, 1991; *Hobart Mercury*, 26 September 1884; *Tasmanian Mail*, 27 September 1884) either reflect a persistence of the incorrect belief that female thylacines possessed six teats in their pouch and could successfully raise six young (Bischoff, 1832; Mudie, 1829; Widowson, 1829), or potentially represent non-violent encounters at territorial boundaries, or neutral watering sources, between two neighbouring family groups, possibly with a shared sibling in each adult pair. For example, J. W. Slebin, a road and shelter builder working at the mouth of the Hibbs River on the west coast in 1914, saw 'eight small tiger pups about four weeks old playing on the beach and at different times I have seen 2 and 3 full grown adults at the same place' (22/4/1937). From successful snaring activities there is one record of multi-generational young. One morning in June 1929 Adye Jordan, snaring on the Arthur River near West Takone, found three thylacines caught in his wallaby snares: 'a female caught by me had its last years cubs with her and [these] were also caught and she threw two mouse size young from her pouch into the mud' (Jordan, 1987, p. 62 – see also p. 33; and *Burnie Advocate*, 15 February 1986). The adult female died shortly after her discovery, the second generation cubs and one of the first generation cubs were already dead, but the other cub was still alive in the snare, and became the last thylacine sold to Melbourne Zoo.

Such a suggested common social structure, in which a stable family group inhabited a fixed home range, while young adults – and probably older, never-mated adults, or ones whose mates had died – led solitary existences, often travelling considerable distances before discovering a mate or suitable unoccu-pied territory in which to live temporarily, effectively defuses one area of vigor-ous debate in the twentieth-century thylacine literature, as to whether or not thylacines were territorial. It appears that both territorial and non-territorial behaviours were characteristic of the species, with the expression of territoriality being dependent upon age and the establishment of a pair bond. Young adults, having recently left the family group, and older unpaired adults appear respon-sible for descriptions of the species as wide-ranging. These nomadic, solitary individuals were the ones most likely to pass by European habitation. Percep-tions that thylacines retired to the mountains to breed (based on Gunn's letter, 12/11/1850, cited in Gould, 1851; and discussed by Dixon, 1989; Guiler, 1985; and Guiler and Godard, 1998) express the reality that the breeding pair were not usually travellers, but largely restricted to a home base and territory in which the family could live – desirably established at some considerable distance from encroaching European settlement (often, through necessity restricted to moun-tainous regions, given the concentration of European settlement around the coast and Midland Plains).

Great variation exists in estimates given for the size of a family group's home range. The smaller of these figures, a suggested one square mile [2.6 km^2] (Wainwright, questionnaire, 10/12/1970), probably represents an estimate of the immediate hunting range around a family camp, consisting of independent but lair-bound young. It has been suggested that, within their home range, the family of thylacines moved between different, semi-permanent campsites, because of their tendency to operate on prey species at above-sustainable levels (Mooney, 12/10/1990; Sharland, 1939), or due to seasonal factors, or the pres-ence and demands of any young. Estimates for the entire range of a family group vary from 10 miles^2 [25.9 km^2] (Cartledge, 17/10/1970), to 100 miles^2 [258.9 km^2] or more (Cooper, 14/10/1970; Ling, questionnaire, 1970; Un-signed, questionnaire, 1970; Willoughby, 15/10/1970). But these higher esti-mates are likely reflections of the extreme rarity of thylacine specimens in the twentieth century, rather than the species' desired social group distancing. On the basis of his detailed knowledge of the Van Diemen's Land Company records of thylacine hunts and captures, Guiler has suggested that, for the coastal and inland region of north-western Tasmania, the average home range of a pair of thylacines was 55 km^2; while his research on the most successful central plateau snarers in the government bounty records suggests a larger average home range of 88 km^2 (Guiler and Godard, 1998, pp. 137, 138). Such differences in esti-mated home range size, based primarily upon nineteenth-century sources, reflect the different climate, biological resources and productivity associated with these contrasting regions.

Some debate exists in the literature about the continued existence of the adult male within the family group, one suggestion being that the essential family group consisted of a territorial female plus young, with only occasional adult male contact, as he moved across a much broader home range, keeping in regular contact with up to three adult females (C. Lord, 1927; Mooney, 12/10/1990). Possibly bound up with this suggestion is an unstated concern that the hypothesised family group structure in thylacines appears atypical when compared with the known social group structures in other extant marsupi-carnivores. While at this distance a definitive answer to this suggestion may not be possible, nevertheless, a number of factors combine to make this suggestion less likely, or at least such expressed social behaviour less common, than the usual stable pair bond.

Parallel evolution, for example, of marsupial moles and placental moles, clearly illustrates that the evolutionary construction of an organism and its behaviour is an interactive product of information in the environment together with information in the genes (Oyama, 1985). In recognition of this, the parallel evolution of the marsupial wolf and placental wolf should come as no particular surprise, nor, in parallel with physical similarity, should surprise be expressed at a social similarity. In placental wolves, while multiple monogamous pairings or polygamy are known within the one extended family pack, the most usual expression of reproductive behaviour – both in the wild and in captivity – is of a single breeding pair within a nuclear family group or extended family pack (Jenks and Ginsburg, 1987). The incidence of monogamy within the Mammalia reaches its peak within the Canidae (the dog family) (Kleiman, 1977; Macdonald and Moehlman, 1982). It should come as no surprise that the com-monest group structure of the marsupial wolf reflects a similar family orienta-tion. It was, indeed, this very social group structure that made thylacines such effective, responsive and rewarding pets.

Despite the wholesale slaughter that had taken place on the species, in the recorded observations by experienced trappers and old-timers of thylacines hunting in the early twentieth century, the number of records for hunting in adult pairs far outweigh the number of reports of hunting thylacines being restricted to a single adult plus young. Of twentieth-century primary source references, supporting the existence of the family group as the typical thylacine social entity, only two old-timers suggest the family group consisted only of the adult female plus young (Jordan, 1987; Ling, questionnaire, 1970).

Anecdotal accounts of the strength of the pair bond, and the attachment between adults and young, may also be used to argue for the stability of a nuclear family group structure within thylacines. McGowan, superintendent of the Launceston City Park Zoo, accepted that if one member of a pair was killed and the body left where it had fallen, its mate would turn up and attempt to drag the carcass away (21/6/1909). Other accounts suggest that, when one member of a pair was captured alive, or killed, and the body or skin carried away, that the other member of the mated pair would tenaciously follow the body or remains for considerable distances (see Bailey, *Derwent Valley Gazette*, 7 May 1997, for an account of an incident in April 1920). In 1914, Dick Rowe and his brothers, having killed the adult female, brought her three cubs back home. Despite the retiring nature of the species, while the cubs were held in captivity back at the homestead, the 'adult male was observed several times in the vicinity at the time' (Rowe, 25/11/1951). A final indication of the strength and importance of the pair bond may be reflected in the reluctance of the species to readily breed in captivity.

With an elliptical pupil in the eye (Gunn, 1850x, 1852w), a phalanx of nocturnal behaviours are to be expected. Certainly, individual thylacines, hunting by stealth or surprise, tended to do so at night. But the requirements of co-ordinated pack hunting – pack in the sense of a small family group of up to six individuals (the adult pair and up to four three-quarters-grown young) – involve either visual or vocal communication, preferably both. A significant repertoire of diurnal behaviours was early noted for the species (Gunn, 1838w; Gould, 1851; Meredith, 1852; *Tasmanian*, 10 September 1887; *Launceston Examiner*, 31 January 1889), and the majority of late nineteenth-century and early twentieth-century observations of diurnal behaviour relate to hunting, usually taking place in the morning or late afternoon, often by multiple thylacines, indicative of the presence of the small family group (D. A. Bell, 1/9/1981; Cartledge, 30/9/1970, 17/10/1970; Hardacre, questionnaire, 1970; Jordan, 1987; Le Souëf and Burrell, 1926; L. Stevenson, 1/12/1972; Wainwright, questionnaire, 10/12/1970; Willoughby, 15/10/1970; Wilson, attrib. *ca* 1922z). Since most of the indigenous mammalian species upon which thylacines fed 'become active in the bush in the early evening . . . and some are still to be seen lazing in the sun and dozing well after sunrise' (Guiler and Godard, 1998, p. 23), significant diurnal hunting behaviours are to be expected. Anatomically, the relatively reduced size of the olfactory lobes in the thylacine's brain, when compared with the olfactory lobes in other marsupi-carnivores (Moeller, 1970), suggests that 'sight is more important than scent in seeking and hunting down prey' (Guiler and Godard, 1998, p. 48). Records of the behaviour of thylacines in zoological gardens certainly support their capacity for diurnal behaviour, and their ability to exist in a standard regime of display during the day (while denied access to their sheltered retreat space) and their use of such artificial dens when access was permitted at night. Finally, the existence of prominent striping down the thylacine's body suggests a significant evolutionary adaptation of the species, in physiology and behaviour, directed towards visual communication and signalling, obviously something not taking place in the dead of night.

It has been suggested that the striking transverse striping in the species was the result of the process of natural selection acting towards effective camouflage (Frauca, 1963; Gardner, June 1972; Grzimek, 1967; Mollison, 14/3/1952). At one level this has a certain superficial plausibility: at a distance such striping could merge with the narrow trunks and dappled lighting found in the understorey scrub and ferns present in sclerophyll woodland. The problem with this explanation is that, at close quarters, rather than disguising the animal, the stripes dramatically accentuate its position and presence. It is far more likely that the selective processes leading to significant striping in the species were designed for such close-up, direct communication.

In order for effective communication to occur, either between different individuals of the same species or between individuals of different species, both the signaller and the receiver need to benefit from it. There is no point in one animal communicating information if a response is unlikely, and there is no point in responding to another's communication unless it might be to the responder's advantage. The recognition of this important interactive relationship, in determining the focus of natural selection upon reliable communication, is referred to as the handicap principle (Zahavi and Zahavi, 1997), and provides the most likely explanation for the thylacine's striping.

> Communication between prey and predator makes sense whenever some of the prey can escape the predator and the latter has to target one of the laggards. Under these conditions, a mechanism for such communication will probably evolve. . . . it is testimony to the power of natural selection to evolve in prey behaviours that will help them escape, and in predators the ability to heed signals that will spare them fruitless chases. (Zahavi and Zahavi, 1997, p. 12)

Examples of such communication signals between predators and prey are to be found in the placental tiger's black and white ears, which make it easy to determine exactly what the tiger is looking at and thus signal its intentions, and the patterning of tigers, leopards and other spotted cats which, while they may be camouflaging at a distance, nevertheless, at closer range, make the predator stand out, frightening the prey and causing them to flee, allowing the predator to chase one of the slower prey individuals – whether due to age, illness, infirmity or inexperience – a pursuit requiring less energy expenditure and a far greater chance of success at capture, than if the choice was merely the closest prey specimen at the time of the predator's appearance (Zahavi and Zahavi, 1997).

The striping on the thylacine's body may thus be seen as of signal importance, primarily designed to announce the presence of a predator to a communal prey species, with the thylacines present able to pick out and pursue one of the weaker individuals in the prey's flight response resulting from the

42

43

Plate 3.1 Richter's lithograph of a male thylacine head. (from Gould, 1851)

deliberate appearance of the predator. The facial markings of the thylacine, including the white upper lip and the whitish area in contact with the large blackish-brown eye, also served to specifically show the orientation and direction of the gaze of an individual thylacine (see Plate 3.1).

Hunting practices

The tendency for one or two members of the small family group of hunting thylacines to first expose themselves to communally grouped prey and start them running, before singling out, chasing and bringing down a single item of prey, has been noted by a number of observers, with respect to predation upon domestic stock: 'They used to come on to the sheep runs at times and kill numbers of them. It would be the two old tigers & sometimes 3 half grown young ones that would be the family. They would get the sheep on a bedding bank [and] kill one . . .' (Wilson, *ca* 1922y). As recalled by the 85-year-old James Dunbabin, whose family caught thylacines for the Launceston City Park Zoo as well as the bounty: 'They would most likely start sheep running from a hill, and, singling out one animal in particular, chase it to lower ground and kill it' (cited in Sharland, 1971x, p. 48). G. Stevenson, whose family provided four thylacine specimens to Launceston City Park Zoo, in his observations on sheep predation, gave an account of the separation and selection of prey: 'I have known two tigers come nightly into a [sheep] paddock . . . and the tigers half circle round them two or three times, and not touch them'. However, on other occasions, having disturbed the sheep, a thylacine 'will trot along beside them, by so doing the strongest sheep will get to the front, then the tiger will cut out at one side and single [out] one . . . and soon catch it' (21/10/41). Stevenson believed thylacines preferentially chose the strongest and fittest prey for capture ('the leaders'). However, the idea that, habitually, thylacines pursued and separated the strongest rather than the weakest prey individuals could only be a possible construction for thylacine families preying upon fenced farming stock limited in their flight response. It was certainly not the experience of the old-timer and friend of Archie Wilson, who with respect to sheep predation commented, thylacines 'seemed to pick weaker ones' (Unsigned, questionnaire, 1970). In the wild, for wild prey species, where movement was only restricted by a common terrain, plus the physical and mental stamina of the quarry and the pursuing thylacines, a maximum return for the thylacine's hunting effort required an ability to identify, separate and chase one of the weaker individuals in any aggregation of prey.

44

The different styles of hunting practised by thylacines were as much related to the social and developmental status of the hungry thylacines as they were to the social status of their prey. Observations made on these different co-operative hunting styles have given rise in the literature to different perceptions on the speed of hunting thylacines. Unpaired adult thylacines, hunting alone, caught their prey mainly by stealth and surprise (Cartledge, 17/10/1970; L. R. Green, *ca* 1975), with a last-minute rush at high speed upon the surprised prey. Paired thylacines, after exposure and selection of a prey item from a group, also required top speed to deliberately run down prey. Accounts of hunting thylacines in these situations gave rise to the perception of the thylacine as a speedy, rapidly moving predator (for example, Cambrian, 1855y; Cooper, 14/10/1970; Hall, 1910; Pocock, 1926; Sharland, 1939, 1941; Wainwright, 10/12/1970, 1/10/1972).

When young thylacines were old enough, they accompanied their parents and took a part in the rapid pursuit and bringing down of selected items of prey. However, the expansion of the hunting base beyond just the two parents meant that alternative methods of hunting by the expanded family group were now possible. When the young were well advanced in development to the point where they could bring down full-sized prey on their own, and the position of the prey and the prevailing terrain made it possible, the most energy-efficient way of capturing socially grouped prey was not pursuit at rapid speed, but the careful driving of prey towards waiting family group members – a process aided by exposure and a more energy-efficient slower pursuit and driving of prey at a distance. Hence this description for the co-operative hunting of macropodid prey: 'The natural food supply of the tiger is kangaroo & wallaby, chiefly in the tea-tree scrubs, there 2 or 3 [tigers] will get together, one will crouch down beside a track, and the others will hunt the wallaby, and when one comes along the [waiting] tiger will pounce on it' (G. Stevenson, 21/10/1941). As recalled by H. Pearce, when macropodid prey was attacked in this way by a waiting thylacine: 'they . . . jump on it. Kangaroos are killed by standing on them and biting through the short rib into the body cavity and ripping the rib cage open' (cited in Guiler and Godard, 1998, p. 25). Observations of pursuing thylacines – driving prey at a distance and at a relatively leisurely pace – have produced the perception (much replicated in the literature for its conformity with placental chauvinism) that thylacines were not particularly rapidly moving predators (for example, Breton, 1835; Guiler, 1958x, 1980, 1985; Guiler and Godard, 1998; Gunn, 1852w; Hardacre, questionnaire, 1970; Lucas and Le Souëf, 1909; Rowe, 26/9/1980), and that, by extension, their only method of success was to wear down their prey through persistent following over lengthy periods of time (for example, Gossage, 1966; Le Souëf and Burrell, 1926; Ling, questionnaire 1970; Regan, 1936; Troughton, 1941, 1944, 1965; West [of Burnie], ca 1953).

45

This latter type of co-operative family group hunting was dependent upon the presence of mature young, and was obviously of less common occurrence, given a limited time interval before the adults bred once again, and the departure of first generation young from the family group to set out on their own. Reflecting this, in all his observations on hunting thylacines, Cooper (14/10/1970) estimated that 90 per cent of the time he saw hunting thylacines they were openly chasing prey at rapid speed, the other 10 per cent of occasions saw thylacines in pursuit of prey at a more relaxed and leisurely pace.

Observation of photographs of live and dead thylacines demonstrates that enormous variability existed in banding patterns, concerning the total number of stripes, how far down the body and thighs they extended, and whether they were straight or curved, tapering to a single point or ending in a fork. Moeller (1968, pp. 370–8) provides detailed dorsal photography and analysis of banding patterns in museum specimens. Obviously a secondary effect of striping as a signal between predator and prey, given the individuality present in striping patterns, was the ready ability of family group members to pick each other out over distance – a particularly vital ability during co-ordinated hunting activities. In such activities, visual recognition between individual family group members was also accompanied by specific vocalisations.

It has been suggested that both size and background colour were associated with thylacine habitat preference, and hence particular sub-populations of the species. Dick Rowe has suggested (Lane, 21/11/1996) that thylacines from high rainfall areas were of a darker background colour, and Nibbs

(9/4/1928) that thylacines from around the shore of the Great Lake were smaller than those found in the heights of the Western Tiers. Such suggestions, that rainforest specimens were darker than those living in the drier, more open central plateau or coast, and that a larger body mass was associated with living in colder, more extreme environments, are not unusual for any mammal, and are expressed, for example, in size and colouration in the populations of brushtail possum and eastern grey kangaroo on the Australian mainland and in Tasmania. As more detailed knowledge of the locality of collection for zoological gardens and museum specimens becomes available, such suggestions may be open to empirical investigation.

Such differences between sub-populations of thylacines in background colour (and hence possible level of contrast with stripes) may explain contradictory reports on striping in juveniles. Lord and Scott (1924, p. 264) noted: 'The young have more pronounced stripes and a distinct crest on the tail'. Cotton, on the other hand (interview, 1980), with reference to the cubs he reared by hand, considered that the contrast in markings on his cubs was much fainter than that usually present in adults.

While the tail crest may well have been more prominent in juveniles, it certainly persisted in a less marked degree, as longer blackish hairs on the dorsal and ventral surfaces at the end of the tail, in adults (Gould, 1851; Lydekker, 1894; Thomas, 1888; Waterhouse, 1841). L. Stevenson (1/12/1972) recalls that a three-quarters-grown captive thylacine 'would stick its bristles up and snarl' at the approach of a stranger. The possibility that the crest of hair at the end of the tail was used to signal emotional arousal in thylacines has also been suggested by Lewis (1977) on the basis of the frequent representation of a brushed distal portion of thylacine tails preserved in Aboriginal illustration.

Of scent and sex

It is possible that young adult females and males established a pair bond prior to the location and establishment of an home range and territory. Certainly for any newly chosen area, at the edge of the home ranges of already existing thylacine family groups, two thylacines would be better equipped to deal with any territorial dispute than one. However, it is also possible that the young adult female first located and established a potential home range and territory, and then made her choice in partner from competitive interactions taking place between unattached males in the vicinity when she was ready to commence breeding.

Apparent sexual behaviour was recorded by Robinson: 'on ascending a small hill, observed a bitch hyena, and a full grown pup . . . their tails stood erect . . . suddenly . . . close by . . . two other full grown cubs . . . ran about scenting the ground . . . At last they got scent of the old bitch and ran off . . .' (Robinson, 14/7/1832). Robinson, aware of the importance of the small family group in thylacine social behaviour, described these four individuals as representing a

47

Plate 3.2 Portion of trophy photograph, from Waratah, *ca* 1901. (Jo Calaby)

family group. What prompts an alternative suggestion is the recording of the elevated tails. The musculature to allow this to happen was certainly present in thylacine anatomy, and trophy photographs of killed thylacines suspended by the hind feet typically show the tail presented in elevated fashion (see Plate 3.2 and Plate 3.3). Additionally, while the majority of Aboriginal illustrations have the tail effectively placed horizontally, nevertheless there are numerous Aboriginal engravings known in which the thylacine's tail is unequivocally positioned vertically, elevated above the animal's back (Plate 2.1b). In other marsupi-carnivores, such as the Tasmanian devil (*Sarcophilus harrisii*) and tiger cat (*Dasyurus maculatus*), elevated tails are read as signs of arousal and excitement (Buchman and Guiler, 1977), often sexual, and in the early stages of courtship, females with elevated tails are followed through the bush by one or more interested males (Mooney, 12/10/1990).

The raised tail is suggestive of olfactory as well as visual communication. Certainly, at a non-sexual level, accompanying the visual identification of individuals through their banding pattern was olfactory individual identification. The naked skin between digits two and four, surrounding the pad on the forepaw of thylacines, was highly glandular (Pocock, 1926) resulting in an easily identified spoor. Meredith provides an account of the olfactory identification of both prey and family members:

> saw a brush kangaroo hop past . . . In ten minutes or so, up cantered a she-tiger, with her nose down, exactly on the track, evidently following the scent, and in another quarter of an hour her two cubs came by, also in the precise track. It was a very nice family arrangement. (Meredith, 1880, p. 66)

As well as individual identification, the thylacine was not short of the internal physiological apparatus for olfactory communication of a sexual nature. The 'rectum . . . is everywhere perforated with the minute orifices of Lieberkühn's glands' and the cloacal sphincter 'in both *Thylacinus* and *Cuscus* . . . forms a complete sheath for the anus and numerous glands' (Cunningham, 1882, pp. 160, 168). In the male, four large Cowper's glands entered the urethra, which also possessed prostatic ducts containing 'a brownish viscid fluid' (p. 165). (Unfortunately, the urinogenitals of Cunningham's female thylacine were mistakenly destroyed during dissection [p. 163].) Such intraspecific pheromonal communication may have been but poorly responded to in our own species.

Generally, thylacines and their excreta appear to have had little distinctive odour perceptible to the human olfactory sense, whether confronting thylacines in the wild (Cartledge, 17/10/1970; Cooper, 14/10/1970; Sawford, questionnaire, 1970; Slebin, 1937; Wainwright, questionnaire, 10/12/1970), around their lair (Ling, questionnaire, 1970), or in captivity (Reid, 27/2/1992; M. Turner, 25/10/1992). Thylacine faeces were, however, deposited in the vicinity of lairs containing young (O'Shea, 25/8/1981) and no doubt carried important information to those animals with the ability to detect it. Some

48

Plate 3.3 Trophy photograph of Mr C. Penny's kill, Waratah, January 1924.
(*Launceston Examiner*)

olfactors have described the thylacine as possessing a clean animal smell, like a horse (Wainwright, 10/12/1970), wallaby (Cooper, 14/10/1970), or dog (Cartledge, 17/10/1970). The last-named reveals an important issue. Just as people differ through their physiology, experience, discourse and culture in their evaluation of, and response towards, the smell of a domestic dog, similar variability is to be expected with humans in the bush. A few people have described the thylacine as possessing a strong, distinctive smell: 'the Tiger has an odour of its own . . . not particularly unpleasant but very strong . . . Unlike any other animal's odour . . . difficult to describe, other than say it's like some unknown herb' (Jordan, 1987, pp. 42, 77). Hardacre (questionnaire, 1970) described the animal as having 'a very rank smell' in the wild. Willoughby (4/10/1970, 15/10/1970) recalled the captive, three-quarters-grown male caught by his father in 1930 as having a strong smell. Cooper (14/10/1970) also referred to captive specimens smelling, with A. Turner (1991, p. 24) suggesting it was not just the captive specimen, but 'you couldn't put up with the smell of the yard when they had him in captivity'. While thylacine urine might have been particularly pungent to some people, note that Jordan (p. 35), who found the animal to have a strong, distinctive smell, did not relate this smell to its urine, and indeed suggested that a thylacine did not mark a territory by 'odourising his boundaries with nature's spray' (1987, p. 42).

Explanatory factors involved in reconciling these contrasting points of view need to encompass the different perceptual abilities of humans and the different cleaning regimes and standards between captive specimens on daily public display in zoological gardens, and those privately held. But the possibility also needs to be entertained that, as a result of elevated hormonal levels due to the stress of capture, newly caught thylacines may well have smelt differently, and the first single thylacine released into a new, caged environment, with no other thylacines in residence, may have been prompted to scent mark the cage, with urine, faeces or glandular secretions. One needs also to recognise that there may have been transient reproductive or age-related factors in thylacine pheromonal messages. For example, while the cloacal region of the young adult female killed in Plate 3.3 shows significant staining, typical of the brown waxy secretions of glandular scent organs produced by marsupials (Pocock, 1926), the adult pair killed in Plate 3.2 show little sign of cloacal staining.

Having mentioned tail behaviour, a brief consideration of other tail behaviours is appropriate. Common in the literature is the idea that the thylacine's tail was a fixed, straight, relatively immovable extension of the backbone, often expressed in the idea that if you held a thylacine by the tail it could not turn around and bite you. In reality, being held by the tail did not immobilise a thylacine (Smith, 1862), as a number of people found out to their cost. The attraction in the literature for an hypothesised fused and immovable thylacine backbone and tail is just a further expression of the myth of the primitive, poorly designed thylacine, beloved of placental chauvinists. Generations of zoologists and comparative psychologists, with fixed, primitive tails in mind, have looked at photographs of thylacines and seen just what they wanted to see. While lateral movement of the thylacine's tail was somewhat restricted, particularly in comparison with a standard, straight-tailed non-Spitz dog, in no sense was the positioning of the tail fixed. Photographs of thylacines in captivity often show the tail diverging sideways from the straight line of the backbone (see Plate 3.4; see also the background thylacine of Plate 3.5). Specific lateral tail movement during aggressive arousal was also recorded. After an aggressive approach to within 2 metres of a human, a thylacine stood its ground 'growling, and with his tail wagging backward and forward after the fashion of a cat on the point of catching a bird' (*Launceston Examiner*, 22 March 1899). Milligan (1853, p. 310) also noted: 'The Aborigines report that this animal is a most powerful swimmer; that in swimming he carries his tail extended, moving it as the dog often does'.

Plate 3.4 Upright thylacine, London Zoo, *ca* 1903. (Zoological Society of London)

In addition to sideways postural placement, significant vertical placement has been preserved in many Aboriginal illustrations and thylacine photographs. Standing thylacines held the tail almost vertically downwards, pointing towards or touching the ground (depending on the length of the tail and the digitigrade or plantigrade positioning of the hind feet). Relaxed and confident individuals exhibited a slight but significant upwards curve to the tail. The height at which the tail was carried was indicative of the level of individual arousal (see Plate 4.1). Disturbed thylacines straightened their tails. When such disturbance was interpreted as threatening to the equanimity of the individual or family group, aggressive intent was exhibited by slightly opening the mouth and hissing. As the level of aggressive intent increased, the jaws opened wider in the typical marsupi-carnivore threat-yawn, the tail was raised to a declination of around 45° and finally the head was raised as well. In zoos, while the tail was straightened

Plate 3.5 Thylacine pair, Washington Zoo, *ca* 1906. (photo: Smithsonian Institution Archives)

and raised to the horizontal for defecation, a straight and uncurved horizontal positioning of the tail also appeared typical of newly arrived specimens, particularly juveniles, unfamiliar with the caged environment, or the presence of non-family member thylacines or photographer. With such juveniles, the crest of hair at the tip of the tail was also raised (see Plate 3.6). It would be delightful to be able to read the level of confidence in the cub involved in a dominance display with its mother, captured in Plate 3.9. The mother's head is lowered in response to the cub's head placed upon her neck, and judging from the root, the cub's tail was positioned close to the horizontal. Unfortunately, the photograph does not capture whether the cub attempted this dominance display with curved-tailed confidence or straight-tailed hesitancy.

Like the dog, the thylacine used its tail to signal arousal and emotional state. Unremarkably, while the specifics of tail movements differed between these widely sundered species, in general, their tails were used to communicate the same basic social information.

Plate 3.6 Young thylacine, London Zoo, with raised tail crest, *ca* 1906. (Australian Picture Library)

Returning to reproductive behaviour, one unpublished archival record has been located that purports to describe intraspecific aggression between males in the presence of an adult female. The Archie Wilson manuscript collection at the Tasmaniana Library, Hobart, consists of three handwritten documents, two of them signed by Archie Wilson, the third unsigned, but written in the same hand on the same type of paper, dated to the early 1920s. The manuscript collection also includes an unsigned, roughly typed fragment of a short story on the thylacine, attributed to Archie Wilson by the archivists at the Library (herein referred to as 'Wilson, attrib. *ca* 1922z'). It is unlikely to have been written at the same time as the documents, and the use of the term 'kangaroo-rat' in the typed fragment rather than the less familiar 'wallaby-rat', as used in two of the handwritten notes (Wilson, *ca* 1922w,y), suggests the possibility that it was at least typed by, and may indeed have been written by another member of the family, possibly based on Archie's knowledge of the animal. The aggressive interaction is described thus:

In an open space . . . the . . . two [adult male] thylacines faced one another.
Deep rumbling noises broke from their throats . . . they waited then, with an
explosive screech, the two creatures rushed together. They snapped and clawed;
dodged and twisted. They fell and rolled over and over until they bounded
against a rock. They sprang apart. Once more they glared at one another.
The slight pause was broken by softly uttered har-aaing calls. The bushes on one
side of the clearing parted. A third thylacine appeared on the scene. She stood
still with only her head and shoulders showing. . . The female's calls acted like
a lash upon the rivals. Instantly they became a struggling mass: a yellowish,
chocolate-striped ball of animated fury. The watching thylacine moved a few
steps forward, then stood still again . . . whenever the rivals fury slackened, her
softly uttered cries redoubled their frenzy. (Wilson, attrib. *ca* 1922z)

Alongside this interesting account, of doubtful provenance, from the wild,
lie rare instances of severe intraspecific aggression in captivity. In July 1889 two
male thylacines (of unknown age) caged together in Launceston City Park Zoo
fought to the death (Trot, 1889). Aggressive behaviour was also recorded
between adolescent and early adult males from different families. In 1904, Bart,
living in north-eastern Tasmania:

had two Tigers he kept for a while . . . One was a half grown male . . . The other
was a three-parts grown male . . . Both of them were kept tied up separately, as,
whenever Mr. Bart brought them together, both went frantic; the older one
attempting to attack the other. (Bart, 6/8/1952)

Both these specimens were eventually sold to Launceston City Park Zoo.

One instance is known of significant aggression between the sexes in
adults. At Washington Zoo in late July or early August 1905, a 3-year-old female
thylacine was gradually introduced (on a day-time basis only) to a more recently
arrived adult male. A savage fight erupted at 11 a.m. on 6 October after two
months of contact, with the female requiring veterinary treatment for her
injuries. The female had 'half of one ear bitten off & a bad cut on her head'
(Blackburne, 7/10/1905). In retrospect, nocturnal introduction would have
been preferable to diurnal introduction, preferably in neutral territory. Alternatively, the introduction of the male to the female's cage rather than vice versa,
might have been more successful (as is the case with bandicoots – Joan Dixon,
personal communication).

While on the subject of ears, they were distinct: large, oval and carried
erect (Australian Museum, 1964; Harris, 1808) even when the owner was asleep
(Renshaw, 1938). The internal ear cavity could be closed (Pocock, 1926), an
obvious advantage during swimming, occasional aggressive social interactions
and the closing moments whilst bringing down prey. In addition to the
aggressive encounter at Washington Zoo, Cotton (interview 1980) recalled his
father finally killing an old, one-eared male thylacine who 'probably had his ear
pulled off or something when he was young'. This single male had been

tolerated 'for years on the place', as he left the stock alone, until in his almost toothless old age, he took to preying on lambs. Externally, the ear was capable of some directional movement – having noticed his first captive thylacine prick its ears at the cry of a native hen (*Tribonyx ventralis*), Mafeking Carter thereon fed both his captive thylacines largely on waterfowl (Carter, interview 22/10/1992).

Given the large numbers of thylacines held in captivity, the very rarity of recorded aggressive interactions suggests the social behaviour of the species was characterised by an early stability and permanence in social interactions between individuals. This is further supported by the analysis of thylacine skins preserved in museums, which show little evidence of significant or continual intraspecific aggression within the species. A number of preserved skins exhibit linear lesions on the feet and neck associated with snared capture, but further damage is largely unknown. McOrist, Kitchener and Obendorf (1993) examined the 1875 skin of a female thylacine with numerous puncture and tear wounds, with the possibility of intraspecific aggression in mind; but concluded, on the basis of measured distances between the canine teeth and the type of injuries, that the wounds were most likely caused by a dog during the specimen's capture.

One incident, capable of interpretation as aggression towards unrelated young, is suggested from captivity. Adult male carnivores are rarely to be trusted – either in the wild or in captivity – with young that are not their own. For this reason a newly acquired female thylacine with pouch young was caged separately at Hobart Zoo in 1924. But a ménage à trois was enforced at Adelaide Zoo in November 1898, when a newly arrived female with two dependent pouch young was added to an existing display of two thylacines, at least one of which was a male (see Plate 3.7). After commencing a semi-independent existence, the two cubs died suddenly within a week of each other on 26 January 1899 and 2 February 1899. The cause of death has not been recorded. Rapid and closely associated deaths such as this were not unknown at this time in Launceston and Melbourne Zoos, where an epidemic disease was present. However, no other evidence for the existence of the disease has been located in Adelaide stock, and none of the adults died at this time. For reasons also unrecorded, unusually, the skins of these two young were not deemed worthy of preservation, only the skeletons of these two specimens were saved by the South Australian Museum. There has to be at least a suspicion that these two young met their ends as a result of behavioural rather than physiological causes.

55

Breeding behaviour and early care of young

Some level of breeding in the species may have taken place at any time of the year. Records of specimens received by the Museum of Victoria, Queen Victoria Museum, Tasmanian Museum, Adelaide Zoo, Hobart Zoo, Launceston City Park Zoo, and Melbourne Zoo, as well as police station records establishing the original date of thylacine bounty claims, indicate that adult thylacines with dependent pouch young attached to a teat were obtained between the months of April and December. Given the marked variability in size of the young attached to a teat, it is suggested that the above matings probably took place between March and October. Such fluidity in breeding is found in other marsupi-carnivores. A number of marsupi-carnivores (belonging to *Sminthopsis*) breed throughout the year (Tyndale-Biscoe, 1973). In the Tasmanian devil, females with dependent pouch young have been recorded in the wild between March and October (Guiler, 1970, 1978, 1983) with a peak breeding season firmly concentrated in early autumn. In the thylacine, the peak breeding period generally extended from late winter to mid spring. This is based on a careful analysis of the government bounty record data, the increased incidence of museum and zoological garden specimens obtained with pouch young from September to November, and anecdotal evidence from selected trappers and old-timers (Hardacre, questionnaire, 1970; Unsigned, questionnaire, 1970). Such timing is not unusual within the marsupi-carnivores. Late winter to spring breeding maxima are typical for many small marsupi-carnivores.

Such a peak breeding season suggests the peak time for adolescent thylacine movement out of the family and migration across the countryside would commence in late autumn/early winter – but note that the actual departure times for all the first generation young could be spread over many months, and may not have been completed until shortly before the second generation of young became independent of the pouch. A late autumn/early winter peak for thylacine movement across the country is supported by the records of capture from the Buckland and Spring Bay Tiger and Eagle Extermination Association, from south-eastern Tasmania (see chapter 6), and a similar winter migratory peak has been suggested for the Van Diemen's Land Company property Woolnorth, in north-western Tasmania (Guiler and Godard, 1998, p. 137).

A detailed analysis of sources and comparative data on thylacine reproductive behaviour is given in chapter 9, focussing on the possibility that captive breeding could have saved the species from extinction. For the present, the following ontogeny is suggested principally from specimen records in museums and the raising of young families of thylacines in zoological gardens. After a gestation period of around a month, each young thylacine was born, crawled into the backwards-opening pouch and attached itself to one of the four teats. It is likely that inexperienced females, still familiarising themselves with the physical parameters and resources of a newly settled home range, as well as being in their first breeding season, probably carried only two attached young, increasing to three or four young during their peak breeding seasons, followed by a reduction in the number of attached young as the animal aged (Guiler and Godard, 1998, p. 19). In the marsupium they developed rapidly, releasing the teat at around three months of age, and making their first forays outside the pouch. Over the next month they became more independent of the pouch and teat, with healthy specimens in the wild probably leaving the pouch permanently around four months after birth.

During this month of semi-independent life – sometimes in the pouch; sometimes out and left for a short time in a lair – the deciduous molar teeth, one in each jaw, were shed (Flower, 1867, 1883) as the young cubs started eating a solid diet. Thylacine adults 'fed their pups as any breed of dogs' (Hardacre, questionnaire, 1970), including the adult female regurgitating food for her young. At a kill, the adult female 'will eat so much herself, and then tear off pieces and swallow them, and when she gets home she will retch them up and feed her young ones' (G. Stevenson, 21/10/1941). An Aboriginal rock painting, from Kakadu National Park, Northern Territory (see Plate 2.1c) possibly illustrates this behaviour; with a young cub, raised on its hind feet, muzzle to muzzle with an adult thylacine; a pose that, in placental wolves and domestic dogs, typically precedes regurgitant feeding.

The return of semi-independent young to the pouch was described in detail by Gunn:

> The present one, in giving suck to its young, used to lie down like a dog, the skin of the pouch thrown back so as to admit of the young ones getting easily at the teats. When alarmed, the young ones crawled in with their backs downwards, the mother assisting by lowering her hind quarters to facilitate their getting in; and by also placing her rump against the side of her cage to give the cubs a purchase with their hind legs against the cage, and thus push themselves in. They were so large when they left [doing] this, that when all in the pouch it hung down very low, and seemed almost a deformity. (Gunn, 1863, p. 104)

Another Aboriginal rock painting, from Arnhem Land, Northern Territory, depicts a female thylacine lying down with four back-downwards young (relative to their mother) with their heads in her pouch (see Plate 2.1d, e). An idea of the bodily distortion caused when four semi-independent young returned to the pouch can be gained from the 1898 photograph, taken in Adelaide Zoo, of a female thylacine carrying her litter of just two semi-independent young in her pouch (see Plate 3.7).

Suckling continued beyond the period of pouch dependence and the size limitations of the marsupium to contain the whole litter. In notes accompanying the exhibition, before the Royal Society, of two photographs of her thylacine family, Mary Roberts, the curator of Hobart Zoo, noted that her captive female thylacine 'had extremely strong maternal instincts', as reflected in her treatment of her three 'tame and playful' cubs, over an eight-month period, from their arrival as independent cubs around nine months of age in July 1909 (see Plate 3.8). The female 'nursed them until they were nearly as large as herself, although throughout that time they had also taken raw meat' (P. C. Mitchell, 1910). These two-thirds-grown cubs were around 17 months of age at the time Roberts wrote to Mitchell (see Plate 3.9 for the family photographed in February 1910 [Roberts, 25/2/1910]). It needs to be recognised, however, that such extended nursing may well have been a reflection of the sundering of the adult pair, and may not have been typical behaviour for entire thylacine families in the wild.

57

Plate 3.7 Female thylacine, with two pouched, semi-independent young, Adelaide Zoo, 1898. (photo: *South Australian Naturalist*)

Another aspect of maternal care, shown in captivity, was a tolerance of ill or physically damaged dependent cubs returning to the pouch, or persisting in pouch-dependent behaviour. Such an instance is illustrated in J. M. Gleeson's painting of the thylacine family that arrived at the National Zoological Park, Washington, in September 1902 (see reproductions in Claude, 1996; Guiler and Godard, 1998; or Moeller, 1997), with the sickly cub, that died ten days after its arrival, ensconced alone (back-downwards) within the marsupium while its healthy, fully independent, 5-month-old siblings tumble around outside.

A variety of lairs and resting places for the family as a whole, the cubs alone, or just individual thylacines have been described; including shelters in hollow logs and hollow trees, under protective bushes or in the shelter of tree ferns, between boulders in rocky outcrops, or in caves. A cave shelter is the type of location illustrated in the only known photograph of a thylacine lair (see Plate 3.10), taken by the Director of Melbourne Zoo, Dudley Le Souëf. It is a shallow cave in the Triassic sandstone beside the St Paul's River, from which a thylacine was captured for the zoo in December 1902. There is nesting material on the floor, much of which appears to be made of fern fronds similar to those of the specimen of *Polystichum proliferum* growing outside the lair. Such constructions

Plate 3.8 Thylacine family, Hobart Zoo, July 1909. (Tasmanian Museum)

of nests of grass, fern fronds and leaves in the wild have been commented upon by Bayley (22/8/1981), Cartledge (17/10/1970), Cotton (interview, 1980), Le Souëf and Burrell (1926), Mollison (14/3/1952) and Sharland (1962), and the carrying of hay in the mouth for the construction of nests in their sleeping quarters was also observed in captivity (*Washington Post*, 8 March 1903). Mollison provided a description of an established thylacine base, ideally situated for hunting:

Plate 3.9 Thylacine family, Hobart Zoo, February 1910. (Tasmanian Museum)

Coming to Fitzgerald one time, climbed on to a limestone cliff & found sapling chewed by tiger pups . . . cubs [had] bit bark & played around a sapling on the ledge . . . & along base of ledge were 5 moss-lined nests . . . belonging to tigers, facing North. Nests [were] on a ledge leading to 5 spurs down which ran trails to various parts of game country. (Mollison, interview 14/3/1952)

In captivity, young thylacines also played with objects, as G. Stevenson (September/October 1941) discovered: 'I . . . soon found that they were of a playful nature so I used to hang a piece of rope to the ceiling . . . [of their] cage about 5 feet high . . . they would get hold of it like a little pup and play with it for a long time'. But primarily, both in the wild (Cartledge, 17/10/1970; Cooper, 14/10/1970) and in captivity (L. Stevenson, interview 1/12/1972; *Washington Post*, 8 March 1903), the cubs tended towards rough and tumble play with each other.

Independent young remained in the lair for a further two to three months until old enough to learn to take on adult roles, and be more of an help than an hindrance during hunting. During this time the adults returned to the lair carrying significant amounts of food for the young in their mouths. Bailey (*Derwent Valley Gazette*, 28 September 1994; letter 29/9/1996) gives an account of predation upon geese in October 1907, with both the adult male and adult female involved in taking killed birds back to the two half-grown cubs at the lair.

Plate 3.10 Thylacine lair, St Paul's River, December 1902. (La Trobe Collection, State Library of Victoria)

While the recorded history of thylacine–human interactions from the nineteenth century often appears to be largely one reflecting ignorance and brutality, it needs to be recognised that there were also many sympathetic observers of thylacines in the wild, and carers of thylacines in captivity, who approached the species with interest and sensitivity. A number of individuals camped in the bush encouraged the interest and curiosity of thylacines, and deliberately left food and food scraps out to be scavenged by thylacines on a regular basis: Bayley (22/8/1981); Eaves (1989); Patman (27/11/1949); *Westerner* (7 October 1982). A significant example of this behaviour, relating to the feeding of young at the lair, is part of the family history recorded by Ruby Lorkin:

> Over one hundred years ago, during a depression, my father, the late James
> Price, was forced to leave his wife, & family, . . . & went prospecting down
> [the] Savage River, in the rugged [North] West Coast. . . . They lived on native
> animals . . .
>
> One evening, near dark, a striped Tasmanian Tiger, came creeping around
> near their tent, & ate the scraps of meat, bones etc from the scrap heap, where
> they threw out their waste food, etc. The men watched from inside the tent as
> she came every evening.
>
> After about a week, they noticed she was taking the food away in her mouth.
> They ventured outside, but could not get close to her, or tame her, they followed
> her, keeping a safe distance behind. She led them to a large fallen tree, &
> underneath this saw she had three young pups. She glared, & growled at them
> as she saw them coming near her puppies, so they decided not to approach any
> further, but she did not attempt to rush them, or bite them. . . . [She] continued
> to come, each night to the camp-site, for food for her young ones.
> (Lorkin, 20/8/1981)

During the months in which the independent cubs were left in the lair, it
is likely that, through playful and semi-playful interactions, they established
working and dominance relationships with their siblings. Overt aggression was
not apparent in the litter mates raised for three months by Dick Rowe (interview, 25/11/1951). 'Bill' Jackson Cotton (aka William Cotton jnr) first noticed
hierarchical behaviour in his hand-reared thylacine cubs, when he changed
from individually bottle-feeding them, to providing them with milk in a bowl:

> they used to sit around a bowl of milk like kittens, instead of like dogs tearing
> all in together, one we had . . . none of the others would dare come near, they
> would sit around in a circle, when he had finished he walked off and then
> another one would go. (Cotton, interview 1980)

These same cubs showed similar behaviour with solid food: 'If given a piece of
meat the pups would not fight over it, like dogs. The one that grabbed it was
allowed to eat it in peace. One squeaky snarl from this one, and the others
would retreat, sitting round him in a semi-circle, on their haunches' (Sharland,
1971y).

Cotton's cubs accepted him in his role of surrogate parent, and although
their movement was highly restricted, once released from their cage, like adult
thylacines taking their cubs to water (Bailey, *Derwent Valley Gazette*, 22 March
1995, 29 March 1995), Cotton 'used to take them down to a creek we had . . .
they used to love the water' (interview 1980). Thylacines were not afraid of
water: suggestions to the contrary and the idea that they were, at best, reluctant
swimmers (Martin Duncan, 1884; Sharland, 1939; W, 1857), principally stem
from inappropriate interpretations of Gunn's critical arguments against marine
predation. Individual thylacines would follow watercourses in search of lone
items of prey (L. Stevenson, 1/12/1972) and they swam readily across rivers and

62

streams in pursuit of prey or simply to reach the other side (Blackwell, 27/11/1951; Fleay, 1946x; A. S. Le Souëf and Burrell, 1926; J. C. Le Souëf, 1934; Ling, questionnaire 1970; Milligan, 1853; Oscar, 1882). However, if there was a handy fallen tree or other bridging structure, they were more likely to choose that path, rather than increase their energy loss through swimming (Le Fevre, 9/6/1938; Sharland, 1937, 1939, 1956; Spurling, 1943; Stanfield, 31/8/1981). Within the confines of a cage, however – and in marked contrast to her Tasmanian devils – Mary Roberts noted that her thylacines were not given to upsetting or splashing in their water container (19/11/1909).

Consistent with the concept of a basic family group structure, associated with co-operative hunting and an established home range or territory, lie the records for a variety of vocalisations expressed in different social contexts.

Vocal expression

Various verbal descriptions have been attached to thylacine vocalisations, and occasional inelegant attempts have been made to parsimoniously apply Occam's razor, and run as many of these as possible into the one description of a single vocalisation: a 'coughing, wheezing . . . creaking kind of cry something between a dogs bark and a cats call' (Anon, 1970). A fabulous tendency appeared to exist, in the minds of both naïve and experienced bush persons, that was likely to interpret any unexpected and unfamiliar shriek heard in the Tasmanian bush, as originating from the marsupial wolf. Sharland's suggestion (1971x, p. 37) that many of the howls and screams heard in the bush and attributed to the thylacine were the likely product of owls or the highly vocal Tasmanian devil seems eminently reasonable. A further problem is that the verbal descriptions that appear to be genuine records relate to a range of distinct types of vocalisation. In the absence of any sound recordings of the thylacine, finding orthogonal factors amongst the legitimate data is, like any correlational or factor analytic approach, a highly subjective process, requiring confirmation outside the process itself by the placement and measurement of these behaviours within an environmental context, that, in the case of the thylacine, can only be historically obtained, defined and delineated.

As a first attempt in dealing with historical records of thylacine communication, five different vocalisations, and their contexts, are suggested. Four of these vocalisations were recorded from thylacines in zoological gardens, and these captive records are reflected in similar types of vocalisations recorded from individuals in the wild or privately held pets; the fifth vocalisation is suggested solely from wild accounts.

1 *A coughing bark*, identified as a social location and identifying call with attractive properties, directed towards immediate family members and used in co-operative, intraspecific behaviour, usually associated with the closing stages of active hunting.

63

Written descriptions for thylacine vocalisations tend to be either onomato-
poeic or else described through the use of analogy. With respect to the coughing
bark, various authors have been confronted with the problem of how to express
the sound of a cough in written English. The coughing bark of the thylacine has
been independently described by both Dick Rowe (26/9/1980) and Alison Reid
(27/2/1992), the last curator at Hobart Zoo, as 'ah-ah-ah-ah'. With a cough-like
expression, while the top pitch of the call itself may be described as high, the
coughing nature of the call produces a rough, smothered, low-sounding edge
to the vocalisation, hence Dick Rowe's alternative description of the call on
another occasion as a 'low-throated yap' (Rowe, 25/11/1951). Walter C. and
R. J. Barwick of Penguin described thylacine vocalisations to Jean Welsh in 1952.
Walter Barwick successfully snared a female thylacine (Shaw, 1972), for which
he received a government bounty (payment approved on 31 May 1906). He
described its call as 'a coughing bark' (Welsh, 1952). Harry Wainwright, son of
George Wainwright, the last so-called 'tiger man' employed by the Van Diemen's
Land Company, who in his own right snared a thylacine in 1922 and sold it to
James Harrison, described the call as 'a little sharp bark' and 'mimicked this by
yelling "hop-hop-hop" at high pitch' (Wainwright, interview 1/10/1972).

64
The suggested context of the coughing bark was as a non-threatening
social call with attractive properties, expressed most commonly during co-
operative hunting in small family groups. R. C. Gunn was the first to record this:
'a male and female . . . with two or three half-grown pups . . . hunting by night,
their exquisite sense of smell enables them to . . . follow the track with untiring
perseverance, occasionally uttering a kind of low smothered bark' (1852w,
p. 245). The accounts of many experienced trappers and bushmen with
demonstrable knowledge of the thylacine confirm the hunting context. Dick
Rowe (25/11/1951) identified the coughing call as used during the 'trailing' of
prey, and Wainwright (10/12/1970) suggested the sharp bark occurred during
hunting. Ingram Cartledge described the call as 'a continuous sharp yapping
. . . when chasing game' (17/10/1970). There is a suggestion that the pitch of
the call varied with the age of the specimen, younger specimens calling at a
higher pitch: 'the mother Tiger and grown pups hunt with a system . . . [the]
mother working the high ridge and the pups working the gully . . . when they
are around in numbers they would be heard hunting for hours and into the
early hours of morning the yap and yelp could be faintly heard in the distance'
(Jordan, 1987, p. 23). An additional context for the call's attractive properties
lies in its suggested use by the female in attracting potential males and
encouraging their rivalry in her presence (Wilson, attrib. *ca* 1922z).

The coughing bark was also widely recorded by curators and zoological
garden staff. At Melbourne Zoo it was described as 'a wheezing bark' (W. H. D.
Le Souëf, 1907, p. 176) or 'cough' or 'coughing grunt' (J. C. Le Souëf, unpublished manuscripts, 1934; 1970). At Taronga Park Zoo, Sydney, it was described
as 'a coughing sound' (A. S. Le Souëf, 1926, p. 937) and 'it makes a series of
husky, coughing barks, the breath being indrawn with a wheeze' (A. S. Le Souëf
and Burrell, 1926, p. 318). The socially attractive nature of the call was also

demonstrated interspecifically, in its use, on more than one occasion, to attract the attention of the curator to a caging problem. Occasionally the staff at Hobart Zoo would leave for home in the afternoon without opening the door to the thylacine sleeping quarters. The thylacines:

> were quiet in the daytime, and then of a night if they were left shut out, then they'd start running around, very agitated. We'd be having tea and it would be getting dark and all of a sudden we'd hear this 'ah-ah-ah-ah-ah-ah' and father would say 'Oh, the men have forgotten to let the tigers in,' and so down he would go, and I went down with him . . . and they were all running around very distressed in their pen, very worried there, and he simply opened the door of their sleeping quarters: silence! (Reid, 27/2/1992)

2 *An audible aspiration*, serving as a social contact call with attractive properties, between family group members, whilst following a scent trail.

The context of this second vocalisation, presented alone while following a scent trail, is suggested from four different observations upon thylacines in the wild. P. Smith (1968, p. 55) gives an account of a female thylacine 'sniffing and sneezing along her mate's pug marks'; Cotton (1980) suggests 'they usually made . . . a snuffly sort of noise as they were following a trail, "snuffle, snuffle, snuffle," that's what they made'; Rowcroft (1843) referred to the 'snuffing' of a thylacine as it approached a flock of sheep; and Martin (11/1/1953) suggested thylacines would 'often follow a person, making a short "clopping" noise, like a pig'. Such vocalisation on a scent trail must have had a social context, in communication between nearby family group members, or possibly, when the trail being followed was laid by another thylacine, in communication with the leading thylacine itself. When the trail being followed was a single item of prey (either from an asocial species, or an already separated individual from its social group) then such communication was likely only amongst those family members bringing up the rear in the pursuit of prey. Once these following thylacines no longer needed the trail, but could both visually and aurally locate the prey, one would expect them to break into the coughing bark already described. As they approached this ideal situation, however, as a following family group member drew close to the leading hunter and the prey, they may well have combined the aspiration with the cough. Amongst coughing bark records four accounts (A. S. Le Souëf and Burrell, 1926; Stivens 1973; Troughton 1941; Welsh 1952) refer to the coughing bark being preceded by an audible, wheezing, in-drawn breath: R. J. Barwick described their cry as 'a long drawn out "shsh" followed by a sharp "coff-off"' (Welsh, 1952). Unsurprisingly, there are no records for this vocalisation in captivity.

3 *An undulating cry*, delivered at any intensity from a whimper to a scream, suggested to be a social contact call, of individual identification, with non-attractive properties.

It is hypothesised that this vocalisation represented a social identifier call, involving primarily a non-attractive statement of identity and position, and, potentially, sex, age, and territory held. A lone thylacine at London Zoo 'often utters a prolonged and very loud undulating cry' (Seth-Smith, 1926) and the undulating cry was also described as like 'the bleating of a lamb' (Breton, 1846, p. 125), and a 'whistling weening sound, something like the neigh of a stallion' (Doherty, 1972). As yet no records of this vocalisation have been discovered from two or more captive thylacines kept together in the one enclosure, hence the possibility of this vocalisation, when delivered at volume, being relevant only to a situation involving attempted communication between unrelated thylacines (or non-family group members) over distance; indicating the presence of an adult thylacine within a claimed territory, that, if necessary, was ready to be defended.

From the wild it has been described as a mournful cry, howling noise, or screech, and variously designated as representing either communication between individuals, most appropriately expressed as thylacines 'cooeeing to each other' (Eaves, 1989) in the bush; or alternatively as a hunting call. While a non-attractive call during co-operative hunting may have occasionally been a reality in some specific situations and stages and places of a hunt, it is suggested that most sources designating the undulating cry as a hunting call are probably in error, and that its usual context for delivery was a stationary one. Answering calls by thylacines from other sites may have been interpreted as rapid movement by the one thylacine, likely to be indicative of hunting behaviour; and some 'experienced' bush persons (as recalled by Doherty, 1972) remained committed to the belief that a single thylacine could 'throw its voice', apparently from one hill top to another, the purpose of this miraculous feat being interpreted as designed to confuse the faster prey, so that they would about-face, and run straight back to the pursuing thylacine!

Delivered at high volume, it represented a plateau of non-attractiveness, with strongly aggressive overtones. The undulating cry, described as a screech or scream, has been suggested in intraspecific fighting (Wilson, attrib. *ca* 1922z) and was certainly given during interspecific fighting between thylacines and dogs (Cotton 1980; Sharland, 1971x). Delivered at low volume, as a whine or whimper, and thus directed towards family group members (Guiler, 1958x), it likely represented non-attractiveness with a nadir of aggression. The only specific context available from the wild for the low-intensity expression of the undulating cry was an injured adult retreating from a human interaction in which it received a seriously damaged foreleg. Its 'whimpering' was translated to be a statement of being in pain, 'like a wounded dog', but, as there was a young 'tiger cub in the vicinity' (*Launceston Weekly Courier*, 17 January 1924) the situational context of the vocalisation was one in which a non-aggressive, individual identifying call with distinctly non-attractive properties, as a warning to 'keep away', would have been an appropriate one.

4 *A warning hiss*, indicative of significant asocial desires. Of low aggressive intent if delivered through a slightly opened mouth, of higher aggressive intent if coupled with a full marsupi-carnivore threat-yawn.

A warning hiss, indicative of the desire to be left alone, was frequently recorded from specimens in captivity. Most commonly it was expressed by newly caught, or recently transferred specimens, which after a time habituated to human presence and activity outside and around the cage. Described as hissing or spitting 'whssh', like a blue-tongued lizard (a large Australian skink, *Tiliqua nigrolutea*), the animal would take 'a deep breath and send the breath out through the teeth' (Rowe, 26/9/1980), the mouth only partially open, indicative of relatively low aggressive intent.

Referring to his caged, captive thylacines (the female having been held in captivity for over six months, the male captured only a month ago) Gunn noted: 'I have purposely kept their cage close to the side of a path where many of my servants pass daily, and where my children are in the habit of playing, and I find that beyond a hissing noise made by the male, they do not seem at all disturbed by any one going close to them' (1850w, p. 90). Specimens in Regent's Park Zoo, London, were also noted as hissing when approached (Aflalo, 1896; Renshaw, 1938). Although specimens in captivity eventually habituated to the presence of humans, as shown by Gunn's female, Rowe (26/9/1980) noted that it was not unusual for caged specimens to continue to respond with a hiss to the first visit of the day: 'they always make it when you go up to them in the morning'. Hissing was also recorded in captive specimens when the animals were physically disturbed during cage cleaning: 'they didn't make any sound at home, just a "haaa" ... when my brother got in ... with them ... and cleaned out' (D. Gould, 23/6/1994). 67

When significantly annoyed or threatened by a stranger, if given the freedom to move, the thylacine would face that person (straighten its tail) and express a hiss with jaws wide open (Rowe, 25/11/1951), in the characteristic threat-yawn, common to other marsupi-carnivores such as the Tasmanian devil *Sarcophilus harrisii* (Dixon, 1989). Such a warning preceded the biting of David Fleay photographing the last thylacine in its cage at Hobart Zoo in 1933 (Fleay, 1979). The yawn was also observed as a threat display in unexpected encounters with thylacines in the wild (G. Stevenson, 13/11/1941). Also from the wild, Jordan (1987, p. 33) refers to both uninjured animals caught in pit-fall traps as 'continuously hissing', in a situation where 'if carelessly handled one could expect to get bitten', and the same behaviour from recently snared or steel-trapped specimens that were not fatally injured.

5 *A warning short, low growl*, indicative of a willingness to physically engage with an issue, the level of aggression, when produced by lone individuals or adults, matched by the intensity of its repetition.

Another warning call, indicating a willingness to fight with another, and expressive of high aggressive intent when produced by a lone individual or adult, was the short, low growl, frequently recorded from encounters in the wild, but also known from captivity, and described as a grunt, growl, bark, snarl, or in Harris' (1808) much replicated description, a 'short guttural cry'. It was the most commonly expressed vocalisation at an unexpected encounter between thylacines and humans in the bush. A thylacine surprised as it stalked around a campsite 'gave vent to a low angry snarl' (Oscar, 1882). Repetitive growling or snarling, accompanied by a deliberate approach, was referred to by Baldcock (1/5/1953) and G. Stevenson (13/11/1941) and described in detail in the *Launceston Examiner* (22 March 1899): 'a large native tiger . . . growling very fiercely . . . came up a step at a time to within 6ft or 8ft. Then he stood growling'.

While it is easy to conceive of Europeans blundering through the bush with little interest or intent upon other quadrupeds, it is more difficult to explain why a thylacine should choose direct confrontation over quiet retreat or observation. Three records provide a context for situations in which a thylacine chose to display itself and vocalise before human intruders. In the winter of 1922 or 1923 Linda Frankcombe passed by 'a large cave' in which it was known that a family of adult and young 'Tigers lived while it was snowing . . . when I got level with the cave my Horse stopped at the sound of a growl & there on my right was a full grown Tiger showing its teeth & still growling, it was 2 to 3 feet from me . . . I was very lucky to have been so close to a tiger' (O'Shea, unpublished letter, 25/8/1981). On another occasion, a 'muffled bark' was associated with 'two native tigers [a male and female] . . . advancing rapidly' towards an approaching human. 'The female tiger was in the lead and was crawling forward. When about 6ft. away she sprang.' The female was killed, the male escaped with injury. 'After the encounter the bushman saw a tiger cub in the vicinity, and this may have been the cause of the attack by the adult animals' (*Launceston Weekly Courier*, 17 January 1924). The growl was also expressed at the close approach of James Price to a wild litter of cubs (Lorkin, 20/8/1981).

This vocalisation was also recorded in a situation of self-defence, rather than just in the above cases of defence of young. A thylacine at bay, fighting for its life against the simultaneous attack of a dog and its human companion wielding a knife and club, expressed itself in 'angry growling' (Oscar, 1882). The warning growl was also recorded in the wild, when a thylacine was disturbed during the killing or consumption of freshly caught prey. 'As a small boy camping in . . . September 1919' J. Cowburn disturbed a 'snarling and growling' thylacine 'as it tore apart a helpless wallaby caught in a springer snare' (Cowburn, 30/8/1981). G. Stevenson (21/10/1941) also records a warning 'growl' from a thylacine reluctant to leave the carcass of a sheep it had just killed. Similar behaviour was recorded from specimens in captivity. 'My mother well remembers many years ago [1900] when her brothers brought home a young tiger . . . & chained it in the corner of a room in the house for about a fortnight. They fed it on kangaroo & wallaby meat, & she says it used to growl & snarl if anyone went near it while it was eating' (Simmons, 31/8/1981). This specimen was sold to Launceston City Park Zoo.

Captive thylacines kept on dog chains on farms or in the bush frequently gave a short low growl (Le Fevre, interview 1976) or started 'grunting' (Semens, 24/10/1992) to warn of a distantly approaching stranger – an ability for which they were prized and held to be markedly superior to domestic dogs. These pet thylacines, while tolerant of sympathetic or non-interfering humans with whom they lived, would nevertheless growl aggressively at strangers in their immediate presence. Arthur Murray's friendly, three-quarters-grown female, that he eventually sold to Hobart Zoo, 'would "squark"... when a stranger came near it' (*Burnie Advocate*, 22 February 1974). L. Stevenson (1/12/1972) remembers the four live thylacines caught by his father and sold to the Launceston City Park Zoo between 1900 and 1906, which would 'snarl' at the presence of strangers as well as horses. G. Stevenson (September/October 1941) also noted that thylacines would arch their back and produce a savage snarl at the presence of strangers. Kathleen Griffiths (1980) recalled her siblings and herself, teasing a chained thylacine by poking it with sticks from a safe distance to torment it, and the 'nasty growl' it would respond with from the length of its chain.

Specimens at London Zoo were recorded as giving a 'grunt' (Renshaw, 1938) or 'a short guttural cry resembling a bark' (J. Gould, 1851; and W, 1855), when seriously alarmed, or disturbed. Harris' (1808, p. 175) mortally injured specimen 'from time to time uttered a short guttural cry' and Bruce Walker (20/8/1981) accompanying his surgeon father to James Harrison's menagerie in the late 1920s, remembers a thylacine with a broken foot 'snarling' in its cage prior to being chloroformed and splinted.

The only occasions known when the short growl was given without overt aggressive intent was in the play fighting of young thylacine cubs. Cotton (interview, 1980) who raised an orphaned family of cubs noted that 'they used to yap like little pups when they were youngsters'. On a previous occasion he had similarly suggested: 'Tiger puppies were very vocal, snarling and yapping very much like dog pups' (Sharland, 1971x).

This first attempt to describe and categorise thylacine vocalisations according to context combined historical records of vocalisations from captivity with those recorded in the wild. While most published records of thylacine calls are associated with specimens held in zoological gardens, nevertheless, a considerable body of data used has come from private records of thylacines kept as companion animals. It is time to consider both the instances and purposes behind the keeping of thylacines as pets.

Thylacine–human interactions

The similarity of thylacine social structure to that most commonly expressed in the Canidae – a single breeding pair within a nuclear family group or extended family pack – has already been commented upon. Derived from this social structure was a need for individual juvenile thylacines to specifically recognise, establish relationships with, and communicate amongst family group members. Once the animal developed to sexual maturity, its further social development needed to encompass the potential ability to enter into a stable relationship with another, unrelated individual, and to transfer its established social skills to this new individual and any resulting family members. It is these same, or at least very similar, social and developmental characteristics that make the placental wolf/domestic dog such an effective and rewarding companion to our own species. It should come as no surprise that, early on in European contact with the species, people started raising thylacines and keeping them in captivity. These thylacine–human relationships proved so rewarding that keeping thylacines in captivity was persisted with, amongst a small sub-population of Tasmanians, until the population dynamics of the species rendered it no longer possible.

It is not easy to locate nineteenth-century records of individually held captive thylacines, as pet-keeping was only rarely translated into publication. While it is still possible to interview people who have memories of pet thylacines held in the early twentieth century, nineteenth-century records of pet thylacines must be constructed from occasional published references, plus the serendipitous recovery of comment while reading archival collections of personal letters and diary entries of the time. Necessarily, the numbers of thylacines kept as pets considered here significantly under-estimates the actual numbers kept in the Tasmanian community at any time. But, starting from 1826, the first known occasion when thylacines were kept in captivity, in the following twenty-five years at least twenty-two thylacines are known to have been kept as companion animals in captivity.

The first person known to have kept thylacines was Edward Abbott of 'Russel's Falls', who in November 1826 raised four 'alive and healthy' thylacine cubs (*Hobart Town Gazette*, 2 December 1826). Robert Lawrence, while at Evandale, noted in his personal diary on 19 November 1829: 'Dined at Captn. Stewarts where I had an opportunity of examining the *Didelphus Cynocephalus* more nearly than hitherto'. George Marsden, who operated the first 'pet shop' in Tasmania from his livery stables in Hobart, had a live thylacine for sale in 1831 (*Hobart Town Courier*, 17 September 1831). William Breton, from his first contact with a captive thylacine, in which he noted that it 'can be tamed with . . . facility' (1835, p. 358) was enamoured of the species. He wrote of five other captive thylacines owned by three different individuals, and on this basis, defended its captive personality from assumptions about the species in the wild: 'It is said to be stupid and indolent; but this is a mistake' (1846, p. 125). Finally, fourteen years after meeting his first pet thylacine, he was able to obtain one himself and proudly brought it along to the Royal Society meeting of 4 August 1847 (Breton, 1847).

Meredith (1852) refers to two captive adult thylacines, as well as 'several instances in which young ones have been kept and reared up kindly' (1852, p. 265). One of these adults became the first of three thylacines kept in the private menagerie of the Tasmanian Governor, Eardley-Wilmot, in the 1840s (Guiler, 1985, p. 54). Meredith had little affection for the species, considering them 'truly untameable', a judgement formed at her initial close introduction to the species, a two-thirds-grown specimen shown to her by a shepherd:

He had the animal secured by a chain and collar, and when it was to be carried off, slipped a strong bag very adroitly over its head and shoulders, pushed the hind legs in, and fastened it. I pitied the unhappy beast most heartily, and would fain have begged more gentle useage for him; but I was compelled to acknowledge some coercion as necessary, as, when I softly stroked his back (after taking the precaution of engaging his teeth in the discussion of a piece of meat), I was in danger of having my hand snapped off. (Meredith, 1852, pp. 264–5)

Except for the most sycophantic breeds, any domestic dog worth its salt, when approached and touched by a complete stranger while eating, would produce exactly the same response. Meredith's description of this particular specimen as possessing an 'untameable ferocity and savageness' (p. 265) is merely a reflection of her foolish approach to a feeding carnivore in the first place.

While Meredith was not attracted towards the species, this was not true for Gunn, who during this time period, to 1851, kept three thylacines in captivity (1850w) – and more thereafter – noting on 9 July 1851, with regards to his new arrival that: 'My living Thylacine is becoming tamer: it seems far from being a vicious animal at its worst, and the name Tiger or Hyaena gives a most unjust idea of its fierceness' (1852z, pp. 156–7).

The extent to which thylacines were not only tameable, but responsive and trainable companions in captivity, has been commented on by a number of individuals. For example:

I was shown a very fine specimen of the native tiger today . . . The animal has been recently caught by Mr. Percy Tucker, who has him chained up in the stable, just as one would chain up a dog. He tells me he is undecided whether to sell him or break him in to work sheep, though he thinks he might prove a little 'hard' for this class of work. The animal seems to take his captivity kindly enough and to be in blooming condition. (Tasman, 1884)

Nobody knows what became of this specimen, but whatever the malleability of captive individuals, Tucker was asking a lot to expect a wild-caught adult thylacine to transform itself into a sheepdog. There has to be a suspicion that his statement was directed at the gullibility of his audience.

Tasman was not the only one to remark upon the similarity of captive thylacines to dogs. Arthur Murray had a pet thylacine, eventually sold to Hobart Zoo on 21 July 1925 for £25 (Reserves Committee, 1925) that 'was extremely tame' (*Hobart Mercury*, 6 December 1983): 'the tiger, a three-quarter grown female, became quite affectionate. "It behaved just like a dog and it got very friendly" he recalls' (*Burnie Advocate*, 22 February 1974).

Rather than being generalists in their attitude towards humans *per se*, captive thylacines appeared highly selective in their interspecific relationships. Similar characteristics are shown by 'all strongly wolf-blooded dog breeds'

71

(Lorenz, 1952, p. 119) such as the chow, which early fix an immutable affection on one mistress or master, and often show, in both sexes, a strong monogamous fidelity as well (Lorenz, 1952, 1954; Simpson, 14/11/1999). It is likely once a thylacine established a relationship bond with a human, that, as in its intra-specific social relations, it was long-lasting and persistent. Suggestive of this were the markedly different responses identified in the behaviour of different captive thylacines to different curatorial staff in the National Zoological Park, Washington. While to most of the staff (and visitors) thylacines appeared dis-engaged, morose and uncommunicative, nevertheless 'with some of the keepers they are as tame as pet dogs' (*Washington Post*, 8 March 1903). Similar behaviour, of a friendly bond established with a single keeper by a new arrival, accom-panied by a tendency to 'snap somewhat promiscuously' at all others, was also recorded at Regent's Park Zoo, London (Cornish, 1917, p. 337).

Far more specific details are known of the treatment and behaviour of privately held pet thylacines in the twentieth century.

Captive thylacines were treated just like dogs: they wore collars and were walked on a lead. During the winter of 1915 the young Almer Saward caught 'a three-quarters grown Tasmanian Tiger' by its front foot in a wallaby snare. He muzzled it, tied its feet together and carried it home. 'Father and I took him down to the barn where we put a dog collar around his neck and tied him up with a trace chain to a beam of the shed' (Saward, 1990, p. 24). The thylacine was purchased by the animal dealer, James Harrison of Wynyard, for £10.0.3, who then sold it to Mary Roberts of Beaumaris Zoo, Hobart (Saward, 1990), for £17.10.0. on 23 October 1915 (Guiler, 1986). It arrived at the zoo with collar still attached (see Plate 3.11).

Thylacines readily adjusted to being walked on a leash. Early last century:

> Mr William Cotton [snr] came into the town of Swansea leading a Tasmanian Tiger, most people at the time were scared of the animal, and were amazed to see a person doing such a thing. . . . Cotton . . . had snares set about 4 miles [6.4 km] west of Swansea . . . and . . . one morning found he had caught a tiger . . . After some consideration he cut a short pole about five feet long, and to the end attached a piece of rope . . . and with a noose made on the end slipped it over the tigers neck, held him at bay, cut the snare, and set of [*sic*] to Swansea leading the tiger with him. He had great trouble to get the animal to travel, but after going a few hundred yards the animal started to act just like as if it was a dog, and followed along beside him for the rest of the way to Swansea with the least of trouble. (Graham, 1/9/1981)

Distance was no problem to the thylacine (or its owner). In the late 'teens or early 1920s the snarer Alex Nickols ('Old Black Alex') walked his pet thylacine on a leash the 10 kilometres along the main road between Fitzgerald and National Park (Larkins, 2/9/1981).

Newly caught captives soon gave up aggressive responses towards their primary care-giver, accepting the parameters of the power relationship newly entered into:

Plate 3.11 New arrival,
Hobart Zoo, with collar
attached, 1915.
(Tasmanian Museum)

> my grandfather caught . . . them and took them . . . down to Hobart into the
> zoo. . . . He took one home and he . . . had him tied up on a dog-chain and he
> used to feed him rabbits, in an old blacksmith's shop. And he walked in one
> night and thought he was back from him – struck a match before he was going
> to give him the rabbit, like – and he was standing up against the old tiger. He'd
> walked further over towards him than he'd thought, and the old tiger's standing
> there, wagging his tail, he said, looking up at him, waiting for him to give him
> the rabbit. (Miles, 1991, pp. 11–12)

But such ready acceptance was not extended to other strange, adult
humans.

The response by captive thylacines to the close approach of strangers was
usually aggressive and vocal. A further illustration of a vocal aggressive response
to a stranger ('AARRR!!'), accompanied by a verbal command from the owner to
his thylacine, which was appropriately responded to, involved a practical joke
played upon a visiting and naïve Alf Walters, sent unsuspectingly into some
stables, to collect a cow's hide:

> I opened the door. AARRR!! There's a damn tiger in there, chained up. I shut
> the door again. . . . [Jack] laughed like blazes . . . He said, 'He can't hurt you'.
> He said [to the tiger], 'sit down and behave, boy!' And . . . he opened the door
> and [we] walked straight in. (Walters, 1991, p. 27)

Exceptions to this aggressive response to strangers were noted, indicative of the sensitivity of captive thylacines to the social and developmental characteristics of its captor species, when the stranger was obviously a non-aggressive juvenile. As a child in the late 1920s, Irene Semmens made friends with one of the Johnson girls at Goulds Country. The Johnsons had a tame, pet thylacine, kept on a very long rope, that used, for preference, to lie on the verandah. (It was kept by the Johnsons as a watchdog, par excellence, and would give vocal warning of the approach of strangers well before the family dogs did so.) If, during the course of a game, a ball landed on or near the thylacine, Irene would just walk up to the animal, pick up the ball and continue playing. The thylacine made no aggressive response or vocalisation towards her approach at all. Irene treated it as if it were a well-trained domestic dog kept about the house – and indeed, that was how the animal behaved (Semmens, 24/10/1992).

To conclude this consideration of thylacines held privately in captivity is this account of a tame thylacine released back into the wild. Col Bailey has provided details of this captive in two different articles published in *Derwent Valley Gazette* (25 May 1994, 22 January 1997), based on information provided by the thylacine's owner. In the Depression of the late 1920s Reg Trigg built himself a bark hut and commenced snaring in the Great Western Tiers near the Walls of Jerusalem. In autumn, he found a young female thylacine in one of his snares. He trussed her up, carried her home in a hempen sack, nursed her back to health, and called her Lucy:

> able to feed Lucy by hand, the young tiger finally responded to his kindness by allowing the bushman to gently stroke her head, seeming to enjoy the experience immensely. As the winter began to take hold . . . Lucy became increasingly restless and . . . Reg became concerned about her unsettled mood. He strongly suspected her of having a male caller while [he was] absent from the camp and finally, out of deep affection and feeling for his precious Lucy he reluctantly released her back into the wild . . . Over two years were to pass, before one morning in early spring as the trapper was starting out to service his snare line . . . he came upon a tiger with two cubs patiently awaiting him beside one of his well-worn trails. He . . . stopped short of the trio, and for some minutes man and beast faced each other, . . . until at length Lucy turned, and together with her cubs walked slowly away into the bush. Reg Trigg continued to trap the area for many more years, but sadly it was to be the last time that he would lay eyes upon Lucy the young Thylacine, or for that matter any others of her kind; for the ominous clouds of extinction were already gathering, the tragic year of 1936 was fast approaching. (Bailey, *Derwent Valley Gazette*, 25 May 1994)

As a point for further reflection, 1936 not only marked the extinction of the species, it also closed forever the possibility of shared empathy and companionship between this species and our own.

Lessons from scrotal sacs

There are people who, without ever having read a single book about comparative psychology, evolution or palaeontology, just 'know' from their personal
experience, that the history of life on earth has naturally progressed towards the
human, producing a species whose intellectual capacities are superior to all
other species, either alive now, or in the past. Such endemic anti-intellectualism
is reflected in a cultural perception of the world entirely bound by the human,
and a measurement of the consequences of human actions expressed solely in
terms of their effects upon our own species. The cultural and educational
limitations involved in closing peoples' minds to an intellectual and aesthetic
appreciation of what it means to be just another twig on the branching bush
representing the 3.9 billion-year history of life on this planet are regrettable. But
even for these one-dimensional people, within Harris' (1808) scientific description of the species there was the indication why, for purely selfish, intraspecific
reasons, the thylacine should have been saved from extinction at all costs.

The very designation of a mammal as a marsupial involves reference to the
pouch, or marsupium, possessed by female marsupials for the suckling,
development and care of young after their birth. The existence of a pouch in 75
adult male marsupials, a scrotal pouch designed to sometimes hold and possibly
protect the scrotum, is unusual. Harris noted it in the marsupial wolf straight
away: 'scrotum pendulous, but partly enclosed in a small cavity or pouch in the
abdomen' (1808, p. 175).

The existence of a pouch in the male thylacine relates, not just to the
evolution of thylacines in particular, but to the evolution of marsupials in
general. If it was such a good thing for adult male thylacines to have a pouch to
protect the scrotum, why do not all adult male marsupials have pouches? In fact,
Owen (1834) theorised that it was likely to be a common male characteristic of
most marsupial species. Yet it is hardly common. Of the 320 or so marsupial
species currently known to science, only two, the marsupial wolf and the South
American water opossum *Chironectes minimus,* are known to have had scrotal
pouches in the adult male (Tyndale-Biscoe and Renfree, 1987). The scrotal
pouch is usually a transitory factor of male marsupial foetal development, with
occasional persistence into the dependent young stage in a handful of species
(Beddard, 1891). The adult male Tasmanian devil protects the scrotum in
ventrolateral folds of the abdominal skin, forming a 'pseudopouch' (Guiler and
Heddle, 1970, p. 882), but the male thylacine was unique amongst large,
terrestrial marsupials in possessing a fully functioning pouch to protect the
scrotum (Guiler, 1985; Dixon, 1989; Hickman, 1955).

At this point it might be tempting to fall into the explanatory arms
of placental chauvinism. If we consider marsupials as inferior, second-rate
mammals, it may come as no surprise that, despite what might be described as
'common sense' advantages in all male marsupials possessing a pouch to protect
the scrotum, nevertheless, only two of these under-evolved animals managed to

do so. Alternatively, since it appears that adult male marsupial pouches were not important structural adaptations, perhaps adult male marsupial wolf pouches represented structural non-adaptations, the simple persistence of the embryological remnants of female marsupial secondary sexual characteristics into male adulthood?

There is nothing unusual in such persistence – as the existence of male nipples in our own species, or for that matter, all male placental mammals, underscores. They exist, but in terms of function are structural non-adaptations. The same basic building plan occurs in all placental embryos, with precursory nipple tissues formed in the placental embryo before the development and differentiation of specific, primary sexual characteristics (S. J. Gould, 1991).

We live with structural non-adaptations that, if functional, would also be adaptive and useful. Why do not male placental mammals breast-feed their young? One would imagine that the selective pressures to encourage male breast-feeding would be particularly high amongst those groups of mammals in which the incidence of monogamy and the family group as a basic social group structure most commonly occur: namely the Canidae, the dogs and their relatives, in which over 50 per cent of known species appear primarily monogamous (Kleiman, 1977), and our own group, the primates, in which a third of all species appear to live in small family groups (Paddle, 1989, 1991). How useful it would be in our own species if male humans could breast-feed their young. But no placental species has, as yet, been able to overcome the embryonic suppression of fully functioning nipples and mammary glands in normal male development. Rather than scoff at two marsupial species with adult scrotal pouches, as placentals, we should rather be impressed that two marsupial species made the significant developmental leap to overcome the suppression of the pouch which normally occurs in the early development of male marsupials. It is tragic that one of these species no longer exists. While it was alive, the thylacine represented the best genetic and developmental model to our own species of a large, terrestrial mammal able to overcome the suppression of female secondary sexual characteristics during embryological and early development. Taking the human perspective as the sole reference point in a value system, we know ourselves less well with the loss of the thylacine; our capacity for understanding, and even altering, the suppressive factors in our own ontogenetic development has been markedly damaged with the loss of this species. From just a selfish, human perspective alone, this extinction should never have been allowed to happen.

The existence of the scrotal pouch in the male marsupial wolf also prompts a small excursion into the minds of scientists and the way in which they operate. In addition to Harris (1808), the scrotal pouch was also described in the nineteenth century by Owen (1841, 1842) and Wright (1888). Despite these verbal descriptions it was not until 1891 that an illustration of the scrotal pouch in the male marsupial wolf was published in the scientific literature by Beddard, based on his dissection of a specimen that died in Regent's Park Zoo, London.

Part of the stereotype involved in the dominant cultural construction of the word 'scientist' involves the perception of a white-coated male figure, in a

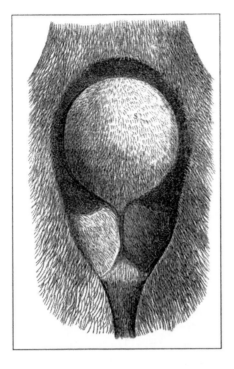

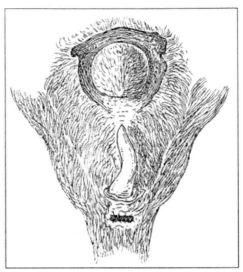

77

Plate 3.12 Scrotal sac illustrations: (a) Smit's illustration (from Beddard, 1891, p. 139); (b) Pocock's illustration (from Pocock, 1926, p. 1064).

laboratory, often accompanied by acolytes, and surrounded by equipment designed to aid objective measurement (S. J. Gould, 1989). While unrepresentative of the diversity of scientists as individuals, or their chosen fields of operation, nevertheless it is a powerful paradigm, and Frank Beddard, in the Zoological Society's prosectorium in February 1890 fits it perfectly. I do not doubt that, as a scientist, during the process of description and dissection of the pouch and brain of a male thylacine, that Beddard behaved as objectively as possible; that he made comments and notes of what he exposed and saw, and shared them with his assistant, the scientific illustrator, 'Mr. Smit', to aid in the production of a plate (see Plate 3.12a) to accompany the verbal description of the paper. It was in this process that a fundamental, but highly instructive, error intruded.

What Smit had available to him, as objective evidence, and what he actually illustrated appear as two different realities. Bound by a dominant placental discourse, Smit worked on the assumption that all male mammals are built in the same way, the way with which he was most familiar, with the penis positioned anteriorly (closer to the head) than the scrotum. While possibly for reasons of propriety the thylacine's bifid penis was not included in the illustration, it obviously belongs at the bottom of the picture. Smit, attempting a little permitted verisimilitude, indicated the positioning of the tail at the top of the plate. Unfortunately, not all male mammals are built the same way. In marsupials it is the scrotum which lies anteriorly, with the penis positioned closer to the tail.

The essential error propounded by Smit was his placement of the tail at the wrong end of the illustration. What is interesting and instructive about this error, and the power of placental chauvinism within the minds of biologists, is that nobody noticed: neither Beddard himself, nor the editorial committee of the *Proceedings of the Zoological Society of London*. Beddard had ample opportunity to correct this error in his future publications on the thylacine (1902, 1903, 1907, 1908) but he failed to do so. He was not alone. It was thirty-five years before this illustrative error was corrected in the scientific literature. In just the same way that hundreds of preconceptually constrained zoologists and naturalists have looked at photographs of thylacines and seen only fixed, straight tails, a whole generation of scientists appear to have stopped, looked at Smit's scrotal sac illustration, seen just what they expected to see, and passed on.

It was finally corrected in 1926 by Pocock, who, at the time, had no scientific superior when it came to the description and illustration of mammalian genitalia. Pocock suggested that 'the figure published by Beddard is difficult to reconcile with his description' and with Pocock's own investigation of the hair surrounding the pouch. Pocock concluded that 'The tail and the hair ... appear to have been added to it from the artist's imagination' (1926, p. 1063), and included a new illustration of his own of the thylacine scrotal pouch, this time complete with penis (see Plate 3.12b).

For all their stereotyped status as pillars of objectivity, when dealing with the supposedly obvious, or under the influence of 'common sense', scientists are as prone to protracted error and as fallible as any other group of human beings. It is for this, and similar reasons, that the previous chapter concentrated not so much on thylacines, but on the behaviour of humans – both non-professional and professional – who observed and interacted with them. An initial reflection on how the concepts of colonial inferiority and placental chauvinism, coupled at times with blatant racism, have played, and continue to play, a part in contemporary behavioural constructions on the species, paves the way for an acknowledgment, arising from a detailed consideration of the history of European interactions with the thylacine, of how all three of these characteristics ultimately contributed to the final scenario of extinction.

Chapter 4

A Predatory Entertainment

Endemic prey species

In Harris' original description of the thylacine he included the results of a dissection of one of the two male specimens that had been captured: 'The stomach contained the partly digested remains of a porcupine ant-eater, *Myrmecophaga* [*Tachyglossus*] *aculeata*' (Harris, 1808, p. 175). This is the only known recorded instance of a thylacine preying upon echidnas, but it has been frequently repeated in the literature. By extension, without needing any additional colonial data, Temminck (1824) and Cuvier (1827), from their European perspectives, suggested that thylacines also fed upon the other Tasmanian monotreme, another essentially asocial species, the platypus, *Ornithorhynchus anatinus*. No supporting Tasmanian observations of platypus predation have been discovered, but the platypus (of adult size up to 55 cm in length, Watts [1993]) may well have been in the mind of those who have simply described the thylacine as feeding on undefined 'small mammals'.

Amongst marsupials, the most commonly recorded prey came from the socially grouped species in the superfamily Macropodoidea. The largest Tasmanian marsupial, the Forester kangaroo or Eastern grey kangaroo (*Macropus giganteus*), exterminated from most of the Midland Plains and now restricted to north-eastern Tasmania, was an accepted item of prey. Captive thylacines were occasionally fed on kangaroos in private collections.

But by far the most commonly recorded marsupial species of prey were the wallabies. The principal species referred to is Bennett's wallaby or the red-necked wallaby (*Macropus rufogriseus*), although it is obvious that some records relate to predation on the smaller rufous wallaby or Tasmanian pademelon (*Thylogale billardierii*). Wallabies may also have been in the forefront of Carpenter's (1848) mind when he recorded the thylacine's diet as consisting of medium-sized mammals. Wallabies were commonly fed whole to captive thylacine specimens held privately, outside zoological gardens. Only one record has been found of a thylacine that refused to eat wallaby, a captive specimen in the late 1920s that refused both wallaby and rabbit carcasses, surviving on a diet of live fed birds (Carter, 22/10/1992). This illustrates a variability in behaviour expressed by individual members of the species, (a not unknown characteristic of other species in captivity), and also a preference for the stimuli of live prey that is understandable in restricted captive environments.

Two smaller macropodids, referred to as potoroos, rat-kangaroos (or, formerly, kangaroo-rats) are also said to have formed part of the thylacine's diet: the long-nosed potoroo (*Potorous tridactylus*) and the Tasmanian bettong (*Bettongia gaimardi*).

Of the five species of possum in Tasmania, only one, the common brushtail possum (*Trichosurus vulpecula fuliginosus*) frequently forages on the ground and was likely to form part of the thylacine's diet.

Various authors have also recorded the thylacine as preying upon the common wombat (*Vombatus ursinus*), often colloquially referred to as a 'badger', Carter (22/10/1992) suggesting they ate a considerable number of them in the wild. Fleming (1939) even suggested that thylacines were responsible for the absence of wombats along Tasmania's west coast in the middle to late 1930s! Loone (1928), from interviews with the old prospector Alexander Coplestone, recorded several experiences with thylacines in the St Leonards area in the 1870s and 1880s, including one mentioning wombat predation: 'Coplestone says it was pitiable to hear the poor badgers squealing when the tigers were killing them at night' (Loone, 1928, p. 105). Weindorfer and Francis (1920) considered the thylacine the only likely predator of a full-grown wombat. In contrast to the above, two nineteenth-century authors questioned the idea that the thylacine ate wombats. In the notes accompanying a female thylacine that had been caught in the St Patrick's River area and was sent to the Zoological Society of London, Gunn noted: 'I sent a trustworthy person up for her . . . He . . . informed me that the Thylacine will not eat the *Wombat*, an animal exceedingly abundant on the St. Patrick's River, and with which they attempted to feed it during the month it was there' (Gunn, 1850w, p. 90). While Gunn makes it clear he was commenting on the behaviour of a single specimen in captivity, it did not stop an exaggeration and expansion of his comments into a general statement in a popular European publication, that in the wild 'nothing will induce it to prey on the Wombat' (W, 1855, p. 248). Unlike some other general expansions and exaggerations about the thylacine's feeding behaviour, this was one that was not taken up by later authors.

Two species of bandicoot are found in Tasmania, both essentially asocial species, the southern brown bandicoot (*Isoodon obesulus*) and the eastern barred bandicoot (*Perameles gunnii*), and both species are likely to have represented significant items in the diet of single, unpaired thylacines. In addition to European accounts, the hunting of bandicoots has been recorded in a Tasmanian Aboriginal legend about the thylacine (J. Cotton, 1979).

80

Predation upon its fellow marsupi-carnivore species – 'le kanguro Dasyure' – was suggested in European publication by Temminck (1824, p. 65), but no Tasmanian records of marsupi-carnivore predation have been located that are contemporaneous with the thylacine's existence. Marsupi-carnivore predation has been suggested by three sources since the thylacine's extinction: Bryden, Director of the Tasmanian Museum, suggested that it fed on tiger cats (*Dasyurus maculatus*) and native cats (*D. viverrinus*) (28/8/1959); Leahy, acknowledging Bryden's input, noted it fed on native cats (1960, p. 17) and P. Smith (1968) suggested that it fed on 'marsupial rats', presumably referring to the larger specimens of dusky antechinus (*Antechinus swainsonii*) and swamp antechinus (*A. minimus*) rather than the smaller-sized white-footed dunnart (*Sminthopsis leucopus*). Predation on native cats, however, was vigorously denied by Isobel Lowther (*Launceston Examiner*, 8 May 1958).

So far as the indigenous, terrestrial placental mammals are concerned, W. H. D. Le Souëf (1907) suggested that thylacines fed on bush or native rats. Certainly the water rat (*Hydromys chrysogaster*) was common and large enough (with a weight up to 1.3 kg [Olsen, 1983]) to be a significant and deliberately sought-after addition to a thylacine's diet. A. S. Le Souëf and Burrell (1926) did not specify the indigenous nature, but recorded the thylacine as feeding on 'small rodents', and they may well have had some of the smaller native species in mind, in addition to the introduced black and brown rats (*Rattus rattus* and *R. norvegicus*).

Birds, in general, are regularly considered to have been common prey. Amongst recent native birds, the largest and potentially most significant prey for the thylacine may originally have been the Tasmanian emu (*Dromaius diemenensis*) that was driven to extinction by the mid-nineteenth century (Crawford, 1952w). While the European invasion of Tasmania rapidly exterminated the emu from the Midland Plains, as a source of food for the expanding settlements, it was never common, despite an abundant habitat, and was obviously much reduced in numbers and heading for extinction in response to Aboriginal predation. While parallels have been noted in the literature between the impending or actual extinction of the thylacine and the prior extinction of the Tasmanian emu – Button (1909), C. E. Lord (21/8/1928) and Stivens (1973) – an emphasis on the important inter-relatedness of the two events, with the emu considered a significant item of prey for a thylacine family, has not been made previously. But the destruction of this large prey species needs to be included in any discussion on the long-term factors potentially involved in the extinction of the thylacine.

The most commonly referred to avian items of prey were the native water birds. Predation on ducks such as the black duck and teals (*Anas* sp.) was recorded from the wild, while the Turners (25/10/1992) fed thylacines on ducks in captivity. Other water birds, such as bald coots (*Fulica atra*), Tasmanian native hens (*Tribonyx ventralis*), red-heads (*Porphyrio porphyrio*), herons (*Ardea* sp.) and adult swans, young cygnets and their eggs (*Cygnus atratus*), were also acceptable items of prey.

81

Specific predation on lizards has been suggested in some secondary sources, on the basis of information provided by experienced bushmen, but no accounts have been discovered of predation on Tasmanian snakes. Two individuals have reported predation on amphibians. Adye Jordan, recalling an incident in his youth, suggests 'A desperately hungry old tiger . . . will take dying tadpoles from a semi-dried pool' (Jordan, 1987, p. 85). Additionally, Bill Cotton, who kept thylacine puppies as a child, in interview recalled that tigers:

> were good swimmers . . . they could swim alright. They could catch fish, mountain trout and frogs, [I've] seen them eating frogs. I used to take them [his pet thylacine puppies] down to a creek we had and . . . I never saw [the] little pups actually catch the fish but they used to chase them and try and catch them with their . . . paws. (W. J. Cotton, interview 1980, pp. 14–15)

It is worth noting that Cotton's is the only primary account of fish predation produced in the twentieth century (and that it is fresh-water, not marine).

An overview of the known and suggested native prey of the thylacine suggests that it was a generalist: individual thylacines feeding on a wide variety of small to medium-sized species, sometimes opportunistically, taking specimens that crossed their paths, at other times deliberately, by stalking and individual pursuit; while small family groups hunted larger, socially grouped prey.

Introduced prey species

Given the thylacine's suggested status as a generalist feeder, conflict between thylacines and European farmers with their introduced domestic stock was inevitable, provided thylacines were prepared to approach the expanding European settlements. As it has been claimed that thylacines consistently did this – to such an extent it was said that they almost destroyed the sheep industry – the recorded level of thylacine predation on other imported farm stock should be an important yardstick by which to measure the claims concerning thylacine predation on sheep.

While cattle farmers and the beef industry certainly provided a significant proportion of the dietary needs of thylacines kept in zoological gardens, outside captivity reports of the consumption of beef and dairy cattle are far more problematical. Only one nineteenth-century source has been located that seriously suggests the possibility that in extreme cases thylacines, 'if driven by hunger', may prey on cows (Krefft, 1871, p. 6). The possibility was considered in an early draft of the manuscript on *Tasmanian Zoology* by Gunn, when he suggested that thylacines have 'never been known to attack . . . Cattle . . . although it is probable young calves occasionally suffer' (Gunn, 1850x, p. 7), but this reference to calf predation was not included in the published text (1852w). Only one twentieth-century source suggesting cattle predation was published while the species was still extant. H. W. Stewart suggested in 1919 the recent discovery of a thylacine den containing the bones of a half-grown calf. It needs to be noted,

82

however, that Stewart's claim that thylacines preyed on cattle appeared in a politically motivated statement that imputed an ignorance, other-worldliness and lack of common sense in those scientists who, in 1919, were calling for the thylacine's immediate protection. Stewart's implication was that thylacines not only remained a present threat to the sheep industry but also represented a future potential threat to the cattle industry as well. Hard evidence that thylacines even occasionally preyed on calves, or were in any way or at any time attracted to the cattle industry, is extremely thin.

After beef, the most commonly recorded alternative food included in the diet of captive thylacines in zoological gardens was horseflesh. No specific instances of predation on horses in the wild have been recorded in the literature, although Krefft again (1871, p. 6) considered the thylacine 'a most ferocious and formidable animal, which will soon overpower even a . . . Horse'. For some people, however, such predation was a real possibility to be feared. Gossage recalls a story from her father, Charles Marriot, working with the Great Western Railway survey party in 1908: the surveyors 'kept a fire going all night' to prevent thylacines from coming 'too close to the camp where there were horses . . . that could become prey' (Gossage, 24/8/1981).

A correspondent to the *Burnie Advocate* (Reader, 1971) reported his grandfather's 'stories of the depredations of the "hyenas"', including the suggestion that 'they were especially fond of young pigs'. Fred Kinane, when in his seventies, was interviewed by Hodgkinson (1982) and also suggested that pigs were fair game to the thylacine. However, only one nineteenth-century record exists of predation on swine.[1] Guiler's detailed work on the station diaries of the Surrey Hills property of the Van Diemen's Land Company (1985, p. 109) has located a single incident in 1844 of a pig apparently killed by a thylacine. It would appear that, while thylacines may have occasionally killed and eaten domestic pigs, these were unusual, isolated events.

By far the most obvious parallels to the hypothesised depredations of thylacines on sheep should be records of their predations on goats, a similarly sized, behaved and imported domesticated species. Thylacines held captive by Monty Turner jnr's family were fed on goat (Turner interview, 25/10/1992), and Turner's father and grandfather would sometimes use the vocalisations of a separated doe and kid in the bush to attract a thylacine for live capture. Murray (1957) recalled sighting a thylacine in January 1906 north of Mount Bischoff that he believed was probably setting off to feed on wild goats, but gave no account of an actual goat kill. Olive Williams (10/9/1981) reported successful predation on 'young kids (goats)' at the Cape Sorell lighthouse in 1932. Possibly on a knowledge of the above, Andrews (1985) suggested significant predation on goats by thylacines, but he provided no references to support the claim. Again, it seems likely that an occasional thylacine took an occasional goat from an occasional farming property, but that this behaviour was in no way common.

In summary, there appears nothing in the recorded behaviour of thylacines towards other farm animals to prepare the ground for the idea that thylacines almost successfully destroyed the sheep farming industry.

83

Two other imported mammalian species require some mention as potential items of prey. Breton (1846, p. 126) noted that: 'A *Thylacinus* that I saw in possession of a gentleman had a puppy six weeks old thrown to it, which it immediately tore to pieces and devoured'. Dick Rowe claimed that 'a day-old litter of pups is an infallible bait for attracting tigers. They have been known to take litters from farms' (Rowe, 25/11/1951). However, while many accounts exist of thylacine and adult dog encounters and fights in the wild with varying results, no record has so far been discovered claiming that the victor proceeded to eat the vanquished.

The other imported mammalian prey was the ubiquitous pest, the rabbit. As rabbits multiplied and spread across Tasmania they became significant grazing competitors for sheep, but also a common prey species for the thylacine. Records exist of old-timers recalling from their youth incidents of thylacines hunting and killing rabbits: Wilf Batty (*Kentish Times*, June 1984); E. A. Bell (unpublished manuscript 1975); G. M. Collins (21/8/1981); Trevor Hall (*Derwent Valley Gazette*, 3 April 1996); Adye Jordan (1987, p. 86); S. Mitchell (25/8/1981); and Monty Turner snr (P. Smith, 1968).

The consumption of rabbits was also recorded in captivity. Renshaw
84 recalled the pair of thylacines that arrived at London Zoo in November 1884 as 'flourishing in excellent condition on a diet of rabbits' (1938, p. 49), and Moeller (1990) suggests rabbits were a significant dietary item in continental European zoos. L. D. Crawford (1952x) referred to a captive thylacine kept by the old bushman Whiteley as being fed principally on rabbits, and Miles (1991) suggests that his grandfather's captive specimens were also fed on rabbits, but Peter Carter when interviewed (22/10/1992) noted that his family's two captive thylacines refused to eat rabbit. Once again, a certain individuality appears in the feeding behaviour of captive specimens; in the wild, the growth and spread of the rabbit in Tasmania at least provided a potential small prey item for opportunistic consumption by the thylacine.

The introduction of bird species into Tasmania – poultry for farming and the acclimatisation of European and Indian songbirds to 'improve' the aesthetic nature of the Australian environment – provided additional potential prey species for the thylacine.

Since its extinction, the suggestion that the thylacine was a significant predator on domestic poultry has received frequent and considerable support in the literature (see Paddle, November 1997, for a list of forty-two sources produced since 1936). These records range from simple, outwardly acceptable statements of the commonness of poultry predation – including Crawford's suggestion (1952w) that one of the problems with keeping thylacines as pets was the tendency of captive specimens to eat their owner's poultry – to the outrageous and obviously incorrect: 'the world's only carnivorous marsupial was sentenced to death as a chicken thief' (Ziswiler, 1967, p. 24).

Only one modern comment has been uncovered that tempers the above popular suggestion of predation on poultry. Noel Thomas (1977) wrote to the *Hobart Mercury* about his early experiences with, and knowledge of, the thylacine, commenting: 'Tigers were never a predatory threat to . . . poultry'.

It is worth considering the history behind the ready scientific acceptance of the commonness of poultry predation, for any light it may cast upon the ready scientific acceptance of the commonness of predation against sheep.

Two things are of concern about this recent emphasis in the literature on poultry predation. First, there is very little support in the pre-extinction literature to suggest that thylacines were ever more than rare and occasional predators on poultry (only six original Tasmanian accounts of poultry predation have been uncovered in the nineteenth century). Secondly, given that there were already two species of medium-sized Tasmanian marsupi-carnivores – the tiger cat, or quoll (*Dasyurus maculatus*) and the Tasmanian devil (*Sarcophilus harrisii*) – that by their own size constraints specialised in the predation of medium-sized birds and mammals, it is hard to envisage a third, larger marsupi-carnivore surviving in Tasmania with a natural and consistent predatory response also deliberately directed to medium-sized birds. (It is worth noting that, to this day, it is argued that the tiger cat and devil still significantly affect poultry production in Tasmania.)

Two of the six nineteenth-century references to thylacine predation on poultry are definitely questionable, given an uncertainty about the species actually being referred to. An adult tiger cat and a young thylacine may be confused under less than ideal observational conditions, and in his diary Robinson noted an early misunderstanding of this kind (26/6/1834). Additionally, in the absence of a specimen's scientific name, confusion is possible since the word 'tiger' is used for both the tiger cat, or quoll (*Dasyurus maculatus*) and the Tasmanian tiger. Less commonly, but just as unhelpfully, while the thylacine was also referred to as the 'native' or 'striped hyena', the tiger cat was also sometimes called the 'spotted hyena' (*Launceston Examiner*, 14 March 1868). The ending of the newspaper articles referring to the 'capture of a native tiger' near Carrick on 11 April 1886 – 'Mr. Hopkins's hen-roost will be all the safer in future' (*Launceston Examiner*, 17 April 1886; replicated in the *Tasmanian*, 24 April 1886) – appears unequivocal. However, this statement is somewhat weakened as a positive indication of poultry predation, as the newspaper reports themselves use the words 'tiger cat' and 'Tasmanian tiger' interchangeably throughout the text as if they were synonymous and referred to just the one species. Similarly, the *Launceston Examiner* (10 July 1891) report of poultry predation is also questionable. Although identified as a 'native tiger', the small size recorded for this specimen suggests it was probably a large tiger cat (which would make sense of publicising its capture) rather than a juvenile thylacine.

Of the four remaining Tasmanian nineteenth-century references to poultry predation, the earliest is from Mudie in 1829, who notes that the '*dog-faced dasyuris (cynocephalus)* . . . commits depredations . . . sometimes . . . upon the poultry yards' (pp. 175–6). With respect to poultry predation and the construction of the behaviour of the species, there are no ecological arguments against the idea that the thylacine 'sometimes' preyed upon poultry in an opportunistic manner when their paths crossed.

The next author to comment on poultry predation was John Gould who recorded that the thylacine: 'commits sad havoc among the smaller quadrupeds

of the country, and among the poultry and other domestic animals of the settler
. . . The destruction it deals around has, as a matter of course, called forth the
enmity of the settler' (Gould, 1851). It needs to be noted that the representation
of the thylacine as causing 'sad havoc' is a description of its predatory behaviour
in general, not a descriptor specifically attached to poultry predation.

Without any pretence of possessing additional knowledge of wild thylacines
W (1855), writing in the popular, London-based magazine *Excelsior: Helps to
Progress in Religion, Science and Literature*, repeated Gould's ideas: 'it commits
great havoc among the smaller quadrupeds of the country; and to the settler it
is a great object of dread, as his poultry and other domestic animals are never
safe from its attacks' (W, 1855). Note, however, that W did not simply repeat
Gould's point of view. There are subtle changes to Gould's description, seen in
the upgrading in the epithet for predation from 'sad' to 'great havoc', and the
now implied emotional response of the settler, changing Gould's 'enmity' – a
quite reasonable response to an occasional poultry predator – into what is now
referred to as 'dread'. Unlike Gould, who wrote and published for the scientific
and intellectual community, the unidentified 'W' wrote the first popular
account published on the thylacine – four pages in length, accompanied by a
full-page illustration. *Excelsior* was available to a wide British readership, with
copies duly exported to the Australian colonies. While the popular audience
may well have encouraged an increased emotionality of writing style, its
dramatic effect by contributing to popular misconceptions about the thylacine
was unfortunate.

The fifth Tasmanian-based account of poultry predation comes from
Willoughby's (1886) book on his travels around Australia. In the chapter on
'Some specimens of Australian fauna and flora' he considered Tasmania's
marsupi-carnivores, emphasising the Tasmanian devil, in both text and illustration. The thylacine received lesser mention in the text and was not
figured. Willoughby did not distinguish between the prey and predatory behaviour of both species, merely providing a sentence which combined their
predatory talents: 'Both "tiger" and "devil" . . . have been so hunted and trapped by the settlers, whose sheep and poultry they killed, as now to be very
scarce' (p. 182).

The last nineteenth-century Tasmanian-based reference is Aflalo's: 'The
Thylacine is a great despoiler of the . . . poultry yard' (1896, p. 67). However,
Aflalo's credibility as an observer of thylacines, while hunting in Tasmania or
visiting London Zoo, is open to serious question. Briefly, concern is expressed
over his inaccurate reporting on the thylacine collection of Regent's Park Zoo
(in terms of the numbers of specimens and the dates of their display), his
unique and divergent views about thylacine physiology (in terms of the
placement and structure of the outer ears, and his denial of the existence of the
male scrotal pouch), and his fanciful behavioural constructions: Aflalo implied
that the thylacine was such an efficient arboreal predator as to be responsible
for the extermination of koalas from Tasmania! With so much of his writing

86

fictional rather than factual, Aflalo's support of significant poultry predation does the argument little good.

In summary, of the six different nineteenth-century Tasmanian-based sources referring to poultry predation, only two of them, Mudie (1829) and Gould (1851), stand as original, unchallenged records of thylacines that 'sometimes' preyed on poultry. The most reasonable conclusion from the nineteenth-century data available is that the thylacine was, at best, an occasional, opportunistic predator on domestic poultry; that it was never a consistent, habitual killer of fowls; and that it did not represent a significant danger to the establishment of domestic or commercial poultry production.[2]

From this conclusion, the question obviously arises: what caused the recent blossoming of the construction that the thylacine was a significant poultry predator? I suggest that three events, occurring between 1921 and 1930, were overly influential in creating this impression.

In 1921 Harry Burrell took a series of impressive photographs of a captive thylacine dismembering a fowl in James Harrison's private zoo. Harrison's cages emphasised natural environments (Plate 4.1), but the photographs, from their first publication in magazine and book formats (Burrell, 1921; A. S. Le Souëf and Burrell, 1926), have consistently been significantly cropped, emphasising the thylacine and fowl and disguising the cage environment. These photographs have been much replicated, with one or more accompanying most of the major publications on the thylacine since its extinction, accompanied by comment on the thylacine's supposed tendency to prey on poultry. The naturalness of the caged environment has even led some authors to exaggerate the significance of the photographs by suggesting that they were taken of a wild thylacine that came out of the bush to raid the fowls on a farm (Joines, 1983; Veitch, 1979).

The second event was the publication in 1926 of the first edition of *The Australian Encyclopædia*. Haswell's entry on the 'Tasmanian Wolf' describes it as 'a formidable enemy of poultry keepers' (1926, p. 541). Professor Haswell, from the University of Sydney, was familiar with the species, had donated a specimen to the Australian Museum, and had written about the thylacine on four previous occasions (Haswell, 1914; Parker and Haswell, 1897, 1910, 1921), without mentioning the existence of poultry predation. To what extent he was influenced to do so in 1926 by the recently published Burrell photographs is unknown. But the effects of the statement were always going to be great, given the popularity of the encyclopædia, whose two volumes made their way not only into reference and school libraries but into many homes as well.

The third event, originally a small report in the *Burnie Advocate* in May 1930, has achieved far greater prominence over time. It is an account of Wilf Batty's killing of a thylacine on his family's farm at Mawbanna. The newspaper report suggests that the thylacine was observed 'prowling about the yard, apparently after fowls' (*Burnie Advocate*, 14 May 1930). This kill and the newspaper report about it have been referred to in almost all significant post-extinction publications on the species, since it represents the last known,

87

Plate 4.1 A captive thylacine with a captive food source, Wynyard, 1921. (Norman Laird Collection, Archives Office of Tasmania)

A PREDATORY ENTERTAINMENT

authenticated kill of a thylacine in the wild. The observation that one thylacine attempted to prey on poultry in the Mawbanna area in 1930 is not questioned. By itself, this record of attempted poultry predation is a welcome twentieth-century addition to the few reliable reports from the nineteenth century. But for reasons unrelated to the poultry predation, its significance has been magnified over time. It has been referred to in publications so many times as to literally force the unrepresentative notion that the thylacine was a significant poultry predator on post-extinction constructions of the species.

The manner in which this concept of poultry predation has grown to be accepted in the literature serves as an important lesson as to how a rare preda-tory event against domestic stock, occurring on just a handful of occasions, can nevertheless, through the selective effects of repeated, popular publication, be magnified out of all proportion in scientific constructions of the species' behaviour. This needs to be borne in mind for the forthcoming analysis of sheep predation.

The consumption of other bird species imported into Tasmania has only been recorded from captivity. When interviewed, Monty Turner jnr recalled that his family's thylacines, destined for the Launceston City Park Zoo, were some-times fed on Indian mynas, starlings, sparrows and pigeons (25/10/1992). The capture and consumption of pigeons by thylacines in outdoor enclosures in European zoos was also recorded (Moeller, 1990; Sprent, 1972).

Sprent's record is an interesting one as it demonstrates how observations by scientists may be distorted by preconceived theoretical positions. Sprent dis-covered a vial in the London School of Hygiene and Tropical Medicine contain-ing four ascaridoid nematodes (Sprent, 1971x), having been passed in the faeces of the last captive thylacine held in Regent's Park Zoo (Sprent, 1972). Though none of the four was male, given his inability to obtain further specimens Sprent described the four he had and proposed a new genus and species for the nematode, calling it *Cotylascaris thylacini* (Sprent, 1971x). Sprent also noted its difference to the ascaridoids he had examined in other Australian marsupi-carnivores (Sprent, 1970), noting that this new species, in the structure of its lips and the position of its vulva, bore affinities with the genus *Lagochilascaris* recorded from the didelphid marsupials of South America. The suggestion of a separate origin for the thylacine from the other Australian marsupi-carnivores has been entertained frequently in the literature, the argument being that the thylacine had distinct affinities with the extinct South America marsupi-carnivore *Borhyaena*. Sprent was aware of these theoretical positions (Sprent, 1970, 1971w), and allowed them to influence his microscopical investigation and consequent suggestions about the origins of Australia's marsupials (Sprent, 1971x). Embar-rassingly, upon reanalysis, Sprent was forced to admit that his supposed new genus of nematode with South American affinities was none other than a female *Ascaridia columbae*, a common avian parasite carried by 'the many pigeons which frequent these [zoological] gardens' (Sprent, 1972, p. 332).

Aside from sheep, one other potential prey species was introduced by the Europeans and needs to be considered: the Europeans themselves.

Consumptive responses towards (and around) humans

Interactions between thylacines and humans had been taking place in Australia for about 60 000 years, to the decided detriment of thylacines. Within a very short time, like all competing carnivores, thylacines would have learnt, no matter how emu-like or kangaroo-like these new bi-pedal arrivals appeared, that to openly attack them in a group was to invite failure and the likely loss of family group members as well. However, picking off individual, stray, unwary humans was another matter. It is in this area of predation that respect for and fear of the species is evident in Aboriginal records. The Tasmanian Quaker missionary, James Backhouse, writing in January 1833, was the first European to report such incidents: 'The . . . Tiger . . . is the size of a large dog, has a wolf-like head, is striped across the back, and carries its young in a pouch. The animal is said sometimes to have carried off the children of the natives, when left alone by the fire' (Backhouse, 1843, p. 122).

Similar incidents are preserved in the oral traditions of thylacines (marrukurli) by the Adnyamathanha people of the Flinders Ranges, South Australia:

90

> In one Dreaming account, called 'Vapapa', they are presented as a danger to human life around the southern part of Lake Callabonna. In the story, a young boy gets killed and fairly thoroughly devoured by one.
>
> In another Dreaming story, 'Marrukurli', they come down the plains on the western side of the ranges, from Lyndhurst to Edeowie. They are ferocious, killing a child on their route southwards. (Tunbridge, 1991, p. 48)

As recalled by his granddaughter, the last Adnyamathanhan who claimed to have seen a live thylacine, the centenarian, Mount Serle Bob:

> warned adults to keep their children inside at night, because there was a danger of being taken by one. They were reportedly heard around the camp at night. The **marrukurli** had a special camp, and there was the danger that they would carry children away to it, and there kill them and eat them. (Tunbridge, 1991, p. 48)

One of the more common behavioural characteristics recorded for the thylacine was its tendency to follow humans as they walked in the bush. Various explanations are possible, ranging from curiosity to casual scavenging or deliberate predatory intentions. Records of the successful scavenging upon the bodies of dead Europeans exist in the literature. Following the wreck of the *Acacia* on the south-west coast in 1905, on arrival the rescue party noted that: 'The bodies had been gnawed by Tasmanian tigers, one of which was found dead near the corpses' (*Hobart Mercury*, 23 May 1905). While such scavenging by thylacines would have been less common than that shown by adult Tasmanian devils with their consistent consumption of carrion, nevertheless the fear of thylacines feasting on the bodies of dead Europeans persisted well into the twentieth century. In 1930 Sergeant E. J. Butler, stationed at Stanley, was called to retrieve the body of James Gaffney, killed in a tree felling accident:

> Owing to the lateness and the possibility of not finding the body in the dark,
> I intended at first to leave it till daylight, but . . . [as] there were a number of
> Tasmanian Tigers back in that part, I decided to go back that night and try and
> find the body. (Butler, 6/8/1934)

The records of aggression, attack and unsuccessful predation on Europeans obviously fed the emotional and mythical response of Tasmanians to the animal. While previous authors have attempted to discuss predatory and aggressive behaviour of thylacines by considering all accounts of attacks upon humans together, I have chosen to separate captive from wild instances, and distinguish situations where the interaction was initiated by the thylacine from those where contact was initiated by the human. Finally, a distinction between aggressive and non-aggressive encounters needs to be made. Warning and aggressive responses from wild thylacines were elicited in certain specific situations when thylacines desired to make their presence obvious. From an intraspecific perspective, predatory acts are non-aggressive and, for lone thylacines, unlikely to be preceded by any warning of intent. Records of thylacines attempting to or actually attacking humans are considered below under four different categories, only one of which genuinely relates to predation.

The first category of interaction to be considered was when a human initiated the contact in an attempt to kill or capture a thylacine in the wild. The idea that thylacines would not attack an adult human unless cornered, snared or attacked, was noted in the nineteenth century (Angas, 1862; Cambrian, 1855y; Silver, 1874; *Tasmanian*, 16 September 1871). But detailed contemporary accounts of such attacks play down the stupidity of the human, and play up the ferocity of the thylacine: *Launceston Examiner*, 14 March 1868; Oscar, 1882; Wayback, 1911.

The second type of interaction, also initiated by humans, consisted of unexpected or unwise approaches towards thylacines in captivity. None of these incidents was published at the time they occurred. From the earliest stages of captivity comes this record from G. Stevenson in the 1890s:

> I had to carry him home about 4 miles on horse back, and by some means I let
> the corn-sack he was in hang down a bit low, and he got his feet on the horse's
> rump and turned in the bag. I had on a flannel, shirt, cardigan, vest, coat, and
> a thick over coat, he bit through the lot, and drove his four tusks into my
> shoulder, the consequence was I had my arm in a sling for a few days.
> (G. Stevenson, 21/10/1941)

Paddy Hartnet lost the top of his thumb to his captive thylacine when he grabbed it by the tail:

they always say, the tiger's got a stiff back; he can't turn. And old Paddy . . . he said he was always led to believe that, but . . . 'Don't ever,' he said, 'believe that!' He said, 'They can turn their back.' He grabbed . . . [a thylacine] by the tail . . . and . . . [it] took half his thumb-nail and half the end of his thumb – took it clean off. Well, old Paddy died . . . with . . . only part of his thumb. The tiger took it clean off. They . . . say . . . the tiger's got a stiff back; he can't turn. But . . . old Paddy reckoned not to believe that. (Steers, 1991, p. 16)

Finally, David Fleay, ignoring a series of threat-yawns, got bitten on the buttock while filming the last thylacine in captivity in Hobart Zoo in 1933 (Fleay, 1956, 1979).

The third type of interaction consisted of an aggressive warning approach by a thylacine directed towards humans walking through the bush. A number of these incidents received contemporary publication (and republication): Backhouse, 1843; *Hobart Mercury*, 8 May 1872; Le Souëf and Burrell, 1926; *Launceston Examiner*, 9 May 1872, 22 March 1899; *Launceston Weekly Courier*, 17 January 1924; *Sydney Morning Herald*, 22 May 1872. Le Souëf and Burrell (1926) and *Launceston Weekly Courier* (17 January 1924) identify the presence of young cubs in the vicinity of the aggressive warning and attack, and such may well be the explanation for most of the above incidents. One other context may be suggested for aggressive reactions to humans in the bush; that associated with the discovery of a thylacine at a fresh kill. Aggressive vocalisations of thylacines disturbed at kills and reluctant to leave them have previously been noted (Cowburn, 30/8/1981; G. Stevenson, 21/10/1941), and Guiler provides an additional record (Le Fevre, 1953, quoted in Guiler, 1985, p. 127).

The fourth and final kind of interaction to be considered consists of unsuccessful, non-aggressive, apparently deliberate predatory attempts on Europeans carried out by thylacines.

The first published account of attempted predation on Europeans involved a child in 1830 when a thylacine 'boldly entered the cottage . . . and attempted to seize one of the little children' (H. W. Parker, 1833, p. 181). An 'old tiger . . . weak and apparently half-starved' attacked an adult working in the bush (Meredith, 1852, pp. 267–8), and without apparent warning a thylacine attached itself to a Mr Littlejohn walking through the bush in 1882 (*Launceston Daily Telegraph*, 4 August 1882).

The most consistently quoted attack on a person was the one carried out upon Miss P. Murray. Unfortunately, the original story has been replicated so many times, without referring to the original newspaper report, that confusion exists as to when the attack occurred, how old Miss Murray was at the time, what she was doing when attacked and the extent of her injuries. There are even inconsistencies in the reporting of her given name. It has been suggested that the original newspaper report appeared in, or around, 1900, in either the *Launceston Examiner* (H. Collins, 6/8/1952) or the *Scottsdale North-Eastern Advertiser* (Troughton, 1944), but it has not yet been located.

Constant repetition and embellishment in various reports has led to the following confused picture: Miss Patricia (H. Collins, 6/8/1952), Pricilla

(R. Brown, 1973), Priscilla (Grzimek, 1976) or Prucilla (Simpson, 1980) Murray; either as a young girl in the 1860s (H. Collins, 6/8/1952), young girl in 1900 (Salvadori, 1978) or mature woman in 1900 (Beresford and Bailey, 1981; Jordan, 1987; Troughton, 1944); was outside, in the winter (R. Brown, 1983; Pizzey, 1968; Salvadori, 1978; S. Smith, 1981), or in the summer (M. Smith, 1982; Troughton, 1944); either doing the laundry – by washing clothes at a bench (M. Smith, 1982; S. Smith, 1981), washing clothes in a river (Grzimek, 1976), or hanging clothes out to dry (Griffith, 1971) – or, alternatively, peeling potatoes (Jordan, 1987); either at Springfield, near Scottsdale (R. Brown, 1983; H. Collins, 6/8/1952; Jordan, 1987; M. Smith, 1982), in a suburb of Hobart (Griffith, 1971), or at a west coast creek (Guiler, 1985); when she was bitten, either on an arm (Beresford and Bailey, 1981; H. Collins, 6/8/1952; Guiler, 1985; Simpson, 1980; S. Smith, 1981), the right arm and left hand (R. Brown, 1973, 1983; Grzimek, 1967; Salvadori, 1978; M. Smith, 1982; Troughton, 1944), shoulder (Jordan, 1987), or leg (Griffith, 1971); by a one-eyed thylacine that was either toothless (as apparently was suggested in the original newspaper report: R. Brown, 1973; Jordan, 1987), or well-fanged (as suggested by the permanent scars left from several teeth: H. Collins, 6/8/1952; Griffith, 1971; Guiler, 1985; M. Smith, 1982; S. Smith, 1981; Troughton, 1944). With the thylacine attached to a part of her anatomy Miss Murray tried to protect herself by reaching for a garden hoe (R. Brown, 1973; Jordan, 1987; Simpson, 1980; Troughton, 1944), hose (R. Brown, 1983) or rake (Grzimek, 1976). In doing so she trod on the thylacine's tail and the thylacine let go. Threatened with the hoe/hose/rake the animal either successfully 'slunk away' (R. Brown, 1973; Simpson, 1980; Troughton, 1944), or ran or fled (Grzimek, 1976; Guiler, 1985, S. Smith, 1981), or unsuccessfully retreated, before being killed (H. Collins, 6/8/1952; Jordan, 1987).

93

From the above, it would appear not unreasonable to suggest that a Miss Murray did exist, who, at some stage in her life was bitten by a thylacine. Both Guiler (1960) and Sharland (1957) have complained about conflicting data and consistency problems in the reports about the thylacine. This is undeniable. What needs emphasis here, however, is that, while there is, at times, evidence of great confusion and contention in the reports, these have not, by and large, been engendered by unreliable or incompetent observations of nineteenth-century naturalists and scientists, as has been suggested by Frauca (1963, 1965); Joines (1983); Pearse (May 1976); Sharland (1966) and Yendall (1982). Rather, more often than not, confusion and contention have been introduced into the reports by twentieth-century authorities through a too casual approach to historical research skills and the interrogation of source material.

It can be concluded that occasional acts of predation were carried out against both Aborigines and Europeans, with a probable preference, associated with a greater degree of success, for the juvenile members of the species rather than adults. More usually, humans were left alone, particularly those with guns, as the east coast convict Henry Tingley wrote to his parents: 'I can go out hunting or shooting of kangaroo ... ducks or swans, tigers, tiger-cats or native cats; there is nothing that will hurt a man but a snake' (Tingley, 1835, quoted in

Nyman, 1976, p. 25). Some thylacines were obviously attracted to humans and often followed them in the bush, but recorded instances of predation are relatively rare.

One other source of food was obtained from humans, not through predation but by scavenging. Some contention surrounds this claim. Guiler has argued that:

> 'Thylacines eat only freshly killed food' appears time and time again in the literature and in the wild state this is undoubtedly correct. Trappers agree on this point, indeed it is one of the few on which there is total agreement. In captivity thylacines were known to eat all sorts of dead material, and this might be interpreted as indicating that they would do so in the wild but there is no evidence of any such habit. (Guiler, 1985, pp. 121–2)

It must be admitted, however, that thylacine trappers had a vested interest in attempting to make their work appear as difficult and dangerous as possible, thereby increasing both their personal, machismo status, and the likely level of remuneration provided by an employer, or other source of private bounty funds, for a successful thylacine kill.

94

With due respect to Guiler's argument, while thylacines no doubt preferred live food, many were not averse to consuming jointed or prepared carcasses as one of their last opportunistic acts in the wild. The first thylacine specimen scientifically described and illustrated 'was caught in a trap baited with kangaroo flesh' (Harris, 1808, p. 175), and the same type of bait was used in the 'Sketch of a Tyger Trap. intended for Mount Morriston. 1823' by Thomas Scott (see Plate 4.2). Contradicting Guiler's claim that 'there is no evidence for any such habit' (1985, p. 122), other nineteenth-century records also commonly refer to wild carrion consumption: *Agricultural Gazette* (1897); Backhouse (1843); Breton (1835); Cuvier (1827); Figuier (1870); Gunn (1838w, 1863); Henderson (1832); *Hobart Mercury* (23 August 1866); Mudie (1829); and Oscar (1882).

Contrasting claims that thylacines never ate carrion or returned to a kill only enter the published record with the bounty debates in the Tasmanian Parliament in the 1880s, part and parcel of the claim that individual thylacines would supposedly kill up to twenty sheep per night merely for amusement (see chapter 6). To be sure, in the wild the opportunity to return to a kill and consume carrion may well have been limited in any area supporting a significant population of scavenging Tasmanian devils (*Sarcophilus harrisii*), but given the opportunity of a carcass being available, and the knowledge that the species was successfully trapped and maintained in captivity, a surprising number of modern authors have still subscribed to this non-carrion consumption position, blithely unaware of its political generation.

It behoves any predator to examine new populations invading its environment. Consequently, it is not surprising that the thylacine was described as inquisitive about Europeans by a number of people known to have been familiar with the species. The frequent absence of aggressive warning and predatory

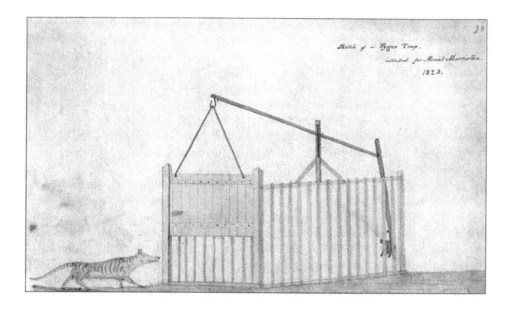

95

Plate 4.2 Thomas Scott's 'Sketch of a Tyger Trap intended for Mount Morriston. 1823.'
(Mitchell Library, State Library of New South Wales)

attack suggests that many instances of thylacines following Europeans as they travelled through the bush may have been motivated by 'curiosity' and/or the opportunity to scavenge and pick up odd and interesting-tasting food.

While not suggesting it was universal behaviour, 'some wolves, however, were known to prowl around bushmen's camps and would take meat scraps and bacon' (Frauca, 1965). 'They would keep circling around & around closing to get a bone or some scraps of food' (Patman, 27/11/1949). These thylacines were 'very fond of roast meat of any kind, but especially pork or bacon. Wombats or possums cooked until $1/2$ burnt, [were] also very attractive' (R. Martin, 11/1/1953). Other original observations of thylacines with a predilection for bacon have been recorded in unpublished letters by Dransfield (25/8/1981) and Tubb (27/8/1981). Original reports of the general scavenging of the remains of casseroles, uncleaned pans and scraps of meat and bone from around European camps have been provided by Wilf Batty (cited in the *Kentish Times*, June 1984); Branagan (18/6/1992); and in unpublished letters by Chaplin (6/11/1956); Gardner (June 1972) and Lorkin (20/8/1981); and Cuvier (1827) suggested similar behaviour around Aboriginal dwellings and encampments.

In addition to the previously mentioned deliberate feeding of a female thylacine with scraps by James Price and his companions (Lorkin, 20/8/1981), the intentional feeding of thylacines in the bush was also recalled by Angela James, whose husband Jack, deliberately 'threw out scraps for them' (*Westerner*, 7 October 1982) and Roy Delphin noted 'they'd have a bit of a lap at the bowls of milk we left near the camp' (Eaves, 1989).[3] George Ransley recalled that his father, who lived in Uxbridge in the Derwent Valley in his youth, 'said in those days the tigers used to come into the people's backyards looking for food' (Reid, 10/6/1992). Deliberate feeding of thylacines also occurred in the early days of these settlements: 'A woman at Marrawah told her husband she had given a "pretty dog" some scraps'. The 'pretty dog' on being tracked, proved to be a thylacine (H. Wainwright, 1/10/1972).

It is in just this scavenging context that placental wolves are likely to have made the first overtures to our own ancestors over 100 000 years ago – leading us both down the path towards domestification. Similar behaviour was present in the thylacine, and fortunately, as is obvious from the intimate life-style details preserved in some Aboriginal illustrations and in some European historical records, there were occasional individual humans, of sufficient sensitivity and intelligence, who responded appropriately to these overtures, developed a close relationship with, and were close observers of, the species.

Some mention has been made previously, under records of predation on wild and domestic species, of the consumption of these same species when offered alive, or more usually dead and jointed, to thylacines in captivity. The former acting curator of Hobart Zoo has indicated that they fed their thylacines at 4 p.m. on beef, never horsemeat, and preferentially on entire day-old calves (Reid, 27/2/1992), reckoning on 3 lb of meat per adult thylacine per day (Reserves Committee, 17/7/1935). In contrast, the staple diet of all the thylacines at Melbourne Zoo was horseflesh (Osborn, 1917, p. 232). 'Beside a ration of meat' – either horse, beef, goat or mutton – thylacines under the care of the Royal Zoological Society of New South Wales were 'given some small game occasionally', most probably rabbits (A. S. Le Souëf and Burrell, 1926, p. 319). While thylacines at London Zoo in the 1880s were mainly fed on rabbits (Renshaw, 1938), by early this century they were usually fed on mutton (Sclater, 1904). Moeller (1990) reports that thylacines in continental European and American zoos were fed horse and goat meat and rabbit bodies. Thylacines at New York Zoo were obviously highly valued specimens, their normal diet of beef and horseflesh (Crandall, 1964) occasionally being supplemented with chickens, milk and the bodies of still-born ruminants, such as elk and hog deer fawns (Ditmars, 29/12/1902, 22/5/1903, 31/12/1917, 21/10/1918). In short, the catholicity of diet in captive thylacines matched the broad range of species on which the thylacine was known to feed in the wild.

One potential item of prey has been deliberately excluded from this chapter. Having mentioned the consumption of mutton in zoological gardens, it is now time to examine the evidence for predation on sheep and lambs in the wild.

96

Notes

1 The newspaper report of a 'spotted hyena' attacking piglets (*Launceston Examiner*, 14 March 1868) is a reference to predation by the smaller-sized, spotted tiger cat (*Dasyurus maculatus*).

2 I have uncovered only one record of domestic poultry predation other than that on the chicken. Col Bailey has published details of two thylacine attacks upon geese in 1907 (*Derwent Valley Gazette*, 28 September 1994, letter 29/9/1996).

3 Roy Delphin's sister, Dorothy Gould, remembered giving their two pet thylacines milk to drink as well (23/6/1994).

Ovisceral Exploitation

98   The concept of the thylacine as a blood-sucking vampire (see chapter 2) is just one instance where popular mythology meets modern scientific construction. The concept of the thylacine as a significant killer of the domestic sheep, *Ovis aries*, may well be another. By the mid-nineteenth century the thylacine had been suggested as 'a great object of dread' (W, 1855, p. 247) for its popularly accepted incursions against sheep farming; to such an extent that the construction of the thylacine as a significant sheep-killer became cemented in popular mythology. Some evidence of its strength is seen in the passage of this construct back to Europe. Recording the folk songs of Ireland, O'Lochlainn (1946) noted that the following verse, referring to two different predators, was to be found in renditions of the popular ballad *The Poacher* up until the early years of the twentieth century:

> Our cots we fence with firing
> and slumber while we can,
> To keep the wolves and tigers from us
> in Van Diemen's Land.

The concept of the thylacine as a significant killer of sheep is also a prominent part of most scientific constructions of the species. However, conformity in opinion from popular and scientific sources should not be, in itself, a recommendation for the blind acceptance of a proposition. If there is good historical evidence for significant thylacine predation on sheep, then the genesis of popular opinion and its incorporation into scientific construction have been a reasonable and sensible process. If, however, there is little evidence for significant thylacine sheep predation, then the acceptance of this idea into scientific constructions of the animal represents as great an error as the scientific acceptance of its supposed vampire-like behaviour.

One starting point for a consideration of the relationship between thylacines and sheep may be the conclusions reached in the previous chapter about thylacine predation levels on other introduced European livestock. With only one specifically recorded incident of predation on goats (in 1932) and only one specifically recorded incident of predation on pigs (in 1844), it would appear that thylacines showed little interest in other introduced medium-sized mammalian species. Despite the frequency of post-extinction claims that thylacines commonly approached farms and were significant predators on poultry, an examination of the evidence for this behaviour suggests this construction is also largely unfounded. There appears to be very little evidence that thylacines became significant predators on any other introduced European livestock – apart from the occasional, casual, opportunistic predatory encounter. The suggestion, therefore, of a significant shift in the predatory behaviour of the species when it encountered sheep would appear, from the start, to be somewhat anomalous.

However, the notion that the thylacine was a significant predator on sheep has received general support. In this chapter I challenge this construction of the thylacine's behaviour. This is not to deny that the occasional thylacine, or family of thylacines, took the occasional sheep from the occasional farm. But the idea that the species as a whole was a threat to the sheep industry, and that thylacines in general were significant predators on sheep, is vigorously challenged. 99

To prevent any imputation of arguing against straw men on this issue, in constructing a bibliography of the thylacine (Paddle, November 1997) I came across in excess of 250 references to the thylacine as a significant predator on sheep. As a concept, sheep predation has captured the scientific debate over the thylacine's predatory behaviour – to the detriment of other predatory options, as well as to the detriment of the species.

With the notion of the Australian economy 'riding on the sheep's back' (Anderson, 1967), any suggestion of predation on this tasty, woolly ruminant was going to do a lot of no good to the species: 'Because of its habit of killing the holiest animal on the Australian scene, the sheep, the Thylacine . . . was very extensively hunted in Tasmania throughout the last century' (Sadlier, 1970, p. 114). From the moment that sheep were introduced into Van Diemen's Land – and some thirty-odd came with the original settlement to Risdon on the Derwent River in 1803 – it has been said that 'thylacines quickly learnt that sheep were easy prey' (Guiler, 1985, p. 15) and 'speedily acquired a predilection for the imported flocks of the settlers' (Cornish, 1917, p. 336).

To the thylacine, 'the sheep was just pennies from heaven' (Morrison, 1957) and they killed 'hand over fist' (L. Stevenson, 1/12/1972), at first just for food. But then they began 'to hunt sheep not merely for food, but also for sport' (C. Gould, letter, 21/10/1861, cited in Sharland, 1962, p. 5), 'working in packs, killing 20 sheep a night' (Thorne, 1970). At an individual level it was estimated that '100 sheep were destroyed' per adult thylacine per year (*Hobart Mercury*, 5 November 1886). The thylacine was 'very destructive to sheep' (Harrower, 26/4/1928) and became 'a serious sheep killer' (Pizzey, 1968) and 'a menace to sheep-rearing' (Harman, 1949) with its 'destructive ravages on the numerous flocks of sheep' (Owen, 1841, p. 258). In order to preserve the sheep industry, bounties on the thylacine were introduced. With a price on its head, the thylacine became locally extinct, a characteristic that was soon to spread throughout Tasmania as a whole.

The idea of sheep predation has not been universally supported by authors writing about the thylacine. After the thylacine's extinction a small number of modern authors, writing without first-hand experience of the living animal, have questioned the claim, without articulating the reasons for their concern. For instance, the thylacine was 'usually considered' (Dixon, 1989) or 'became generally known as a sheep killer' (Clucas, 1978). Slightly stronger criticism of the construction of sheep predation has been expressed through the use of such modifiers as 'supposed to be' (Frauca, 1963), 'reputed' (R. H. Green, 1973) and the more popular 'alleged' (Burton, Burton and Pearson, 1987; Guiler, 1973w; Sharland, 1941). In a shoot-the-messenger approach Small (1985) commented: 'From early Tasmanian settlement the tiger became more and more of a threat to farmers, often the result of sensational, ill-informed journalism'. The sources for such journalistic comment require investigation for their motivation and validity, a point that will be emphasised in the consideration of parliamentary debates on the thylacine in the 1880s (see next chapter).

100

Sceptics have found some support in the pre-extinction literature. During the second half of the nineteenth century there was a handful of authors who argued that, while thylacines did not appear to be attacking sheep at that time, for example, Meredith (1880, p. 66), nevertheless, they had been 'formerly' a problem to sheep farmers in the early days of settlement (Hull, 1871; *Launceston Examiner*, 22 March 1899; Silver, 1874).

Significant numbers of authors have described the thylacine not as an habitual killer of sheep but as merely an occasional predator. These include the initial, early nineteenth-century records of sheep predation, which contain no reference to the behaviour of the species as a whole; pre-extinction comment on sheep losses as occurring only rarely and in remote areas; and post-extinction comment involving an acceptance of the early recorded incidents, with a certain cynicism directed towards later claims. For example, Lloyd (1862, p. 76) reported that 'The native tiger occasionally makes an attack upon the neighbouring flocks', while Farmer (1887) argued that: 'There are comparatively few sheepowners troubled with those vermin'. These points of view continued into the twentieth century, the thylacine 'occasionally causes loss to sheep-owners' (Leach, 1929, p. 409).

Aside from these general arguments are recorded instances where thylacines and sheep lived together in one locality, with little or no recorded predation. Trooper Nibbs, posted to the Great Lake police station in the late 'teens and early 1920s, recorded that in his time in the district he personally saw four thylacines, none of which appeared interested in domestic stock as items of prey: 'I did not have any complaints from the Sheep Owners whilst at the Lake' (Nibbs, 9/4/1928). In the same locality, some fifteen or twenty years previously, Patman noted the rarity of sheep predation by thylacines, and that genuine 'anti-thylacine' feelings were only reasonable amongst the snarers:

the tiger was blamed for a lot of sheep & lambs killed by dogs. The Hunters at the Lakes did not like the tiger as he scared the kangaroos away from his snares. I . . . seen . . . many of thes[e] tigers . . . In 1900 when camped on Dunrobin Estate I seen as many as 8 in one summer & I only seen three sheep killed by the tigers. (Patman, 27/11/1949)

Blaming thylacines for kills by dogs has played no small part in constructions of the thylacine's behaviour, and in the destruction of the species.

There is a final category of author opposed to the idea that thylacines preyed on sheep, namely those who deny that sheep predation ever took place at all! Two types of comment are included here. First, there are those from the early nineteenth century that deny the existence of any native animal that represents a danger to animal husbandry. Secondly, there is a handful of recent records obtained from old-timers with established knowledge of the species, who deny that sheep predation ever took place. S. Mitchell (b. *ca* 1902) has commented: 'they done no harm to any stock that I can recall . . . I have heard they killed sheep but I had no reason to think that they did' (25/8/1981). Similarly, Albert Turner (b. *ca* 1923) suggested: 'They reckon they used to kill their sheep and that, but I'm a bit doubtful about that' (A. Turner, 1991, p. 25). No doubt at all was expressed by D. A. Bell (1/9/1981): 'tigers do not kill sheep'. Similarly, K. M. Crawford (2/9/1981) wrote: 'sheep . . . were never taken or attacked by the (hyina)'. In a letter published in the *Hobart Mercury*, Noel Thomas (1977) argued that 'the presence of stray dogs' was the most likely cause for stock losses attributed to the thylacine.

Thus, there were significant differences of opinion regarding the predatory behaviour of the thylacine. These differences are not a product of recent research but stretch back to the early nineteenth century. It might appear to be a reasonable approach to adopt a middle position and conclude that the thylacine was an occasional predator on sheep, as is known to have been the case with other introduced European livestock. But this would not explain the apparent need for bounties against the species, or the idea that the thylacine was a consistent, exploitative predator on the sheep industry. To understand the genesis of these conflicting positions it is necessary to re-examine in detail the historical relationship between thylacines and sheep.

1803–1829: All quiet on the thylacine front

Neither of the early published descriptions of the thylacine that originated in Van Diemen's Land made any reference to it as a potential sheep-killer (Paterson, 1805; Harris, 1808). In addition, while the personal diary of the Reverend Robert Knopwood refers to two encounters with 'tygers' neither entry makes any suggestion of the likelihood of the species becoming a problem to the new settlement (20/8/1803, 18/6/1805). In 1810, in a report written for the Home Office, Oxley directly referred to the benign nature of the native fauna as experienced at both the Derwent River and Port Dalrymple settlements:

both [settlements] are free from that destructive animal to Sheep, the Native Dog [dingo], the dread of the Stock Holders in New South Wales. The only Animal unknown on the Continent is the Hyaena Opossum [thylacine], but even here they are rarely seen . . . it flies at the approach of Man, and has not been known to do any Mischief.[1] (Oxley, 1810)

By 1819 the number of sheep in Tasmania had grown from the approximately thirty introduced in 1803 to the staggering total of 172 000, more than double the number of sheep present at that time in the founding Australian colony of New South Wales (Heazlewood, 1992). Only two records of thylacine predation on sheep have been found in the sixteen years during which this explosion in sheep numbers occurred, both attacks being close to the original Risdon Cove/Hobart settlements. The first, from April 1817, suggests:

A few weeks ago a male animal of the tyger species was killed on the premises of Edward Lord, Esq. at Orielton Park; it measured 6 feet 4 inches from the tip of the nose to the extremity of the tail. It was long a terror to the numerous flocks in the neighbourhood, and had at different times destroyed a number of sheep. It required the joint exertions of two dogs, and the stock-keeper, before it was killed. (*Hobart Town Gazette*, 5 April 1817)

102

How much reliability may be granted to the suggestion that the thylacine had 'destroyed a number of sheep' needs to be viewed against Lord's reputation as an inefficient stockholder. Considered by other east coast settlers as 'an arrogant, land-grabbing troublemaker' (Nyman, 1990, p. 7), instead of his farming Lord is best remembered for building his own racecourse and organising (possibly in more ways than one) race meetings and associated betting activities (*Hobart Town Gazette*, 5 October 1816). In 1826 the Land Commissioner, Roderic O'Connor, found Lord's 30 000 acres to be largely unimproved and his stock overseen by out-of-control ruffians (S. Morgan, 1992, pp. 39, 45). He considered Lord 'evil', corrupt and 'one of the greatest destroyers of the prosperity of this Colony' (O'Connor, 21/6/1826, cited in McKay, 1962, p. 12). While I am happy to accept that a thylacine was killed at Orielton Park, the suggestion that it 'was long a terror to . . . the flocks' may just as likely be read as a convenient excuse for obvious mismanagement from inefficient stock keepers.

The second record also dates from 1817. Another male thylacine was killed, this time at Bagdad, on the property of George Evans, the Deputy Surveyor-General: 'It was attacked by seven dogs, and made a stout resistance, till at length it was killed with an axe by the stock keeper' (*Hobart Town Gazette*, 6 December 1817). While it was initially considered by Evans that this specimen 'had at different times within a week killed thirty sheep', later he apparently changed his mind. Granted an occasional thylacine took the occasional sheep, but even an hypothesised rogue thylacine specialising in sheep-killing was going to be hard-pressed to find reason to kill more than three or four sheep per week for its own energy requirements. The mass destruction associated with killing well in excess of energy requirements every couple of days, apparently solely for the enjoyment of killing, is atypical thylacine behaviour, but very typical of the behaviour of feral dogs, which were soon to be singled out as the major problem for sheep farmers close to established settlements. That Evans soon concluded that the loss of sheep had been caused by feral dogs rather than the thylacine is suggested by his failure to ascribe sheep-killing behaviour to the 'opossum-hyena' [thylacine] in his description of colonial animals (1822, p. 56). Evans did accept that there might be a problem native predator – of which he had no personal experience. He reproduced a second-hand story, written by a non-Tasmanian, suggesting the existence of a native 'panther' that supposedly 'commits dreadful havoc among the flocks' (1822, p. 57). These two newspaper reports represent a very mild introduction to the concept of occasional predation, and provide little basis for the vitriolic attack that was shortly to be mounted against the species by an influential individual with personal economic interests threatened by the success of sheep farming in Tasmania.

In 1819 reference to the thylacine was finally made in book format, published in London and available to a wider audience. Written by W. C. Wentworth who described himself as 'a native of the colony' of New South Wales, the title unashamedly announced its political and economic purpose: *Statistical, Historical and Political Description of The Colony of New South Wales and Its dependent Settlements in Van Diemen's Land: with a particular enumeration of the advantages which these colonies offer for emigration, and their superiority in many respects over those possessed by the United States of America*. With its largest, boldest typeface reserved for the words 'New South Wales', the book represented a political gesture to gain primary investment interest in the colony of New South Wales. It was an influential comment and publication, running to three editions in its first five years (Meston, 1958). Wentworth's description of the thylacine's predatory behaviour was much repeated by other authors. It was regrettable that, demonstrably, it bore so little resemblance to reality.

The differential success stories of sheep farming between the New South Wales and Tasmanian colonies was obviously a source of some concern to Wentworth and his fellow New South Welshmen. The much older, original Australian settlement, established in 1788, possessed only 80 000 head of sheep in 1819, less than half Tasmania's total. The New South Wales coastal fringe tended to be an unsuitable environment for sheep production, and the presence of the dingo did little to encourage the industry. The origin of Wentworth's fabricated description of a 'panther' and its supposed effects upon sheep farming is unknown. He may have been merely reporting hopeful rumours around Sydney at the time, alternatively, he may have deliberately set out to exaggerate the two reported incidents of thylacine sheep predation in a manner disadvantageous to investment in sheep farming in Tasmania.

> The native dog [dingo], indeed, is unknown here [Tasmania]; but there is an
> animal of the panther tribe in its stead, which . . . commits dreadful havoc
> among the flocks. It is true that its ravages are not so frequent [as the dingo's];

103

but when they happen they are more extensive. This animal is of considerable size, and has been known in some few instances, to measure six feet and a half from the tip of the nose to the extremity of the tail; still it is cowardly, and by no means formidable to man: unless, indeed, when taken by surprise, it invariably flies his approach. (Wentworth, 1819, pp. 118–19)

Wentworth had obviously read Oxley (1810) and the *Hobart Town Gazette* (5 April 1817), but at this point all contact with reality in his description stops. No evidence can be found to justify his extravagant claims that the thylacine's destructive effects on the sheep industry in Tasmania 'are more extensive' than the dingo destruction on flocks in New South Wales. Whatever the origins of the idea, Wentworth's description represents a distortion of the known thylacine data that just happens to suit a particular Sydney-based economic outcome.[2]

Wentworth's claim achieved a wide readership and an often uncritical acceptance, leading to its frequent repetition in the work of other authors, but his suggestion that a 'panther' existed in Van Diemen's Land was confusing to those with a better knowledge of Tasmanian animals, who were aware of the thylacine and the economic unimportance of its predatory behaviour (Evans, 1822; Lycett, 1824). These authors were forced to suggest that there must be two different large carnivorous species in Van Diemen's Land, a panther (of concern to the sheep industry) and the thylacine (of no concern to the industry).[3] Wentworth's unsubstantiated suggestion that the thylacine caused 'havoc' and committed 'ravages' amongst the sheep struck a responsive chord in the minds of many readers. A significant number of nineteenth-century authors borrowed Wentworth's description directly, referring to 'havoc' caused amongst the flocks. The phrase that Wentworth borrowed from Oxley (1810) also struck a chord in his readership. The statement that the thylacine 'flies' or 'flees' from the presence or approach of man was picked up and repeated by Lycett (1824), Parker (1833), Lloyd (1862) and Ireland (1865).

Van Diemen's Land colonists needed to counter Wentworth's construction of thylacine predation if they were to maintain and increase investment and expansion in the colony's sheep industry. Jeffreys (1820) argued the colony's suitability for sheep farming, noted the presence of the thylacine amongst the native fauna and merely pointed out the precise, available, known data on the species: 'The wild animals consist of . . . some few of a species of hyena, but . . . there have not been seen more than four since the island was first discovered'[4] (Jeffreys, 1820, p. 108). As far as animal husbandry was concerned, to Jeffreys the thylacine presented no danger.

Further, more positive reports followed that of Jeffreys, presenting Van Diemen's Land as an ideal environment for sheep. James Dixon, captain of the ship *Skelton*, reported that 'The country is peculiarly adapted for sheep; and that animal thrives uncommonly well, and increases astonishingly' (1822, p. 35). With only two thylacines reported to have attacked and killed sheep since 1803, Dixon gave no credence whatsoever to the idea that the thylacine was a danger to the developing sheep industry.

September 1823 marked the twentieth anniversary of the colony. One month before this, the third recorded instance of thylacine sheep predation occurred:

> A few nights ago, a hyena tiger, an animal so rarely seen in this Colony . . . was found in the sheep-fold of G.W. Gunning . . . Four kangaroo dogs, which were thrown in upon him, refused to fight, and he had seized a lamb, when a small terrier of the Scotch breed was put in and instantly seized the animal, and, after a severe fight, to the astonishment of every one present, the terrier succeeded in killing his adversary. (*Hobart Town Gazette*, 2 August 1823)

This represents the earliest account in the literature of the reluctance of most dogs to approach a thylacine. The fact that the thylacine was killed by a small terrier should not be used to infer that it did not know how to defend itself, as implied in the comment by Beresford and Bailey: 'So much for the tiger's alleged ferocity' (1981, p. 17). Surrounded by an unspecified number of people and five dogs the lone thylacine was hardly fighting on equal terms. In this situation, the thylacine would be likely to attend to the whereabouts and behaviour of the large humans and the four larger dogs in its immediate vicinity, rather than on the small, tiger-cat-sized dog that was also present and ultimately became its downfall.

105

Another guide to Van Diemen's Land was published in the same year (Godwin, 1823). In it Godwin referred to the successful husbandry of the sheep, cattle, horses and other domestic animals imported by the settlers. He borrowed his feline descriptor of the thylacine from Wentworth (1819), including amongst his list of wild animals 'an animal of the panther tribe' (p. 18), but he ignored Wentworth's fabricated claim about the 'dreadful havoc' purportedly caused to sheep farming. After all, the colony was twenty years old and only three thylacines had been recorded as having attacked and killed sheep in that time.

One other human–thylacine interaction from 1823 has been assumed by some authors to reflect upon the thylacine's supposed predilection for sheep. It relates to the sketch, commonly reproduced, by the surveyor Thomas Scott of a 'Tyger Trap. intended for Mount Morriston. 1823' (see Figure 4.2). Von Stieglitz (1966, p. 4) has suggested that 'Tasmanian tigers must have been troublesome at "Mt. Morriston"', and Whitley (1973) agreed with this sentiment. But as is clear from the sketch, the design of the trap was not to kill or injure the thylacine but to capture a live, uninjured specimen. It stretches the bounds of credibility to relate this to the idea of sheep predation, since two far more probable explanations are available. At this time there appear to have been only two thylacines in Europe: the Linnean Society in London had an entire juvenile specimen, and there was an adult skull in the Leyden Museum. By 1827 a further three entire specimens were present in European museum display (Renshaw, 1905; Temminck, 1827). Their source is unknown but, obviously, somewhere in Tasmania, someone was capturing thylacines in undamaged condition, killing them and preparing them for European museums.[5]

Alternatively, the trap may have been designed to capture a live thylacine for a pet.

Edward Curr arrived in the colony in 1820 as partner in a merchant business which included the export of wool. He obtained a land grant of 1500 acres (607 ha) and commenced sheep farming, but in 1823 he returned to England where he published a book on his experience, in which he predicted a bright future for the Tasmanian sheep industry (Curr, 1824). He was interviewed by the directors of the Van Diemen's Land Company in London who were so impressed that they appointed him chief agent and manager (*Hobart Colonial Times*, 4 November 1825). He left England in October 1825 to return to Van Diemen's Land to obtain land and commence farming for the Van Diemen's Land Company, an enterprise which continues to the present day (Heazlewood, 1992).

On the desirability and economic importance of sheep farming Curr commented: 'Van Diemen's Land . . . is blessed . . . with a salubrity of climate which no country can surpass, and which is found to be peculiarly favourable to the rearing of sheep. . . . In the flocks of Van Diemen's Land its true riches will always be found to consist' (Curr, 1824, pp. 65, 83). The concept of 'riding on the sheep's back' arose early! But, in writing positively about the future of the sheep industry in Tasmania, Curr also wrote frankly about his experience of the problems that beset the enterprise. As he was later to play such an influential role in categorising the thylacine as a significant predator on sheep, it is worth considering Curr's original recommendations for the management of sheep in Van Diemen's Land.

Prior to the purchase of sheep the 'proper preparations required' the construction of two huts, and a 'sheep fold . . . made of rough logs or of brushwood'. Curr argued that 'usually' three shepherds (plus the master in the first hut) were required for five to six hundred sheep. He advised every stockholder to 'separate his flocks into smaller bodies, never allowing more than five hundred to run together, and keeping breeding ewes, lambs, and wethers asunder'. A resounding problem faced the sheep farmer amongst his convict labour: 'To find a sufficient number of experienced shepherds is impossible; the nature of their employment at home does not often subject them to the penalty of transportation' (Curr, 1824, pp. 70–1, 75–7). As 'the most important business of the shepherd in Van Diemen's Land is to prevent the plunder of his flock' (p. 73), an overseer not only had to have confidence in his own convicts used as shepherds, he also had to have confidence in the management of convicts on neighbouring holdings. This was not always easy to find:

> Sheep stealing in this island . . . is organized into a most complete system . . . instances have occurred where five or six hundred have been driven off at once, and irrecoverably lost . . . Many persons, in calculating the profits on sheep, allow a deduction of twenty-five per cent. for robberies . . . The practice of yarding the flocks every night cannot be too strongly recommended.
> (Curr, 1824, pp. 35, 38–9)

Six years before the publication of Curr's book, in a typically unsuccessful attempt to act as a deterrent, sheep-stealing had been made a capital offence.[6]

To Curr, there were no significant native predators to sheep. He therefore suggested that 'those who, from the situation of their lands, are not much subjected to plunder, always suffer their sheep to bed themselves at night; and, if they have not been disturbed, the shepherd may be generally certain of finding them in the same place in the morning' (p. 74). So much for Wentworth and his 'dreadful havoc'! Rather than any native predator, Curr suggested 'the scab is their principal enemy' (p. 70).

After nearly four years of practical experience of the sheep industry in Tasmania, in considering the problems facing farmers Curr did not mention the thylacine as a problem to sheep. This omission presumably reflected both his personal experience and the small number of published accounts of thylacines attacking sheep. This number did not change for the next four years. The colony reached the quarter-century mark in 1828 still with only three published incidents of thylacines attacking sheep.

While Curr's personal knowledge and experience of sheep farming and thylacines did not apparently encompass thylacine sheep predation, and the few published accounts do not support it as being common, nevertheless thyla-cines were around at the time new settlements were taking place, and it is possible to read in private comments the early growth of rumour and popular mythology about the species. Nowhere is this developing mythology better illustrated than in the fanciful over-exaggeration of the adventures and difficulties of life in Van Diemen's Land, penned by Adam Amos to a friend back home in Scotland:

> The natives are the very last of the human species . . . they are very dangerous and troublesome . . . convicts have fallen a sacrifice to them on our settlement . . . if they fall in with a white man or two unarmed it is only a miracle if they escape . . . Parties of the crown prisoners are allways breaking away to the woods and rob and plunder the settlers . . . the murders and robberies committed by them were incredible . . . Tigers are plentifull amongst the rocky mountains and destroy many sheep and lambs. (20/4/1826, cited in Barrett, 1944, pp. 132–3)

Before taking these claims at face value, the spirit in which the letter was written needs to be acknowledged. I think it possible to detect touches of David-Temple-around-the-campfire-with-Geoffrey-Smith in the letter's construction. Amos notes 'I have had difficulties to encounter and where my fortitude and patience have been tried', but, of course, true grit, courage and a Christian way of life win through in the end – Amos' descendants were still farming the same land in the 1950s. Realistically, if social interactions in the colony were so anarchistic and nature so antagonistic, there was no way the colony would have survived.

Two other references indicative of a contemporary belief in thylacine predation on sheep were noted in Sharon Morgan's research on land settlement in Tasmania. She read the diaries of George Hobler, a farmer on the banks of the North Esk and neighbour of Robert Lawrence, and discovered a single diary entry about a thylacine that was shot by a shepherd in the bush in April 1828. Although it was not around the sheep at the time, Hobler wrote 'many [sheep] has he destroyed no doubt'. This is an interesting statement of attribution but, as Morgan points out, no evidence was offered to support such a claim, either at the time of entry or elsewhere in the diaries (S. Morgan, 1992, p. 117). Morgan also discovered in the collection of George Meredith's papers an undated note, suggested as '1828?' in which 'Mary Meredith told her husband that thylacines had been very destructive to William Talbot's flock and had even killed his dogs' (pp. 117, 190). As discussed in detail later in this chapter, the multiple killing of sheep (and, in this case the multiple killing of farm dogs) is atypical thylacine behaviour, but characteristic of attacks by packs of wild dogs. Furthermore, in this case it may not have been wild dogs at all. Nyman (1990, p. 23) gives an account of an incident in 1822 when, in a dispute over land grant ownership, George Meredith deliberately set his working dogs upon William Talbot's pure merino flock.

For the first twenty-five years of the colony, the thylacine's attacks upon sheep were so rare and occasional as to be worth little comment by those writing of the colony and sheep farming from a first-hand perspective. It is only in the second-hand replications of Wentworth's fanciful construction where claims of significant sheep predation may be found. Outside the published records, in the realm of popular perceptions, the situation may best be summed up by concluding that, while there was a little predation, there was a lot of rumour, much of which, by its nature, was over-exaggerated, unsubstantiated or misattributed.

108

The glowing reports of the potential of Van Diemen's Land for sheep farming published in the 1820s encouraged the formation of a number of agricultural companies in London for the purpose of investing in the sheep industry in Tasmania. The New South Wales and Van Diemen's Land Establishment (later known as the Cressy Company) was formed in London in 1825. The company managers, together with indentured servants, shepherds and stock arrived at Hobart in May 1826 and bought 1200 acres (485.5 ha) at Cressy, south of Launceston, on which they built 'eight miles of post and rail fencing' in their first year. The Cressy Company never claimed significant predation of their sheep flocks by thylacines. The first manager, Henry Widowson, after his return to England, wrote a book about the colony (Heazlewood, 1992, pp. 25–7). He obviously had limited experience of its marsupi-carnivores. He noted: 'The only animals that can be termed carnivorous are the small hyena, the devil, and the native cat' (Widowson, 1829, p. 179). The thylacine is described as the smallest of these three species:

> The hyena, or as it is sometimes called, the tiger, is about the size of a large terrier; it frequents the wilds of Tasmania, and is scarcely heard of in the located districts. Where sheep run in large flocks near the mountains these animals destroy a great many lambs. The female produces five or six at a birth; the skin resembles the striped hyena. (Widowson, 1829, pp. 179–80)

Widowson's zoological knowledge left a lot to be desired. As well as seriously under-estimating the size of adult thylacines, he over-estimated their reproductive capacity (thylacines possessed only four teats in their pouch). His suggestion that thylacines were preferentially feeding on lambs (rather than sheep) needs to be treated with the same caution applicable to his other statements about the species. It is reasonable to doubt whether Widowson (the manager of one of the largest sheep farming enterprises in Tasmania at that time) had any first-hand knowledge or experience of the thylacine at all. In this case, if his account merely reflects casual, popular perceptions of the animal that he had encountered while in Tasmania, it demonstrates even at this early stage, just how far from reality popular perceptions of the thylacine may have been.

Unfortunately, this lack of knowledge was not obvious to the lay, or equally naïve, reader. Consequently Widowson's claim of the small size of thylacines, with six young and preferential predation upon lambs, was picked up by his contemporaries, and in similar manner to Wentworth's comments (1819) replicated in other publications (Mudie, 1829; and Bischoff, 1832). These replications represent a selective use of Widowson's comments, for Widowson judged the overall effects of this seasonal lamb predation by thylacines as of minor importance in his review of sheep farming in Tasmania and its comparison with New South Wales. Widowson noted an absence of large, significant, native predators in Tasmania, but a necessity of vigilance to prevent sheep-stealing: 'there is no native dog to destroy them [sheep] in Tasmania, the only dog to be feared is the two-legged one' (p. 149).

109

One unpublished event from 1829 balances Widowson's publication. Robert Lawrence, farming near Launceston, noted in his personal diary on 19 November 1829 that he dined with one Captain Stewart, the owner of a captive thylacine. Lawrence's diaries from 1829–31 and excursion notes from 1833 cover his sheep farming and scientific interests. Although he also wrote about encounters with thylacines on other occasions, he never made reference to the thylacine as a problem for the sheep farmer.

At the close of the decade data on the relationship between thylacines and sheep shows that occasional, isolated incidents of predation took place. Two much replicated and influential books suggested otherwise (Wentworth, 1819; and Widowson, 1829) but for all their influence upon popular perceptions, these publications were not based on demonstrable, first-hand knowledge. Wentworth's construction appears economically motivated, and Widowson's image of the significant killer of lambs needs to be set against his other palpably incorrect statements about the thylacine's small size and reproductive behaviour. Available insights into popular perceptions of the thylacine at the time indicate, on the one hand, a developing mythology about the species, little tied to reality, with exaggerated and misattributed claims about the thylacine's reproductive and predatory behaviours, while on the other hand, different individuals had started raising thylacines as pets and companion animals. Up to 1829 only three published first-hand accounts of thylacines being found amongst or preying on sheep (two in 1817 and one in 1823) have been discovered.

The next year, however, this situation changed dramatically. Without warning, it would appear that almost the entire thylacine population of Tasmania in 1829 and 1830 turned its attention to the flocks held on the frontier settlements of the Van Diemen's Land Company (Map 5.1); to such devastating effect that sheep farming on two of their holdings, 150 000 acres (60 700 ha) at Surrey Hills and 10 000 acres (4050 ha) at Hampshire Hills, was discontinued, and a bounty placed on the thylacine's head.

1830–1838: The first private bounty scheme

On 16 April 1830 Curr entered the following details of a bounty scheme in the Van Diemen's Land Company *Letterbook*:

> The superintendent of the Hampshire and Surrey Hills Establishments[7] is authorised to give the following rewards for the destruction of noxious animals in those districts:-
> For every Male Hyena   5/-
> For every Female [Hyena] with or without young   7/-
> Half the above prices for Male and Female Devils and Wild Dogs.
> When 20 hyenas have been destroyed the reward for the next 20 will be increased to 6/- and 8/- respectively and afterwards an additional 1/- per head will be made after every seven killed until the reward makes 10/- for every male and 12/- for every female. (Curr, 16/4/1830)

This memorandum represents an important and influential act in terms of constructions placed upon the thylacine's behaviour. Guiler commented:

> The thrust of this bounty was directed mainly at the thylacine, and the increasing scale of the rewards shows that the Company was determined to get rid of the pests, which obviously were to be found in considerable numbers on its holdings. The Company realized that a major effort would have to be made to reduce the population. The dogs and devils were not regarded as such a serious pest being more numerous and easier to destroy and were worth only half the thylacine rate. (Guiler, 1985, pp. 16–17)

Guiler's response has been to accept the bounty at face value, infer a problem with the thylacine and the sheep, and accept the bounty as a reasonable attempt to solve that problem. The significance of this bounty scheme has been frequently noted by other authors, and represents a major plank in the edifice constructed for the thylacine as a significant predator on sheep.

It has long been automatically assumed that the Van Diemen's Land Company introduced a bounty on the thylacine because the thylacine was destroying sheep. However, the history of the Company and the events that led up to the announcement of the bounty suggest an alternative explanation.

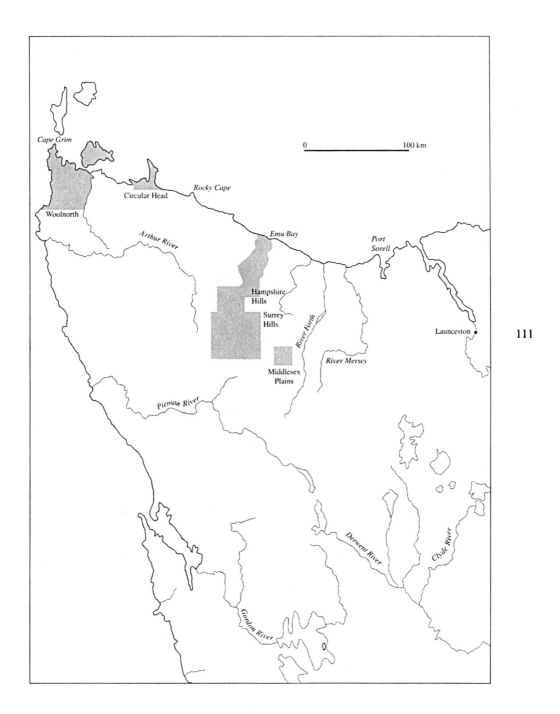

Map 5.1 Holding locations of the Van Diemen's Land Company in north-western Tasmania.

Like the Cressy Company, the Van Diemen's Land Company originated in London in 1825. The Company's intention was to raise sheep and it sought a grant of 500 000 acres (202 350 ha) in northern Tasmania on which to do this. Stephen Adey had been led to expect that the position of chief agent for the Company in Van Diemen's Land would be his, however, after the directors had read Curr's book and interviewed him they decided to appoint Curr instead. Adey was named a joint commissioner with 'equal' status to Curr, but Curr had right of veto over all decisions. Unsurprisingly, the relationship between the two men was never easy (Meston, 1958). The Company's representatives arrived in the colony early in 1826 and a vigorous period of exploration in north-west Tasmania took place, in search of suitable pasture land. Henry Hellyer was placed in charge of the survey department and housing. Adey was given primary control over sheep and wool production. Directly under Adey was Alexander Goldie, whose responsibilities involved clearing and fencing, cultivation of the land, and the general charge of livestock (Meston, 1958, p. 47). Land was quickly taken up at Circular Head and then, with Hellyer's discovery of 'glorious, gently rising, grassy hills' extensive holdings in the highlands were granted to the Company as well.

Adey was sent to Circular Head to plan the settlement and farming there. Fourteen weeks later he returned to Hobart. The young settlement was facing starvation, but rather than send a subordinate to Curr requesting help, Adey quit the settlement himself. Curr sent him back but Adey: 'grew despondent and dissatisfied. To Goldie and Hellyer he was in the habit of saying on all occasions that he did not see how the undertaking was ever going to pay and that it was very probable that it would be abandoned' (Meston, 1958, p. 46). Both Goldie and Adey also proved incapable of managing the mixed workforce of indentured servants and convict labour. The convicts went on strike and refused to work without more rations above the government regulation issue; they instituted their own sick leave, effectively holiday leave, and neither Adey nor Goldie appeared capable of enforcing discipline or productivity amongst them. Curr was compelled to go to Circular Head and sort out the problem himself in January 1827 (Meston, 1958).

Meanwhile, development was proceeding in the central highlands areas. A jetty had been constructed at Emu Bay and a road heading south from the coast put through to the Hampshire Hills and Surrey Hills holdings, and a track put in from Surrey Hills eastwards to Launceston. In October 1827 at the London directors' request, Adey became the Company's agent in Hobart and Curr was relocated to Circular Head (Meston, 1958). Curr visited for the first time the Hampshire Hills and Surrey Hills holdings and 'was dismayed at the poor quality of the country the Company had selected' (Mercer, 1963, p. 20). Hellyer had seen it at the end of a dry, late autumn, now in October the plains were dormant and soggy rather than at a peak of spring growth (Curr, 10/11/1827). No turning back was possible, the Company had applied for these areas of land and road and jetty building had been proceeding for five months. The first stock station was built in March 1828 at Hampshire Hills. Because the winter weather would make the road impassable, cattle were introduced to Hampshire Hills and Surrey Hills in March, even though the stockyards and station at the 150 000-acre (60 700 ha) Surrey Hills holding were non-existent and construction only commenced in May 1828 (Mercer, 1963). Under the care of Goldie, and such workers as he could command, the cattle were incompetently managed; many wandered off and got lost, some died in the winter cold, others were possibly killed by Aborigines or, occasionally, may even have been killed by a family of thylacines.

It is easy to argue that Curr, stationed along the coast at Circular Head, should have taken a more active interest in the Hampshire Hills and Surrey Hills holdings, given the known factors of Goldie's management problems and the premature stocking of the land. However, Curr had far more distracting problems to deal with. The Van Diemen's Land Company had invested its money in the Van Diemen's Land Bank. In January 1828, with true entrepreneurial spirit, Adey, in Hobart, opened his own private bank, called the Derwent Bank, where he was both managing director and cashier. To provide his private bank with a capital base he withdrew the entire Van Diemen's Land Company funds from the Van Diemen's Land Bank and placed them in his own bank. Curr was both worried and incensed – and, when they finally heard about it back in London, so too were the Company's directors. Eventually, Adey was forced to return the Van Diemen's Land Company money to the Van Diemen's Land Bank and Adey then 'resigned' from the Company. Thus the person primarily responsible for sheep and wool production left the Company. Throughout much of 1828 Curr was plagued with uncertainty as to the financial security of the Company and his own ability to readily access and use the Company funds in developing the planned enterprise (Meston, 1958). It is understandable that Curr's supervision of the more remote holdings suffered.

113

In March 1829 Goldie suggested that the Hills holdings were ready for sheep. Soon after, the core of the Van Diemen's Land Company sheep enterprise, the stock imported from the United Kingdom together with others purchased from breeders in Van Diemen's Land, were released at the Hampshire Hills and Surrey Hills holdings. The earliest historical account of the Company succinctly remarked of this 1829 venture that 'the farming did not succeed' (Cattley, 1863). More recent histories of the Company have referred to it as a 'disaster' (Mercer, 1963; Meston, 1958). Of the 5500 finest-quality Saxon merino sheep sent to the Hampshire Hills and Surrey Hills holdings, 'only a few hundred survived' and came down from the mountains when the exercise was finally abandoned (Pink, 1990).

This sheep disaster of 1829 followed the cattle debacle of the previous year and was predictable. Virtually none of the requirements that Curr had outlined in his book in 1824 for successful sheep farming in Tasmania was in place when the stock were released on the Hills holdings.

The preparation of the huts, fencing and separate sheep folds was inadequate and behind schedule. 'Hellyer . . . had to learn by experience the way to build turf huts, log and weatherboard houses' (Meston, 1958, p. 55), and of

Goldie's work Curr commented: 'Even at the Hampshire Hills amidst the best of materials his fences are by no means what they ought to be' (4/1/1830). Instead of separated flocks of only 500 sheep and experienced shepherds, the sheep were 'looked after' *en masse* by inexperienced convicts under an incompetent manager incapable of getting them to work consistently.

Initially, the holdings were too new and too far from other settled districts to fall prey to bushrangers or sheep-stealing, but 'from the earliest days of the Company's establishment the loss of stock through the depredations of the aborigines was a constant source of anxiety' (Meston, 1958, p. 52). In one single incident, 118 sheep were killed by Aborigines at Woolnorth in 1827, and similar attacks were carried out at the Hampshire Hills and Surrey Hills properties during 1829 (Meston, 1958, p. 51), in understandable responses to European invasion.

As if this was not bad enough, the one 'blessed' factor that Curr had identified, 'a salubrity of climate . . . peculiarly favourable to the rearing of sheep' (1824, p. 65) turned against him:

> what must have been, one of the worst winters for many years set in. In late April [1829] the cold became intense. In May and June the infant settlements were battered by almost continuous howling southerly winds bringing driving sleet and blizzards. The winter progressed into July and August . . . The real trouble started when the ewes began to lamb. The cold was so intense that the lambs stood no chance of survival. October came and the winter weather showed no signs of abating.
>
> In November Hellyer wrote, 'The snow is so deep that we are completely hemmed in by it.' . . . Poor Hellyer witnessing the horrors of this severe climate must have realised sadly that his 'glorious, gently rising, grassy hills' were no garden of Eden and certainly not agricultural country. (Mercer, 1963, p. 26)

Cattley's comment that 'the farming did not succeed' (1863) is a masterly understatement.

News of the disaster spread rapidly. In the *Hobart Town Almanack for the Year 1829*, the published comments about sheep production make no mention of any thylacine problem but, in a thinly veiled criticism of what happened to the Van Diemen's Land Company, there are complaints about how poor management, exposure to the 'inclemency of the weather' and the 'crowding of the sheep, particularly in the lambing season' affect the sheep industry (Ross, 1829, p. 142).

Curr's response to the debacle in the Hills holdings, and the necessity to explain such devastation to the Company directors back in London, led him to search for excuses. I suggest that the first that he found to hand was the thylacine. It represented a convenient scapegoat and, importantly, one that did not involve criticism of the management or the running of the properties. The thylacine was a natural part of the environment. As it had not previously been identified by Curr as a significant predator on sheep, no criticism could be directed at the management for not taking precautions against it. Also, the species' vernacular names of 'native tiger' and 'native hyaena' carried valuable imagery to the directors and investors back in London, and problems with 'the Hyenae and native Dogs' were noted in the 1830 Annual Report (Van Diemen's Land Company, 1830, p. 5). Arguably, the proclamation of the bounty against the thylacine in April 1830 had more to do with saving the necks of the management than saving the necks of the sheep.

Curr's second scapegoat, rather more justifiably, was to be Goldie. Curr wrote to the directors about Goldie: 'As a farmer he does not make a proper comparison between the end to be gained and the cost of gaining it. Hence in a great measure the dreadful destruction . . . wherever he has had the management' (4/1/1830). Goldie resigned before the end of the year.

Both of the Company's staff appointed with primary responsibility for the sheep had now resigned. In 1832, as the full extent of the loss from the venture of sheep farming in the Hills holdings – well over 90 per cent of the stock – became known, Hellyer committed suicide (Pink, 1990). On the face of it, Hellyer and Goldie's departures would appear to be rather extreme personal responses to a problem traditionally assumed to be not of their own commission, but supposedly caused by the thylacine.

Few records exist of the operation of this first Van Diemen's Land Company bounty scheme, nor do we know when the bounty actually ceased. It just seemed to fade away for the unimportant issue that it really was. As well as the extensive written archives preserved from the early Van Diemen's Land Company, there are also the independent diary records made by George Augustus Robinson as he passed through the sheep holdings at this time. Consideration of these data sources supports the idea that, even on the very fringes of European settlement, the thylacine was merely an occasional predator on sheep, and most certainly not the major predator, or even the most obvious recipient of a bounty scheme.

Eric Guiler has spent a considerable amount of time over the years working on the station diaries, stock returns and account books of the Van Diemen's Land Company, to establish the levels of stock predation on the Company's holdings (1961w, 1985; Guiler and Godard, 1998). My research on the Van Diemen's Land Company archives has concentrated on the preserved letter-books and correspondence files. I use Guiler's figures on predation levels extensively in this section to support my argument on the lack of importance that thylacine predation had on sheep farming for the Van Diemen's Land Company. Guiler has used the same figures to argue the opposite though not, I believe, without significant difficulty. They are the same data, only the interpretations are different. Guiler believes that 'These [Van Diemen's Land Company] records are the only strictly factual accounts of thylacines during the pre-bounty days and are devoid of any hysterical exaggeration or distortion' (1985, p. 92). I disagree. I hope to be able to show that the manner in which the data on thylacine predation were ignored, distorted and abused at the time by the Company's middle-level management in Tasmania is as good an example as any in the literature of an 'hysterical exaggeration' of the thylacine's behaviour to suit an economic agenda.

The day after Curr's bounty proclamation for the Hampshire and Surrey Hills holdings, Joseph Fossey, superintendent at the Woolnorth holdings, wrote to Curr about being 'visited by a Hyena' and suggesting that he offer a reward for its capture (17/4/1830). Curr responded with a copy of his original bounty proclamation (12/5/1830). This single thylacine continued to cause minor problems to the sheep at Woolnorth (Curr, 5/6/1830), before it was killed, but neither Curr nor Fossey saw it as anything other than a minor irritant. The greatest predatory problem to the flocks at Woolnorth was clearly identified in their correspondence as 'those shepherds' dogs which have taken to hunting alone' (Curr, 20/5/1830). Nevertheless, not wishing to impugn Fossey's management of his shepherds, or the behaviour of the Company's dogs, when writing to the Van Diemen's Land Company directors in London Curr deceitfully blamed all the predatory problems at Woolnorth on the thylacine: 'I am sorry to say that the Hyenas have committed considerable destruction here [Woolnorth]. The shepherds have watched their flocks at night for several weeks, and one Hyena has been killed. At present they seem to have suspended their ravages' (Curr, 18/11/1830).

116

Robinson, on his mission to establish peace, a settlement and treaty with the Aborigines (H. Reynolds, 1995) records two instances of bounty payments. On 22 August 1830 a female thylacine was killed on the Middlesex Plains and presented for payment at Hampshire Hills a few days later: 'Mr. Robson . . . consented for McKay to have 7/- out of the store for a bitch hyena (5/- being allowed for a dog)' (25/8/1830). At this stage the bounty scheme had been operating for four months, but so few had been submitted for bounty payment that the price still remained at the initial, opening level.

In April 1831 Curr wrote to Samuel Reeves, the new superintendent at Woolnorth: 'I recently informed you that 8/- was the price of a hyena. It is raised to 10/-, and both sexes alike' (2/4/1831). The increase in price from 7 to 8 shillings may be an indication that by April 1831 after the first year of the bounty's operation, twenty female thylacines had been caught on the Company holdings. Assuming that there were equal numbers of males and females caught, this still represents only forty specimens killed over the entire holdings of the Company, some 350 000 acres (141 640 ha) in extent (Guiler, 1985, pp. 90–1). The suggestion that the real reason for the bounty's establishment may have been political rather than predatory is more than adequately supported by this interpretation of the data: only one thylacine being killed for every 8750 acres (3540 ha) of the Company's land. Whatever the real reason for the first thylacine bounty, it obviously had little to do with the actual population numbers and presence of thylacines on the Van Diemen's Land Company holdings.

The direct movement from 8 to 10 shillings and removal of sexual discrimination was clearly not in response to the numbers of thylacines caught and presented for bounty payment. It was, however, able to be interpreted this way, and as the reality of the debacle in sheep farming became known in London, it must have been satisfying to see the directors prepared to blame the thylacine as the major source of the Company's woes: 'we hope the Hyenas will gradually disappear and be destroyed' (Court,[8] 3/6/1831).

Curr went out of his way to encourage this belief in communications with London: 'Mr Reeves informs me . . . 3 [lambs] have been carried away, he supposes by Hyaenas' (25/7/1831), and 'Kennedy . . . [the] new Shepherd . . . has been so annoyed by Hyenas that he has been watching his Flock every night for three weeks' (9/3/1832). The directors in London responded appropriately: 'we fear the *Hyaenas* and *Wild Dogs*, more than Climate' (Court, 10/10/1833).

Back in Tasmania, Robinson came across a dead thylacine in the bush at Mt Cameron: 'The New Hollanders skinned the hyena for the purpose of carrying it to Cape Grim to get the ten shillings reward from the Company' (29/8/1832). In addition to these two bounty claims, Robinson commented a number of times on stock losses experienced by the Van Diemen's Land Company. Only on one occasion did he record a predatory event as solely the work of thylacines (17/8/1830). In all other instances where Robinson mentioned thylacine predation on sheep, he also mentioned wild dog predation:

> Saw several carcases of sheep which had been killed by hyaenas or wild dogs. [24/8/1830] . . . Spoke to Mr. C's nephews about the wild dogs; . . . [they] say nine or ten dogs and hyaena together had been caught in a month. [25/1/1834] . . . Mr. Chitty said the Company could not keep sheep here, that if they were to fold them they would soon die, that it was only moving about that kept them alive. Wild dogs and hyaenas are numerous. Watchmen were kept to look after them. (4/6/1834)

117

As well as presenting wild dogs as viable alternatives to thylacines over incidents of sheep predation, Robinson also offered some insight into alternative explanations for the problem of Aboriginal predation: 'discovered the skeletons of eight sheep . . . doubtless killed either by the natives or white men, most probably the latter as one man had told the superintendent he would do some mischief before he left' (14/7/1832). The possibility of other deliberate acts of sabotage by convicts against the Company's sheep should not be discounted.

Problems in sheep farming continued to grow rather than decrease for the Company, for reasons that had nothing to do with thylacines. As the settlements became older they attracted more white transients and settlers who further displaced the indigenous population. In November 1833, forty-eight merino ewes and eighty-eight lambs were believed 'stolen by the sealers', from West Hunter Island where 'No Hyenas, wild dogs, or any other animal capable of committing such destruction' were to be found locally. In addition, elsewhere at Woolnorth, in a typical guerilla strike, 'the natives had appeared . . . and robbed the shepherds hut of all the Provisions it contained, and as they had several dogs with them, the sheep were very much disturbed & scattered about in all directions' (Hutchinson, 16/12/1833). Like the thylacine scapegoated to cover the losses of sheep at Hampshire Hills and Surrey Hills, in communications with London it was politically expedient to blame nearly all of the heavy sheep losses at Woolnorth in 1833 on the Aborigines and their dogs. Later internal investigation – not communicated to London – was to identify the major source of the problem as Europeans and their dogs.

Two years after the disaster of sheep farming at Hampshire Hills and Surrey Hills became known, an attempt was made to filter some of the truth through to London and change the directors' blindness to the rigours of Tasmania's climate. While the view that thylacines were to be feared more than the climate had been deliberately encouraged in the past, to prevent further losses by restocking these holdings with sheep some of the realities of the enterprise had to be made known (Hutchinson, 4/8/1834). This produced the desired response. Sheep farming was to be abandoned at Hampshire Hills and Surrey Hills, which were then to specialise in cattle. The directors decided 'to pursue tillage chiefly at Circular Head' alongside sheep, with Woolnorth to specialise solely in sheep (Court, 26/2/1835). However, the myth of thylacine predation followed all the sheep across to Woolnorth:

> The *Woolnorth* Returns show occasional losses of Sheep by *Hyenas*. This Establishment is now looked upon as our principal one for Sheep, we cannot therefore too strongly impress upon you the absolute necessity of using every means to exterminate these wild animals at once, otherwise they will propagate and in time be as great a scourge as they were in the Surrey Hills.
> (Court, 26/2/1835)

In his published research on the relationship between the Van Diemen's Land Company and the thylacine, Guiler (1985, pp. 90–113) has tried to find evidence of significant thylacine predation in the period 1830–34 that would justify the creation of the thylacine bounty scheme. While Guiler was able to uncover evidence of some thylacine predatory acts at Surrey Hills between 1832 and 1834, he could find no evidence at this time of any thylacine activity against stock on the 20 000-acre holding at Circular Head (p. 110) or the other large 150 000-acre holding at Woolnorth. With concern for his argument about significant sheep predation by thylacines Guiler comments:

> The Company must have had some trouble from thylacines killing sheep at an early stage in the development of the [Woolnorth] property, but I have found no comments or statistics about this in the few available records, and in fact, the only statistics are to the contrary. The tables of stock increases at Woolnorth 1832–1834 give no losses due to thylacines nor do they pass comment on this topic although they record all sorts of other calamities . . . (Guiler, 1985, p. 95)

There was actually no 'must' about it. Not only are Guiler's statistics of the stock returns to the contrary, the written records from Woolnorth from 1830–34 that I have covered above are also to the contrary: one thylacine that killed sheep in 1830 before being killed itself, one or two thylacines present in 1831 that 'supposedly' killed three lambs, and nothing more (Curr, 18/11/1830, 25/7/1831).

From the station diaries preserved for the Surrey Hills holding from 1832 to 1852, Guiler (1985, pp. 108–10) was able to extract data on sheep loss due to predation. At the same time as the Woolnorth data given above, at Surrey Hills between 1832 and 1834, 688 sheep were identified as being lost to predatory acts; only 48 of these (less than 7 per cent) were ascribed to thylacines (Guiler, 1985, p. 105). The political nature of the first thylacine bounty scheme is more than adequately exposed by these figures. While the directors and investors back in England had been encouraged, through 'hysterical exaggeration', to believe that the thylacine was primarily responsible for the great sheep losses at Surrey Hills and Hampshire Hills between 1829 and 1834, the station diaries indicate that thylacines were responsible for less than 7 per cent of the predatory acts against the sheep. Wild dogs at Surrey Hills accounted for 228 kills, and 412 predatory kills were not specifically ascribed. What is unusual about these station diary records is that, for the predation records between 1832 and 1852, of the 903 sheep killed only one instance, a lamb in 1842, is ascribed to the Tasmanian devil. This does not suggest confidence in the assignment process for the cause of kills committed by native predators. Heazlewood, in his mono- 119 graph on the sheep industry in Tasmania, suggested that up to 1880, after the scab, predation by the Tasmanian devil was the most common problem to the sheep farmer, and identified the Van Diemen's Land Company as the only sheep farming enterprise in Tasmania that apparently consistently suffered from the ravages of the thylacine (1992, p. 86). It appears likely that many of these unascribed kills may have been caused, not by wild dogs or thylacines but by Tasmanian devils.

An inability to distinguish clearly between the different marsupi-carnivore species was remarked on previously in chapter 4. It is worth noting that James Bischoff, who was sent to Tasmania to replace Adey, was unable to clearly distinguish between thylacines and Tasmanian devils and any bounty payments he authorised are obviously questionable. In his book on the colony and the Van Diemen's Land Company he used Widowson (1829) to describe the thylacine as a terrier-sized 'small hyena', preying upon lambs, and bearing up to six young at a time (Bischoff, 1832, p. 29). To Bischoff, any sheep-kill by a large native predator such as the thylacine was going to be ascribed to the Tasmanian devil, while any lamb-kill by a smaller native predator, such as the Tasmanian devil, was going to be ascribed to the thylacine.

While the political expediency of the creation of the first Van Diemen's Land Company thylacine bounty has not been noted before, my claim that thylacine predation was unimportant is not new or radical. This viewpoint was tacitly held by all the early authors of histories of the Van Diemen's Land Company. Cattley (1863), writing a brief history of the Company from 1825 to 1863, makes no mention of the thylacine as a problem. Nor is there any mention of thylacine predation in the history of the Company by Edwards and Edwards (1941). In 1958 A. L. Meston's monograph on the Company specifically for the years 1825 to 1841 was published. In his text he also made no mention of the thylacine. However, the book was published posthumously, with an abstract and introduction unseen by Meston. Such has been the power of popular perceptions of the thylacine as a significant sheep-killer, that the anonymous author of this supposed 'Summary' of the text maintains: 'Stock losses were heavy from weather, marsupial wolves and aborigines' (Meston, 1958, p. 3). However, while the monograph deals at length with the Company's sheep problems ascribed to Aborigines and the weather, there is not one actual mention of marsupial wolves in the text at all!

Thylacines do not appear as a suggested predatory problem until the publication of the next work that included a history of the Van Diemen's Land Company: 'the aborigines, Thylacines and Tasmanian Devils slaughtered the sheeps [sic] and lambs . . . Tasmanian Tigers . . . took a terrible toll on the sheep' (Mercer, 1963, pp. 15, 26).

Taking a broader approach than just the early years of the first bounty scheme, for Surrey Hills, Guiler records the loss of 716 sheep to predatory acts between 1832 and 1842. Of these 239 (33.3 per cent) were specifically attributed to dogs, and only 62 (8.7 per cent) were attributed to thylacines (Guiler, 1985, table 6.6, p. 109). The demonstrably far greater problem of feral dogs to the Van Diemen's Land Company requires some background information, before considering the minor placement and status of the feral dog in the Company's bounty scheme.

Dogs first set foot in Tasmania in 1798 with the visit of the explorer George Bass. The sealers who followed Bass to Tasmania and the islands between Tasmania and the mainland brought dogs with them. The early attempts to integrate sealers into Aboriginal economy and society, involving customary marriage negotiations and the traditional exchange of women's sexual services for hunting dogs, flour and other gifts (McGrath, 1995), soon degenerated into the capture, enslavement, rape and prostitution of Aboriginal women by the sealers and later the settlers, sometimes accompanied by the offer of dogs to compensate for the abuse. When the sealers left, the women were sometimes returned to their people, with dogs as pets or companions (Jones, 1970). The early European settlements soon ran out of food and parties of hunting dogs and convicts were sent out into the bush to catch kangaroos and emus. As the European invasion of their territories increased the Aborigines began to attack the hunting parties, killing and injuring both dogs and Europeans, and seizing the captured game. Jones (1970) records several instances from 1806 and 1807 of the killing or wounding of dogs, but notes that it soon stopped, as Aborigines began to appreciate and use the captured canines rather than killing them. Aborigines who were in close contact with the settlers, such as those living on the Van Diemen's Land Company holdings, also received dogs as gifts or rewards for their services (for example, Robinson, 22/7/1832, 27/8/1832).

In 1816, just fourteen years after settlement of Port Dalrymple in the north, James Kelly encountered a 200-strong band of Aborigines accompanied by at least fifty dogs (Bowden, 1964). In his diaries Robinson made numerous references to Aborigines hunting with dogs between 1830 and 1834 and Lawrence (18/1/1833) also recorded dogs accompanying Aborigines whilst hunting.

Dogs became lost or escaped from European hunting parties and farms, and no doubt some dogs became lost and escaped from Aboriginal hunting parties as well. Convicts escaping to the bush sometimes took dogs with them, and as these bushrangers were finally rounded up and captured or killed, their dogs were left to roam the bush and farmlands of Tasmania. They soon joined together in packs and made their presence felt. As early as 1819, the *Hobart Town Gazette*, making reference to sheep farming, suggested that it 'would be found very beneficial if the owners of stock were to limit the number, or get rid entirely of the dogs . . . which induce the men to employ almost all their time in hunting' (25 December 1819). The problem of feral dogs soon spread to urban areas:

> By 1826 it was apparent that dogs were becoming a menace, and the attention of settlers and stock-keepers was called to the danger of multiplying the race . . . they had become wild in the woods, ravaging the flocks and the herds . . . innumerable hordes of them infesting the town [Hobart] . . . often bit passers-by. (Goodrick, 1977, p. 118)

121

Sheep farmers started to take collective action. Two public meetings of sheep-owners were held in the Campbell Town district in eastern Tasmania in August and September 1833, to 'determine the most effectual means . . . to check . . . the alarming increase in the number of Wild Dogs' (*Hobart Town Courier*, 20 September 1833). It became obvious that some government response was needed to deal with the dog problem (Jones, 1970); consequently, starting in Hobart, regulations to restrict the rearing of dogs were promulgated in 1830. Feral dogs were to be destroyed by the police and, when known, owners held responsible for the actions of their dogs. However, the regulations had little effect, and 'by 1834 it was quite obvious that wild dogs were the greatest drawback to the prosperity of the settler' (Goodrick, 1977, pp. 118–19).

The dog problem was exacerbated by the continuing European war against the Aborigines. The Aborigines, in counter attack against the settlers, used their dogs to destroy sheep (Jones, 1970; H. Reynolds, 1995). Specific instances of this occurred on the Van Diemen's Land Company holdings and were noted by Robinson (17/2/1834, 18/2/1834). As Robinson proceeded to convince the Aborigines their best future lay in peaceful negotiation, the establishment of a treaty and their supposedly temporary removal to Flinders Island (H. Reynolds, 1995), the Aborigines left their dogs behind in the bush (Cotton, 1980; S. Smith, 1981, p. 22) allowing them to spread like the dingo on the mainland. Breton warned of the dangers of feral dogs to the farming community (1835, p. 410) and by 1846 had found all his worst fears confirmed (1846, p. 132). In contrast to the predatory activities of thylacines, wild dogs frequently banded together and roamed the countryside in packs and, true to their domestic legacy, frequently killed as much for pleasure as for food.

An indication of the effects of the different predatory styles and comparative loss caused to sheep from feral dogs and thylacines may be obtained from a full consideration of Guiler's analyses (1961w, 1985) of sheep losses from predation on the different Van Diemen's Land Company holdings. From an analysis of the station diaries at Surrey Hills, for the eighteen years between 1832 and 1849, dogs were identified as being responsible for the deaths of 299 sheep; a predatory response double the level of that recorded for thylacines, which in the same time period were held responsible for killing only 146 (Guiler, 1985, p. 109) or 147 (Guiler, 1961w, p. 207) sheep. Guiler also worked on the stock returns at Woolnorth – recorded monthly between 1839 and June 1847, and submitted quarterly after this date to September 1850 – where the primary problem of feral dog predation was even greater than at Surrey Hills. During these twelve years – where distinction was made between predatory loss due to thylacines and dogs, rather than recording the cause as merely 'vermin' – 660 sheep (an average of 55 per year) were recorded as lost to predation by dogs, while only 40 sheep (less than 4 per year) were recorded as being killed by thylacines (Guiler, 1985, table 6.1, p. 96). From these figures, the predatory loss due to feral dogs was 16.5 times greater than the loss caused by thylacines. The tendency for feral dogs to run flocks down and maim and kill more sheep than they required was illustrated in two incidents that occurred at Woolnorth in 1843: sixty-four lambs were slaughtered by dogs during the one night in June, and in August 'no fewer than 78 [sheep] were driven into the sea . . . by dogs' (Guiler, 1985, p. 96). No such mass killings were ever recorded against the thylacine in the stock returns or station diaries of the Van Diemen's Land Company.

The significance of feral dog predation, and the relative unimportance (and even possible misassignment) of thylacine predation (as argued above on the basis of archival statistics), were also common knowledge to the managers of the Van Diemen's Land Company stock at the time. George Robson, completing the August 1832 monthly stock return for the Surrey Hills holding, explicitly noted:

> Dogs notwithstanding the numbers which are destroyed continue to kill but some particularly harass and scatter our flocks in spite of every exertion . . .
> It is worthy of remark that while we are so overrun by Dogs we are comparatively free from Hyenas only four sheep have fallen by this animal and it is probable that even these may have been killed by Dogs as it is a little difficult to discriminate a while after death. (Robson, 30/8/1832)

Essentially, the problem of feral dogs became acute to Tasmania's farmers from 1819 onwards. The Van Diemen's Land Company records of predatory stock loss confirm this, and demonstrate that feral dogs were a far greater menace to the flocks than thylacines, given their tendency to kill at levels far in excess of their energy requirements (Guiler, 1985, p. 109). It is understandable that feral dogs would be part of a bounty scheme set up by the Van Diemen's

122

Land Company. What is not easily explicable is their bounty value being only half of that awarded to thylacines. Guiler comments: 'The dogs . . . were not regarded as such a serious pest being more numerous and easier to destroy and were worth only half the thylacine rate' (Guiler, 1985, p. 17). Guiler matches the apparent illogicality in the bounty structure with an illogicality of his own. Two paragraphs after suggesting that wild dogs were not regarded as the most 'serious pest' he nevertheless admits that 'The VDL Company in fact had more trouble with wild dogs than with thylacines' (p. 17). Feral dogs were certainly the most numerous, consistent and significant predators on sheep at Surrey Hills. If the principal aim of the Van Diemen's Land Company bounty was to protect the sheep, the bounty should have been aimed at the major predator and have provided a maximum reward for the killing of feral dogs, with a reduced reward for the killing of markedly less important predators such as the Tasmanian devil and thylacine. This criticism of the bounty structure is not easy to resolve, particularly if one is committed to a belief that 'exaggeration or distortion' are not present in the Van Diemen's Land Company records about the thylacine (Guiler, 1985, p. 92). If, however, as I have suggested, the principal aim of the bounty was to disguise unfortunate decision-making and protect an incompetent, local management, then the bounty structure, that pretended a significant problem existed with an emotively named native predator, becomes far more explicable. This 'incompetent management', through the escape of dogs from the holdings and the deliberate gift of dogs to the Aborigines, was also largely responsible for the feral dog problem anyway – another political reason why the thylacine and not the wild dog needed to be placed at the head of the first Van Diemen's Land Company bounty scheme.

123

From this analysis of sheep farming by the Van Diemen's Land Company it is suggested that the establishment of the thylacine bounty did not reflect reality. Disastrous losses had occurred to the Company's sheep due to an inappropriate choice of land, the weather, poor management and predation by feral dogs. With thylacines identified in the stock reports (Guiler, 1985) as the cause of death for sheep at levels varying from 0 per cent at Circular Head (1832–38), 0 per cent at Woolnorth (1832–34) to only 7.4 per cent at Surrey Hills (1832–38) of all predatory kills, there is nothing in the Company records to suggest that at this time the thylacine was ever anything more than an occasional, relatively insignificant predator on sheep. The status granted to this first Van Diemen's Land Company bounty in the thylacine literature, and its effects on scientific constructions of the behaviour of the animal, are to be deplored.

The temptation to blame stock losses from feral dogs on the thylacine that began with the Van Diemen's Land Company grew as the years progressed. Popular perceptions of the thylacine as a significant sheep-killer were likely to result in a greater status and importance being granted to the role of the shepherd in the community, and less blame would be attached to a shepherd whose flocks were supposedly decimated by ravaging native tigers rather than by their own, or someone else's, dogs. Consequently, confounding the two predators and increasing the number of kills attributed to the thylacine was common. Owen

House described how, on the family farm in 1919: 'Several sheep and lambs were dead; others had lumps of flesh hanging from their bodies and tears in their throats oozing with blood. Some were tangled in fences and torn to pieces. In all, forty dead and others had to be destroyed' (2/9/1981). His father, who supposedly 'was well acquainted with the Tasmanian tiger' assumed 'this was the murderous work of the tiger that had been sighted on the property a few days before'. The culprits returned, attacked the sheep again and were killed. They were two large wild dogs. House concluded: 'I have often wondered if the tiger was exterminated and wrongfully blamed for deeds it did not commit' (2/9/1981). Serventy criticised the attitude of mind possessed by the average Australian farmer, to blame all stock problems on out-of-control predators: 'By seeking and accepting . . . scapegoats . . . farmers are failing to investigate more deeply and discover what is really killing their stock' (Serventy, 1966, p. 37).

Returning specifically to the written records of sheep predation on Van Diemen's Land Company holdings in the last half of the 1830s, we find that feral dogs were causing such severe problems to the sheep at Woolnorth in 1835 that Schayer's proposal 'to prevent the keeping of dogs at Woolnorth' was approved, but his request for an increase in the bounty payment awarded for the killing of wild dogs was rejected (Curr, 15/3/1835, 6/7/1835). During 1835 some thylacines also appeared at Woolnorth once again and killed some lambs (Court, 9/7/1835). Schayer requested a dog that would specifically kill hyenas, but Curr was initially unable to provide one (21/11/1835, 19/1/1836). Instituting enquiries, Curr discovered there were two dogs belonging to Reeves, now working at Circular Head, that were purported to have killed thylacines, but Curr desired they remain on the Circular Head holding. He suggested that Hewitt, Richardson and Cummings at Woolnorth had, or claimed to know of, 'vermin' dogs that would kill thylacines and suggested these should be used instead (8/2/1836).

Unfortunately, the Richardson and Cummings dogs lived up to their names in an unexpected manner. When released amongst the sheep at night, supposedly to protect them from the thylacines, they turned on the flock instead and started slaughtering sheep with abandon. The dogs were immediately recognised as the same ones that had caused so much destruction to the sheep at Woolnorth in 1833, when their ownership had been attributed to the Aborigines. A Local Board inquiry was set up and found Richardson and Cummings and their dogs responsible for 'the large loss which took place at the South Downs in the Improved flock in the year 1833' (14/4/1836). While Richardson and Cummings were fined accordingly (Curr, 18/4/1836, 30/6/1836, 10/8/1836), it was deemed politic not to correct the perception held in London that the 1833 losses were of indigenous origin.

Still uncaught, the 1835 thylacines hung around the flocks at Woolnorth, 'the Hyenas are more than usually troublesome' (Curr, 7/5/1836) and Curr discovered details of 'a contrivance for *catching Hyenas*' and sent them to Schayer (30/11/1836) but the 'contrivance' was not needed. Presumably the thylacines were finally killed, for their depredations against the flocks ceased in October 1836 (Curr, 13/5/1837, 15/7/1837). Nearly three years were to elapse before thylacines reappeared on the property, and made their presence felt in a minor way on the flocks. While in the past on Woolnorth there had been 'considerable losses of Sheep by Hyenas . . . this year, indeed since October 1836, there have been no losses whatever' (Curr, 15/5/1837). Curr further expressed himself as 'very glad to observe . . . you are not troubled by Hyenas' (16/8/1837) and 'You are most fortunate in being so little troubled with hyenas' (23/7/1838).[9]

During the 1830s Curr did not restrict himself to blaming the economic woes of the Company on either the thylacine or the Aborigines. Given a perception that he was 'the manager of an establishment . . . on which so many thousands have been expended with so small a prospect of a return' (Ross, *Hobart Town Courier*, 27 March 1835), Curr set out in 1835 to scapegoat the convicts, demonstrating how difficult it was to have to rely on a workforce of convicted criminals. Curr claimed to have discovered 'a conspiracy amongst the reprobate parts of the convicts of this establishment to seize upon their officers and the free people, and after helping themselves to all the convertible property they could lay their hands on, make off in the Company's schooner to America' (Curr, 27/3/1835). In an editorial comment answering charges of libel levelled against him by Curr, James Ross (27/3/1835) cynically dismissed this story as a fabrication: 'a mere Will-o'-the-wisp . . . conjured up to frighten the natives and the Directors'. Within the space of a few years all this had changed. By the end of the decade the convicts were being portrayed as model and productive citizens, while it was the tenant farmers and the indentured servants that were now suggested as the cause of the Company's economic woes:

> The Directors, anxious for dividends, continued to send these [indentured servants], in spite of Curr's protests that Irish families sent in 1840 were equal to second or third rate prisoners and that he preferred to hire ticket-of-leave convicts; in 1841 he complained that his convicts were being corrupted by the Irish servants. (Meston, 1958, p. 50)

Returning to the published chronological record of the relationship between thylacines and sheep, the comments on sheep farming in the *1830 Hobart Almanack* again reflected the news from the Van Diemen's Land Company. On this occasion, rather than mismanagement, the problems supposedly caused by the thylacine were emphasised in an anonymous publication, now known to have been penned by Thomas Scott (Von Stieglitz, 1966).

> At three miles from Westbury the Van Diemen's land company have a hut, and occupy a large tract of ground . . . for the purpose of grazing a numerous flock of fine woolled sheep . . . Considerable numbers of the native hyena prowl from the mountains near this, in quest of prey among the flocks at night. The Shepherd is therefore obliged during the lambing season, either to watch his flocks during the night, or to secure them in a fold. One of these animals had just been caught before the party passed. It measured six feet from the snout to

125

the tail. The skin is beautifully striped with black and white on the back, while the belly and sides are of a grey colour. Its mouth resembles that of a wolf, with huge jaws, opening almost to the ears. (Scott 1830, p. 53)

Originally published in *Hobart Town Courier* (28 February 1829), of restricted distribution, its republication in a readily accessible book format (omitting only comment on the suggested military splendour of the thylacine's skin), Scott's comments became much replicated for the next thirty-five years; but again, not everyone writing on sheep farming in Tasmania subscribed to this position. For example, Betts' (1830) overview of Tasmanian sheep farming failed to credit the thylacine as a problem to the industry.

No new records of predation were published in 1831. While Scott's story was republished by Grant (1831), the *1831 Hobart Town Almanack* returned to the position adopted in the 1829 issue, that the thylacine was relatively unimportant as a predator on sheep, and failed to mention this behaviour in an overview of Tasmanian plants and animals (Ross, 1831).[10] The records of sheep predation published in 1832 were not original either: Bischoff (1832) replicated Widowson, and Henderson (1832) reprinted Scott. In January 1833 Robert Lawrence, a young but experienced sheep farmer following the example of his father, mentioned the presence of the 'Hyaena' in his journal during an excursion into western Tasmania but, as previously, he made no reference whatsoever to sheep predation when mentioning the species (17/1/1833). While Prinsep made no reference to thylacine problems in her comments on sheep farming (1833), two direct replications of Scott's 1830 predation account were again published in this year by Melville (1833) and Parker (1833).

Breton (1835), in a publication that warned of the danger of feral dogs to the farming community, also mentioned the thylacine and, influenced by Widowson (1829), suggested that there was some economic importance to the thylacine as it 'destroys lambs'. However, Breton corrected Widowson's small, terrier-sized thylacine with the specific measurements of a tame 5 ft 5 in (1.67 m) specimen in captivity (Breton, 1835, pp. 357–8).

Richard Owen began writing and publishing on the thylacine in 1837. Initially, he considered either the incidence of sheep predation to be so rare as to be unworthy of specific comment, or the sources suggesting regular sheep predation to be so unreliable that, when he wrote about the behaviour of the species he failed to mention sheep predation at all: Owen 1837, 4/1/1837, 1838w,x, 1839w,x, 1840. In 1838 a new record of sheep predation was established courtesy of R. C. Gunn: 'when I was recently at the Hampshire Hills two were caught in one week at the sheep' (Gunn, 1838w, p. 101). No reference, however, was made to the thylacine bounty, and it would appear that it had wound down some years previously (when its political role was no longer necessary).

To sum up, the 1830s commenced with a Van Diemen's Land Company bounty against the thylacine and an attack on the species republished in the *1830 Hobart Town Almanack*. The validity of the bounty has been brought into question, suggesting a political rather than predatory origin. Occasional acts of predation on sheep were recorded at Surrey Hills and even more rarely at Woolnorth. Only one additional published account of a specific incident of sheep predation has been found (Gunn, 1838w). However, while some commentators failed to give credence to sheep predation when discussing the thylacine or sheep farming in Tasmania, a growing number of authors were becoming attracted to that position.

Scott's dramatic account of thylacine predation (1829), republished in the *1830 Almanack*, differed from the positions expressed in the 1829 and 1831 almanacks, but nevertheless, it was the one that was picked up and widely repeated in the 1830s, just as Wentworth's (1819) account was repeated in the 1820s. It is understandable from a publishing point of view how the ideas of these two authors with their dramatic accounts of sheep predation captured attention. Such tales maintained interest and suggested excitement and danger in the wilds of a colony on the other side of the world. It is, however, unfortunate that these reports appear to be so unrepresentative of the behaviour of the species as a whole.

1839–1849: The second private bounty scheme

Yet another republication of Scott's account of the thylacine occurred in Martin (1839).

Australia's first great rural depression commenced in 1840, and the effects were particularly severely felt in Tasmania. The price of wool on the London market was low and all the Australian colonies were overrun with sheep (Heazlewood, 1992). The new Port Phillip Colony (later Victoria), across Bass Strait from Tasmania, proved to have more and better wheat and grazing lands and offered wider prospects to investors and settlers. Consequently there was a massive diversion of capital to this new mainland colony (Fitzpatrick, 1949). Unlike England, which had usury laws to restrict the legal rate of interest to 5 per cent, Van Diemen's Land had none: 'By means of borrowed money on which they were paying ten, fifteen and even twenty per cent interest, the Tasmanian settlers in the 'thirties built themselves a magnificent card-house to live in' (Fitzpatrick, 1949, p. 20). Their fall was sudden and dramatic.

At the same time there was an increase in the number of convicts transported to Tasmania, because transportation to New South Wales ceased in 1840. The greater number of convicts placed greater financial demands on the already stressed Tasmanian economy (Robson, 1985). There was also a change in convict management, from the assignment system (convicts assigned to work individually in either government or settler enterprise, and to be housed at their place of employment) to a probation system (where convicts were to be housed together in separate camps, with discipline and employment related to various stages of rehabilitation). Effectively, the convicts spent one to two years working on government programmes, before merely being 'eligible' to work for nearby settlers. Colonial enterprise established far from the settled areas of the island was then deprived of the high levels of cheap convict labour upon which they had previously relied (Fitzpatrick, 1949, p. 315). Thus 'in 1840 the [Van Diemen's Land] Company lost its convict labour when the assignment system was abolished' (Mercer, 1963, p. 38).

Finally, the rains failed: 'Nature . . . turned her back on the Tasmanians and for three successive years the wheat harvest failed' (Fitzpatrick, 1949, p. 20). The drought was particularly severe on the north-west coast of Tasmania – the location of the Van Diemen's Land Company's best land, the 150 000 acre Woolnorth holding (Mercer, 1963, p. 38).

The Van Diemen's Land Company, set up to specialise in sheep, was faced with enormous financial difficulty. While the local managers of the Company might be blamed for a lack of foresight in not encouraging greater diversification of stock and production, most of the other economic and environmental problems were beyond their control. With a world glut of sheep and sheep products; with the rural and general depression throughout the Australian colonies exacerbated in Tasmania with the collapse of the Tasmanian economy and the loss of investment to Victoria; with the cessation of cheap convict labour and severe drought for the first time in the colony's thirty-eight-year history; the Van Diemen's Land Company was in serious economic disarray. Maintaining an historical precedent of what to do when faced by economic disaster, in 1840 the Van Diemen's Land Company introduced another thylacine bounty.

The first bounty scheme had paid at the rate of 10 shillings per head, irrespective of sex, from April 1831 onwards, until the political expediency (rather than any predatory reality) associated with its existence ceased. Emphasising this is the lack of contiguity between the two bounty schemes. On its introduction the second bounty scheme actually *decreased* the price paid for killing thylacines by 40 per cent! This bounty was obviously not designed to encourage the Company's employees to renew their efforts at capturing the species. Some other purpose must be found to explain its existence. Like the first, the new bounty scheme was also presented on an increasing scale: 'six shillings per scalp for less than 10 scalps, eight shillings each for 10 to 20, and ten shillings each for more than twenty scalps' (Guiler, 1961w, p. 207).

This second bounty scheme ran for seventy-four years, from 1840 to 1914 (Guiler, 1985, p. 26). Knowledge of its establishment has had considerable effect on popular and scientific perceptions of the species. Numerous authors have alluded to the 1840 bounty and drawn conclusions from it about the thylacine's behaviour, one of the most frequent being that thylacines were obviously common and constantly killing significant numbers of sheep:

> This scale of remuneration suggests that thylacines were plentiful and that a considerable number of sheep were being killed. (Clucas, 1978, p. 2)

> It is evident that the Company was having much trouble with thylacines, which must have been relatively abundant for such a bounty to be offered. (Guiler, 1985, p. 17)

128

A second interpretation claims that thylacines had been persistently slaughtered since 1840 but still managed to breed successfully and survive on the Van Diemen's Land Company properties for seventy-five years (Australian National Parks and Wildlife Service, 1978; Guiler, 1961w, 1985). As the species could supposedly withstand persecution for long periods of time, some people are still, today, waiting for the recovery of the species after the government bounty that ceased in 1909! This second Van Diemen's Land Company bounty has been such a powerful image in the construction of recent scientific ideas about the thylacine that one almost needs to apologise for suggesting an examination of the Company records to test whether any of these assumptions is valid.

There is no evidence up to 1838 that thylacines were ever more than very occasional predators on sheep on the Van Diemen's Land Company holdings. If there was a sudden behavioural change in the species that prompted the development of a new bounty scheme in 1840 – and this is not a behavioural impossibility – then one would expect to find evidence of the frequent killing of sheep in 1839, prior to the new bounty's establishment. Fortunately, predatory loss records have been preserved at this time for nearly all of the 350 000 acres controlled by the Van Diemen's Land Company. They record that in 1839 four sheep were lost to thylacines on the 150 000 acres at Surrey Hills (Guiler, 1985, table 6.6, p. 109), one sheep was lost to thylacines on the 150 000 acres at Woolnorth (Guiler, 1985, table 6.1, p. 96), and no sheep were lost to thylacines on the 20 000 acres at Circular Head (Guiler, 1985, p. 110). Five sheep lost to predatory attack by thylacines on 320 000 acres in 1839 provides no rational basis, either for seeing the thylacine as a frequent killer of sheep, or for the establishment of the 1840 bounty scheme. Whatever the real reason for the construction of this second thylacine bounty, it appears to have had little to do with the actual behaviour of thylacines living on the Van Diemen's Land Company holdings. I believe that this second bounty, like the first, was designed to suit a political and economic end: creating a smoke-screen to help excuse the reality of the Company's immanent economic collapse. Whatever the reason, the bounty bore no relationship whatsoever to the Company's own data on the incidence of thylacine predatory attacks on its sheep.

Just as the origin of this second bounty is dubious, its implementation is also open to question. The scheme officially commenced in 1840, but the first record of a bounty paid under this scheme does not occur until some ten years later on 31 August 1850 (Guiler, 1973x, 1985). Guiler analysed the Van Diemen's Land Company account books and cash books available for sixty-four years out of the bounty's seventy-four-year existence, to determine the total number of thylacines submitted for payment under the scheme. In his own words: 'The cash books from 1841–85 and the individual accounts from 1849–1904 show only eight bounties as being paid' (Guiler, 1985, p. 22). Only eight! Truly, the impact of this bounty scheme on contemporary public and recent scientific perceptions of the thylacine's population numbers, predatory behaviour and reproductive behaviour, far exceeds the reality of its operation.

129

Ivan Heazlewood, in his book entitled *Old Sheep for New Pastures* (1992), wrote the history of the Tasmanian branch of the Australian Society of Breeders of British Sheep, covering the early arrival of sheep in Tasmania, their rapid increase, significant individual Tasmanian farmers, and the establishment in the 1820s and 1830s of the large agricultural companies primarily concerned with wool production. From his detailed knowledge of the history and archives of the early sheep farming ventures in Tasmania, and the unique problems they confronted, he singled out the Van Diemen's Land Company alone for comment as the only farming enterprise that apparently 'lost numbers of sheep and lambs to Tasmanian Tigers' (1992, p. 86). While I take issue with Heazlewood about an inference of significant sheep predation arising from the known existence of the Van Diemen's Land Company's bounty schemes, I see no reason to question Heazlewood's inference that other major sheep farming ventures in Tasmania *did not* suffer depredations from thylacines. In fact, I have had the opportunity to test this thesis against an archive collection largely ignored by Heazlewood. He made only scant reference to the Clyde Company, primarily because, for much of the time, its Tasmanian partners were less interested in wool production or preserving the purity of British breeds of sheep since it possessed tendered government contracts for the supply of meat to Hobart and surrounding areas (later extended to Launceston). (An additional factor relates to the fact that most of the Clyde Company archives are held in the State Library of Victoria, rather than Tasmania.)

The Clyde Company origins date back to the arrival in Van Diemen's Land in 1822 of the first of five sons of Philip Russell, of the County of Fife. It was formally dissolved in 1873 (P. L. Brown, 1971). Most of the records of the Clyde Company – correspondence, stockbooks, company returns and some diary comment – were published in a series of eight volumes by P. L. Brown between the years 1935 and 1971. Additional, unpublished archival sources about sheep farming in Tasmania include the day book and ledgers for all the Tasmanian properties between 1838 and 1874 and the private diary comment of William Russell. I have read all the Clyde Company archives relating to the sheep farming ventures in Tasmania by members of the Russell family between 1822 and 1874. During these times the Russell family sheep suffered significantly, and constantly, from predation by wild dogs, and at irregular intervals, from the effects of sheep-stealing and attacks by Aborigines, bushrangers/escaped convicts and own-convict sabotage. While archival comment indicates thylacines were certainly around at the time, I have found no specific instance of thylacine predation on sheep in any of the surviving stock records, letters, diary and day book entries made during this fifty-three-year period.

Initially, contact with thylacines was considered an item of interest, as in William Russell's diary entry of 13 May 1841: 'One of Mr Bisdee's shepherds brought a tiger he had killed on the side line'. No suggestion of sheep predation or payment of a reward was made at this time. However, William made an extended visit to Hobart, Camden and Richmond for five days in July that year. Who he spoke to about thylacines, or what he read during this time is unknown, but on his return to Dennistoun, he offered the shepherd 10 shillings for 'Killing Tiger' (22/7/1841). News of this reward travelled through the company's employees, such that on 2 August 1841 William 'paid James Newman £2 for killing 4 tigers' and again, on 2 October 1841, William paid 'Shepherd for Killing a Tiger 10/-'. Outside of these three occasions, I have found no other record of any reward or bounty payment made to a Russell family employee for the destruction of thylacines in the day books and ledgers of the Tasmanian properties up to their closure in 1874.

This Clyde Company thylacine bounty scheme was remarkably short-lived, existing only between July and October 1841. As it is unaccompanied by any mention of thylacine predation on the flocks in the detailed stock records, it is tempting to suggest that it was set up to reflect a fashion-of-the-times. Alternatively, if there was a thylacine stock problem, it was obviously minimal, given the failure to record such incidents and the transient existence of the bounty. The most one can possibly suggest from the Clyde Company archives is that a thylacine or thylacines may have possibly killed a few sheep during a single four-month period of the Russell family's fifty-three years of sheep farming in Tasmania. Heazlewood's inference that, apart from the Van Diemen's Land Company, the other major Tasmanian sheep farmers appear to have been little troubled by the thylacine survives this investigation of the Clyde Company archives unscathed.   131

One other more fictitious than factual construction has been placed upon the Van Diemen's Land Company records: those associated with the employment of the so-called 'tiger men'. In 1834 Robinson noted that watchmen were employed to guard the flocks at night from wild dogs and, potentially, thylacines (4/6/1834). The Company employed its first official 'tiger man', James Lucas, at Woolnorth in 1836. Its last 'tiger man' was George Wainwright jnr. The designated employee position of 'tiger man' ceased in 1914 (Guiler, 1985, p. 104).

It has been suggested that the position of 'tiger man' involved a full-time commitment to the eradication of the species. Some farmers 'went to the extent of employing full-time trappers on their sheep runs' (Sharland, 1966, p. 55); 'sheepowners . . . had to keep a huntsman for the express purpose of killing these animals' (*Hobart Mercury*, 27 August 1887); and the 'tiger men' were 'the Van Diemen's Land Company's full time tiger hunters' (Pink, 1982, p. 17). It has also been suggested that the importance of these men was such that, in those holdings of the Van Diemen's Land Company unfortunately lacking a 'tiger man', 'stockmen and shepherds were assigned to shoot or snare tigers on at least one day of their working week' (Pink, 1982, pp. 16–17). The existence of these 'tiger men' has long been much commented on in writings about the thylacine. The misleading interpretations placed upon the two Van Diemen's Land Company bounties is enough to suggest caution. An analysis of the history of employment of the first and penultimate 'tiger men' employed by the Company also suggests that we cannot take the role of the Van Diemen's Land Company 'tiger man' at face value.

In a letter to London, Curr outlined both the reasons and conditions for the employment of James Lucas as 'tiger man' at Woolnorth:

> I have recently visited Woolnorth, where I found the sheep generally in satisfactory condition. The Hyenas were troublesome, and chiefly to the Saxon flock. To check their ravages I have sent to the Hills for a man who resides there with a pack of dogs making a living by killing wild animals. (5/8/1836)

Lucas was already an employee of the Company when he was sent to Woolnorth with the expectation that he would be able to survive by working full-time catching thylacines on the remuneration scale of: 'a ration of flour and the half guinea [10/6] per head for such Hyenas as he kills' (5/8/1836). Lucas was unwaged on Curr's expectation that, given flour from the Company's store at the level of 'the ration of a free man' (27/9/1836), he could make a reasonable living (and pay for his other necessities of life) from the rewards for killing thylacines. However, the local overseer, Schayer, thought otherwise.

Although the one or two thylacines that had been around the sheep at Woolnorth since 1835 'were troublesome', nevertheless, they were not common, so Schayer, in a sincere attempt to prevent Lucas from dying of starvation, suggested his employment as a night shepherd. Curr was not enthusiastic:

> If you employ him as a night shepherd, which I do not wish, for my only object in sending him, is that he may destroy wild animals, you will give him wages in addition to his rations at the rate of £15 per year, but if when he is night shepherd he allows any of his sheep to be destroyed by Hyenas you will give him no wages at all. (27/9/1836)

As events were to show, Schayer was correct. When these one or two 'troublesome' thylacines voluntarily, or at the hands of Lucas, departed the property in October 1836 it was to be nearly three years before a thylacine reappeared again to attack the sheep at Woolnorth. Thylacines were held responsible for the death of just one sheep at Woolnorth in 1839, and for the deaths of four sheep in the years 1840 and 1841. Then once again the thylacines disappeared or were destroyed, and there were no instances of sheep predation attributed to the thylacine in 1842 (Guiler, 1985, p. 96).

Lucas survived because, despite being the Van Diemen's Land Company's first, supposedly full-time 'tiger man', he was nothing of the kind. He was actually employed at Woolnorth as just another shepherd. Once again, the information Curr sent to London, excusing the lower-than-expected financial returns by blaming the thylacine and suggesting a new way of dealing with the problem – by employing a full-time 'tiger man' – fails to match the reality of the situation. The constructed position of employment as the 'tiger man' at Woolnorth was a mythical fabrication from the start. It continued in that vein.

In May 1903 a memorandum of agreement was drawn up between Ernest Warde, appointed as 'tiger man' at 'Mount Cameron Woolnorth' and the Van

Diemen's Land Company. He was to be paid £20 p.a. with an additional £1 for each thylacine killed. His duties as 'tiger man' were specified as follows:

> It is hereby agreed that the Snarer shall proceed to Mount Cameron Woolnorth, with as little delay as possible and shall devote his time to the destruction of Tasmanian Tigers, Devils and other vermin and in addition thereto shall tend stock depasturing on the Mount Cameron, Studland Bay, and Swan Bay runs, and also effect any necessary repairs to fences and shall immediately report any serious damage to fences or any mixing of stock to the Overseer & shall assist to muster stock on any of the above runs whenever required to do so & generally to protect the Company's interest & shall also prepare meals for stockmen when engaged on the Mount Cameron Run. (McGaw, letter, 29/5/1903)

Rather than being a full-time position hunting thylacines, or even a position in which a considerable part of employed time was spent hunting thylacines, the duties of the 'tiger man' appointed at Mt Cameron make it clear that it was only a titular position. The responsibility of the 'tiger man' was to act as a stockman, boundary rider and camp cook over the Mount Cameron, Studland Bay and Swan Bay runs. Occasional snaring activities would be appreciated and the capture of a thylacine would reap a £1 reward.

133

Undeniably, the position of 'tiger man' carried a certain masculine status, one that has only appeared to increase over time: 'The name "tiger man" . . . became known throughout the state and with the decline of the thylacine . . . acquired a certain mystique which still exists today' (Guiler, 1985, p. 108). Unfortunately, the name bore little resemblance to the actual duties of the employee concerned. The influence of the known existence of these 'tiger men' on constructions of the thylacine's predatory behaviour are to be regretted. 'Tiger man' was a descriptor with semantic rather than authentic status.

In concluding my analysis of both the Van Diemen's Land Company bounties against the thylacine, and the appointment of the so-called 'tiger men', it is worth pointing out that the argument, for the relative lack of importance of thylacines as predators on Van Diemen's Land Company sheep from the late 1820s to the 1840s, based on my own research, appears to remain true for the rest of the Van Diemen's Land Company archives as investigated by Guiler. With reference to the Company stockbooks and station diaries covering the period 1874 to 1914, Guiler comments for this later time period as well, 'sheep died more from other causes than from thylacine attacks' (Guiler and Godard, 1998, p. 131).

Supposed problems with thylacines on the Van Diemen's Land Company holdings made their way into the Company's annual reports, presented and circulated in London, and were acknowledged in the London-based newspaper designed to attract wealthy investors and immigrants to the antipodes (*South Australian Record and Australasian Chronicle*, 21 March 1840). Such publicity may have been the cause for the fashionable acceptance of sheep predation occurring briefly in British scientific publications on the species. For a two-year period in 1841 and 1842 Richard Owen gave countenance to the idea that the thylacine commits 'destructive ravages on the numerous flocks of sheep' (Owen, 1841, p. 258) and is 'a great pest to the shepherd' (Owen, 1842, p. 70). But he soon reversed this position, returning to his earlier stance of not mentioning sheep predation when he published on the thylacine (Owen, 1843; 1846; 1847). In 1841 Waterhouse published a review of the species, suggesting that thylacines 'attack sheep' (Waterhouse, 1841, p. 128). However, when he published again on the thylacine in his 1846 monograph on the marsupials, he omitted all reference to sheep predation from the text (Waterhouse, 1846, pp. 456–9). Also from 1841, J. E. Gray produced a check list and notes of all known eastern Australian mammals. The thylacine was listed as commonly occurring in only one of Tasmania's nine geographical areas, that in the north-west around 'Circular Head' (p. 400) – thanks, no doubt, to the Van Diemen's Land Company propaganda. Gray, however, did not take the opportunity to refer to any predatory behaviour on sheep, and other works published in the mid- to late 1840s on the characteristics of the thylacine also failed to mention sheep predation: Strzelecki (1845), Grant (1846), Swainson (1846), Breton (1847) and Baudement (1849). It is regrettable that constructions of the thylacine's behaviour by twentieth-century scientists and naturalists often failed to show the same critical evaluation of source material expressed by their forebears in the 1840s.

134

1850–1883: The concept (and the species) fade from scientific sight

The first live thylacines went on public display in 1850 at Regent's Park Zoo, London, and, come 1854, a thylacine was on public display in Propsting's animal collection in Hobart. Melbourne Zoo followed in 1864, and soon there were live thylacines on display in Paris, Berlin, Vienna and Launceston (Paddle, 1996). Thereafter, as thylacines became common in Australian, European and American zoological gardens,[11] a whole body of literature was generated on the species, considering not just its wild behaviour but also concerned with the observation, care and display of captive specimens. It needs to be recognised that, despite the general acceptance in the twentieth century of a common nineteenth-century understanding of the importance of the thylacine as a significant predator on sheep, references to sheep predation in the scientific literature published between 1850 and 1883 represent increasingly minority viewpoints, when considered against the volume of comment generated on the species during this time.

R. C. Gunn was very much aware of the dissonance between popular mythology, expressed in the wholesale slaughter of the species by shepherds, and the reality of sheep farming, in which the thylacine was an insignificant problem when compared to feral dogs and theft, and he expressed himself accordingly (1850w,y, 1852w,x,y). But popularly, particularly in British publication, the ideas and phrasing of Wentworth (1819) and Scott (1829, 1830) were still widely replicated, updated at times into more modern language and, for magazine publication, written in a more spectacular and arresting journalistic style (Angas, 1862; Baird, 1858; Bunce 1857; Howitt, 1855; Ireland, 1865; W, 1855).

The mythology that surrounded the thylacine was only part of the popular mythology that exaggerated the difficulties of life in Tasmania. Louisa Anne Meredith arrived in Tasmania in 1840, and her first published comment on thylacine sheep predation reflected popular mythology: 'The Native Tigers are . . . to be dreaded among sheep' (Meredith, 1852, p. 264). However, she later distanced herself from such popular constructions and considered thylacines to be so rare, retiring, shy or unobtrusive – 'we do not hear of them very often now' – that they no longer represented a potential threat to sheep as they had in the past (Meredith, 1880, p. 66). Nor were they a potential threat to humans. Meredith spoke out against the representation in British publications of the dangers to settlers in Tasmania: 'why our really peaceful lives should be represented at home as invested with such terrors by day and perils by night . . . I am wholly at a loss to conjecture' (Meredith, 1852, p. viii). Meredith's lack of understanding represents a naïvety to the social construction of knowledge typical of many scientists. Popular perceptions are frequently constructed with little recourse to objective data, and are not going to be changed by the mere presentation of objective evidence that negates such construction. A similar scientific naïvety typified the later inadequate response of scientists attempting to preserve the species from extinction.

135

Only two new published first-hand accounts of sheep predation have been found between 1850 and 1883. In 1868, a newspaper report that commenced with the predatory attack against some piglets of a tiger cat (*Dasyurus maculatus*), referred to as 'a spotted hyena', segued into another 'hyena' incident, one that now obviously referred to a thylacine. 'Some years ago' a 'hyena', supposedly previously responsible for killing twenty-five sheep, 'came as soon as it was dark and killed a sheep, but was shot down with coarse slugs of lead' (*Launceston Examiner*, 14 March 1868). In addition, the *Tasmanian* reported on details of the chase and killing of a large 6 ft 3 in (191 cm) thylacine by a shepherd and his dog at Blessington, noting that it 'was in the act of eating a sheep' when it was first discovered (*Tasmanian*, 16 September 1871). Not a particularly impressive total for a thirty-four-year time-span.

Some recent authors, allied to a significant sheep predation position, have chosen to rely on two dubious records for support. One of the more over-rated sources on sheep predation is the bastardised publication of Bunce (1857). Bunce has received uncritical acceptance in the literature by some authors, and has been referred to as a primary source and authority for significant sheep predation continuing on the Van Diemen's Land Company properties into the late 1850s (Guiler, 1985, p. 18; Guiler and Godard, 1998, p. 116; S. Smith, 1981, p. 23). However, Bunce's supposed first-hand account of thylacine predation on sheep on the Van Diemen's Land Company holdings in the late 1850s is totally fraudulent, a direct plagiarism from Scott (1829, 1830) published twenty-eight years previously.

The *Illustrated Melbourne Post* (22 March 1867, republished in *Illustrated Sydney News*, 16 April 1867) gave a picturesque account of a wondrously strong thylacine encountered near Fingal that, supposedly due to 'the Ravages which it commits amongst the flocks', was set upon and finally killed after a horseback chase of one-and-a-quarter hours at 'tremendous speed' and then fifteen minutes of being flogged with a whip. A grand round-the-campfire story of human persistence and ingenuity winning out over brute strength and stubbornness! While Guiler and Godard (1998) accept this account at face value, I consider the veracity of this report as questionable. Whatever the truth behind the original incident, if such an incident occurred it would appear that, in the process of leaving Tasmania, and crossing Bass Strait to be published in mainland newspapers, the deeds and behaviour of both protagonist and antagonist in the story appear to have undergone considerable exaggeration.

By the late 1870s and early 1880s Tasmanian naturalists and scientists were very much aware that whatever the potential threat thylacines might have represented to sheep production it had long since passed (Meredith, 1880). The animal appeared to be so rare as to make obtaining specimens extremely difficult. Advertisements were now placed in the press requesting thylacines. The animal dealer, Jamrach, of London, 'asks for twelve striped wolves (meaning our native tigers), for which he offers £4 each . . . dead, [and] pickled in strong brine of salt' (*Hobart Mercury*, 16 April 1874).

Kendall Broadbent, on a collecting expedition for the Australian Museum, Sydney, spent eight months in Tasmania, but could not purchase any thylacine specimens: 'The people tell me they are all caught' (15/3/1879). Reflecting Broadbent's lack of success, the thylacine was so restricted in distribution and so rarely seen that advertisement was again required in the early 1880s in order to obtain museum specimens. In 1882 Officer began advertising in the *Hobart Mercury*:

> WANTED to purchase a few Tasmanian DEVILS and TIGERS. Highest price given. Apply, S.H. Officer, New Norfolk. (*Hobart Mercury*, 21 February 1882)

How many specimens were obtained, for what price, and where they were sent is unknown. But the supply of specimens certainly failed to meet the demand and Officer was still advertising for specimens a year later (*Hobart Mercury*, 21 February 1883).

Notwithstanding popular perceptions, by 1883 concern is common in scientific circles about the future of the species, its rarity and the difficulty of obtaining specimens, alive or dead. Crowther (1883) described the species as 'not yet extinct' but heading in that direction; the *Launceston Daily Telegraph* (24 August 1883) referred to thylacines as 'these rare animals' and A. A. C. Le Souëf (26/11/1883) commented that the species was 'daily becoming rarer and more difficult to obtain'. Unquestionably, the occasional thylacine took the occasional sheep, but the actual threat that thylacines posed to the industry may best be summarised by recalling that, for the first eighty years of settlement to

136

1883, I have found only six original, published accounts based on 'eye-witness' observation that detail specific incidents in which thylacines preyed upon sheep: *Hobart Town Gazette* (5 April 1817, 6 December 1817, 2 August 1823); Gunn (1838w); *Launceston Examiner* (14 March 1868) and *Tasmanian* (16 September 1871). Popular mythology of the time was certainly not tied to published reality. Suddenly, without warning, the dam burst in 1884.

Notes

1 As a measure of how much greater problem the dingos were to sheep farming in New South Wales than thylacines were to sheep farming in Tasmania, by 1826 the Agricultural Society of New South Wales had already established a state-wide bounty against the dingo (Atkinson, 1826).

2 Morgan (1992, pp. 94–5) has identified a similar bias in Wentworth's discussion of fruit and vegetable production, in which he completely ignored Tasmanian productivity and specifically considered the apple enterprise only in terms of its potential in New South Wales (so much for Tasmania's designation as 'The Apple Isle').

3 It is important not to underestimate the power of seminal texts. Such has been the credence given to Wentworth (1819) that this confusion, suggesting the presence of two large indigenous predatory species in Tasmania, persisted well into the twentieth century. The popular *Everyman's Encyclopædia*, a common component of Australian school and home libraries up to the fourth edition, published in 1958, persisted with this error: 'there are sev. species peculiar to the is. – the T. devil (*Sarcophilus ursinus*) and the tiger (or striped) wolf, which, on account of the damage it wrought to sheep is now extinct. Among other larger indigenous animals are the native hyaena (*Thylacinus cynocephalus*)' (Bozman, 1958).

4 By this time five specimens were already acknowledged in publication. However, the anomalous incident of sheep predation (*Hobart Town Gazette*, 24 July 1819) sometimes attributed to the thylacine (Morgan 1992, p. 190), would appear to relate to the killing of a tiger cat (*Dasyurus maculatus*), scavenging a sheep previously killed by feral dogs.

5 One possible source was Thomas James Lemprière, who arrived in Tasmania in 1822 (Whitley, 1966), and commenced sending zoological specimens back to Swainson in 1827 (Gunther, 1900).

6 In Hobart at 11 a.m. on Thursday 11 June 1818 William Trimm 'was executed ... for sheep-stealing' (*Hobart Town Gazette*, 13 June 1818). The Land Commissioner, Roderic O'Connor, referred to another capital occurrence: 'one Davis, since hanged for sheep stealing' (23 June 1826, cited in McKay, 1962, p. 12). Note that no one was ever hung for the murder of Tasmanian Aborigines!

7 While written and dated from Hampshire Hills, the bounty was not restricted to the Emu Bay, Hampshire Hills, Surrey Hills complex. It was immediately extended to cover the Woolnorth–Cape Grim holding (Curr, 12/5/1830), and

137

Robinson (diary entry, 25/8/1830) records the bounty being paid on a specimen caught on the Middlesex Plains holding.

8 In communications with Curr, the word 'Court' refers to the senior managing executive of the Van Diemen's Land Company in London.

9 A similar pattern of local extermination and later recovery was still evident on the property nearly fifty years later (Guiler and Godard, 1998, p. 137). Sheep farming at Woolnorth, however, had not become plain sailing. Sheep and bullocks were stolen and killed by Company employees (Curr, 30/5/1839) and the remaining Aborigines stepped up their attacks on the sheep (Schayer, 31/1/1842). Feral dogs roaming the settlement built up to almost plague proportions and slaughtered indiscriminately: (S. J. Dyer, 31/3/1843, 30/4/1843, 30/6/1843; Gibson 8/5/1843, 10/5/1843, 24/5/1843, 31/5/1843, 22/9/1843).

10 Guiler (1985, p. 16) suggests an unpublished incident of 'tigers killing sheep' on the east coast in 1831, sourced to John Helder Wedge's diary. Unfortunately, his diary for the period 4/7/1830 to 27/1/1833 is no longer in existence and it is not possible to confirm this claim.

11 Zoological gardens in Antwerp, Cologne, New York and Washington also displayed thylacines. Thylacines were purchased in Australia for live public display in India and South Africa, and they were present in a private animal collection in Canada.

Chapter 6

Mythology Becomes Misology

Prior to the 1880s the Tasmanian Parliament had shown some interest in the preservation of native fauna – principally motivated by a recognition of the need to preserve native game from outright obliteration to allow future generations to share in the destruction. As a result of colonisation, problems with the environment surfaced on the agenda of all Australian colonial governments in the 1870s and 1880s. In 1871 Tasmania became the first Australian colony to pass a Bill for the destruction of rabbits (Rolls, 1969), but parliament was initially reluctant to move from a position of accepting the deliberate destruction of an introduced herbivore, to one accepting the deliberate destruction of a native carnivore. This chapter examines the evidence that contributed to this vital political shift between the years 1884 and 1888.

Local politics: stock protection associations and petitions

While professional scientists, naturalists and collectors were complaining about the rarity of the species, frequently attempting in vain to obtain living or dead specimens; as soon as these professionals left an area, thylacines apparently appeared *en masse* to wreak havoc on the poor pastoralists and their defenceless sheep!

In February 1884 the *Tasmanian* reported that Mr F. Brewer, of Ferney Hill, had 'captured a large tiger on Saturday last in a snare, but unfortunately it had been in the snare too long to procure it alive' (*Tasmanian*, 8 March 1884). Whether Brewer wanted a live thylacine as a pet for himself, or to offer it for sale is unknown. Certainly other north-eastern residents kept thylacines around this time. For example, Percy Tucker had a pet thylacine, kept on a chain, and a pet kangaroo, kept separately in an enclosure (Tasman, 1884). Alternatively,

Melbourne and Launceston City Park Zoos would have readily purchased a live thylacine specimen at this time.

In July of the same year Thomas Dunbabin: 'snared a large native tiger 5 ft 2 in long. Mr. Dunbabin . . . has lately missed a number of sheep, which have, no doubt, been carried off by these animals. A short time ago a tiger was caught, but it only measured 4 ft 11 in long' (*Tasmanian News*, 10 July 1884, reprinted 11 July 1884). The specimen was sent to the Royal Society's museum in Hobart (*Tasmanian News*, 11 July 1884). A report of its arrival at the museum was accompanied by an appeal to the community at large to help the museum obtain further specimens of these rare, hard-to-get marsupi-carnivores:

> The donation is one of a kind that the council of the Royal Society are very desirous to get, and our country friends would be of great assistance to the Museum if they . . . at any time capture any tigers, devils, native cats, etc., even if they were to send specimens that the Museum have already on view, the additions would be valuable for exchanges with other museums. (*Tasmanian Mail*, 12 July 1884)

140   Both the *Hobart Mercury* and *Tasmanian Mail* championed the claims of thylacine captures by the French brothers. The earliest reference I have found to their claims comes from the *Hobart Mercury* (4 July 1884): 'An active family named French, living at Tin Pot Marsh, have captured another native tiger, one of a number caught lately'. Further details followed:

> they have another, a large female, with four young. Exclusive of the young this capture makes twenty-seven of these destructive animals taken by this family during a short time. Their property is fenced high, so that nothing can get over, and apertures are left for snares, of which they set nine hundred, catching various animals. They will, if possible, secure the tiger and young for the Museum. (*Hobart Mercury*, 5 July 1884; *Tasmanian Mail*, 12 July 1884)

Unfortunately, these five thylacine specimens somehow disappeared and were never sent to the museum.

August 1884 saw the establishment of the first of four known 'stock protection associations':

> At a meeting held in Buckland on the 14th inst. . . . it was unanimously decided that an association be formed, to be called the Buckland and Spring Bay Tiger and Eagle Extermination Association ...
>
> Rules were adopted for the working of the association, the principal being, that to raise a sufficient fund every sheepowner pay $1/2$ d. per head for each sheep up to 1,000, and $1/4$ d. per head for all sheep above 1,000; and that a reward of £5 be paid for each full grown tiger caught in the district, and £2 10s. for all cubs equal in size to a full-grown domestic cat, the skins of all the animals caught to become the property of the association. (*Hobart Mercury*, 21 August 1884)

In true and typical fashion of the ruling class squattocracy, note the imbalance inherent in the source of funds for the Association. The small farmer, on a small plot of land with a small number of sheep was to be levied at a half-penny per head. The large sheep-holders, those who actually decided to form the Association, with their extensive holdings and their extensive numbers of sheep beyond their first 1000 sheep were to be levied at only half that rate, at a farthing per head.

The Association considered: 'presenting a testimonial to the Messrs. French for the important services rendered to all sheepowners . . . they having within the last three years caught no less than 33 tigers' (*Hobart Mercury*, 21 August 1884). If the report is to be believed the French brothers, with their 900 snares, had caught an additional six thylacines in the past month, and were now capturing eleven adult thylacines (plus young) per year. Once again, however, for unknown reasons, they were unable to send even one of these specimens to the museum.

Some tension is evident between the difficulty experienced by museums and zoological gardens in obtaining thylacine specimens, and the reason given for the formation of the Buckland and Spring Bay Tiger and Eagle Extermination Association. Some locals, unimpressed by the formation of the Association, apparently even suggested that thylacines were locally extinct in the Spring Bay area:

> The Curator of the Royal Society's Museum has just received . . . a large male specimen of the Tasmanian tiger . . . killed on Mr. Joseph Hodgson's estate, between Sorell and Spring Bay, . . . its discovery removed the belief existing in the minds of many persons that no native tigers are now to be found in that locality. (*Hobart Mercury*, 26 August 1884)

Hodgson was one of the committee members of the Buckland and Spring Bay Tiger and Eagle Extermination Association. The report also requested 'country residents' to continue sending specimens to the museum. This request was soon met with, from the same source:

> This morning the Buckland and Spring Bay Tiger Extermination Association forwarded a fine male tiger to the Museum . . . The Curator states that he hopes the example will be followed by other country folks, for the skins of these animals, if not kept in the Museum, are always useful for exchanging. (*Tasmanian News*, 29 August 1884)

William Turvey, treasurer of the Association, is recorded as the donor of this specimen in the museum records (Cooper-Maitland, *ca* 1968). Within the first fortnight of its existence two thylacines had been presented to the Extermination Association and both had been forwarded to the museum.

The first published statement of dissatisfaction with the operation and purpose of the Buckland and Spring Bay Tiger and Eagle Extermination Association was printed the same month:

141

A sheep-owner writes:- 'I see . . . that there has been a meeting held at Buckland by several sheep-owners, to form an association to exterminate the tiger and eagle, but . . . there are more enemies to sheep and lambs than the tiger and eagle. Why not have all included, and make the reward a little less for the tiger? There is the wild dog, the native devil, and sea hawk, the two latter very destructive among the lambs . . . Those who do not suffer from the attacks of the tiger and wild dog do so from the others more or less. I would not be against paying my share, barring I knew the general body of sheep-owners were deriving some benefit from the movement. (*Hobart Mercury*, 29 August 1884; *Tasmanian Mail*, 30 August 1884)

This disgruntled comment from a sheep-owner recognised that the thylacine was just one of many potential predators on sheep, not necessarily deserving of primary focus and attention. The writer, in fact, suggests that thylacine predation is not omnipresent but local, there are sheep farmers 'who do not suffer from the attacks of the tiger'. While the restricted incidence of predation may be put down to the good fortune of farmers whose stock was overlooked by unpaired hungry thylacines on their travels; it may alternatively be suggested that the species was rare, and apart from the occasional rogue individual, thylacines were not particularly partial either to eating sheep or to living in such drastically altered areas of habitat as farmland.

The following month the Association sent a further donation to the museum: 'Mr. Wm Turvey . . . forwards a female tiger and four young ones which were caught by the recently formed Buckland and Spring Bay Tiger Association . . . This is Mr. Turvey's second donation of the kind, so that he is becoming a very useful contributor to the society' (*Hobart Mercury*, 26 September 1884).

Three adult and four young thylacines were thus captured by the end of September after just six weeks of the Association's existence. The bounties paid would have totalled only £15 for the three adults; the four young being too small to attract bounty payment. This might suggest that perhaps thylacines were locally common and a problem to sheep. However, the Association was presented with no further thylacines for nine months. Rather than being locally common, after these first captures, the species appears to have become locally extinct, until further thylacines migrated from elsewhere. Not until winter of the following year were Association members again able to supply the museum with specimens.

On 28 October 1884 Mr James Gray, member of the House of Assembly as representative for the Sorell district on the south-east coast, presented a petition to parliament on behalf of twenty-six Spring Bay sheep-owners who 'suffer serious losses amongst their stock by the ravages of Native Tigers, &c', requesting 'an enabling Bill . . . giving Local Districts powers . . . to . . . rate upon stock' (J. Gray, 1884). It is worth noting that three days earlier, an article in the *Tasmanian Mail* commenting on sheep farming on the east coast had suggested 'tigers are comparatively scarce' (Special Correspondent, 1884), and that feral dogs were the major nuisance to the east coast sheep farmer.

142

Publication of the *Journals and Printed Papers of the Parliament of Tasmania* commenced in 1884. The annual reports on the status of the sheep industry prepared by the Chief Inspector of Sheep, formerly merely tabled in parliament, were now annually published. While the petitions and bounty debates over the perceived threat of thylacines to the sheep industry were argued over in parliament between 1884 and 1887, resulting in the establishment of a government bounty in 1888, at no time was a thylacine problem ever entertained by the respective Chief Inspectors of Sheep. Their annual reports deal frankly with the problems of the industry over stock reduction and stock loss in the 1880s. Disease, drought and competition with rabbits are cited as the main environmental problems to the industry. At no stage in any of the official annual reports on the state of the industry during these crucial years was the thylacine mentioned even as a minor problem of significance to sheep farmers: Harrison (1884); Tabart (1885–91).

Australian politicians not being ones to let objective facts about the environment interfere with the political process of representing one's economic power base, Gray was not bothered by newspaper comment, industry statistics or government publication and presented the petition anyway.[1] Essentially, the petition sought the granting of legal power to local district municipal governments to levy residents on agricultural matters. At a minimum, this would involve all sheep farmers at rates based on their stock held. This may be seen as an attempt to force universal subscription by local sheep-owners to the Buckland and Spring Bay Tiger and Eagle Extermination Association. Subscriptions to the Association had probably been lower than expected, given the rarity of thylacines, their infrequent attacks on sheep, and the imbalance in subscription rates between small and large stockholders. But, at another level, the petition requested that local government be granted the legal power to tax agricultural production within its municipal boundary.

The suggestion that the House of Assembly should cede some of its powers on agricultural and fiscal matters to local district government on the basis of, at best, a few rogue thylacines, was never going to get much support on the floor of parliament, outside the rural-rump lobby. On consideration, it was decided to print the petition (not including the names) in the *Journals and Papers of Parliament*, thus acknowledging the political gesture made by Gray, but nothing further was done about it. Any pressure to force the issue disappeared as the Extermination Association entered the nine-month period in which no thylacines could be found on which to pay a bounty.

The *Hobart Mercury* (29 October 1884) and *Tasmanian News* (29 October 1884) noted the presentation of the petition to parliament, but provided no editorial comment. However, the *Hobart Mercury* (3 November 1884) published a letter from a disgruntled sheep farmer objecting to the marginalisation of the sheep farmer's stock loss problems on to the thylacine:

> I believe we have a greater evil to contend with . . . [than] tigers . . . my fellow sheepowners should also combine to suppress, and, if possible, effectively

143

eradicate a greater pest, namely the 'Duffer,' [sheep stealer] who travels far and
fast, and cannot be either snared or trapped . . . tigers rarely kill more than one
sheep a night . . . Whereas Mr. Duffer removes sheep in fifties, aye, and in
hundreds, and leaves neither skin, bone or sign behind him. (Money, 1884)

Instead of creating stock protection associations against minor native predators
and involving parliament on the issue, it was alternatively proposed that 'a
Sheepowners' Mutual Protection Union be formed . . . and offer a reward of
£100 for the conviction of any person or persons stealing a member's sheep,
horses, or cattle' (Money, 1884).

In contrast to the position of the Hobart press, the *Launceston Daily
Telegraph* produced an editorial on the petition, opposed to the principles
behind any power to levy, praising the thylacine as the 'harmless . . . monarch
of the forest', and condemning the large sheep-holding squattocracy (the
'wool kings') for their rapacity and attempt to increase their local political
power:

> Mr. JAMES GRAY has presented a petition to the House of Assembly from
> 26 sheepowners, praying that steps may be taken to exterminate the native tiger.
> They are of opinion that if an enabling bill was passed, giving local districts
> power to deal with the evil, either by a rate upon stock, or other means, a great
> good would ensue, and the evil would be ultimately overcome. No doubt these
> 26 sheepowners believe great good would ensue – to whom? Our own opinion is
> that the Legislature interferes too much with matters that ought to be taken in
> hand by the people themselves . . . If the sheepowners want to get rid of tigers
> on their runs we would advise them to do so at their own proper cost, and not
> aim to levy a tax upon people in the towns, and the poor farmers who do not
> suffer from tigers . . . We are glad that Parliament was fooled into running the
> country to the expense of printing Mr. Gray's petition. It has enabled us, on
> behalf of the unlucky residents who own no sheep, to protest against this way of
> raising money for the benefit of the wool kings. (*Launceston Daily Telegraph*,
> 12 November 1884)

The politics behind the petition were more important than the content of the
petition itself. Pity the poor thylacine! Indicative of the actual population
numbers of the species, thylacines still proved so difficult to obtain that further
advertisements were placed in the press by 'Naturalist', New Norfolk, requesting
specimens (*Hobart Mercury*, 11 June 1885).

June 1885 also saw the registration of three thylacine specimens donated to
the Royal Society's museum in Hobart by members of the Buckland and Spring
Bay Association. The French brothers had caught an adult male and female
thylacine, and finally remembered to submit their captures to the Association
for a reward, before sending them along to the museum. An additional adult
thylacine donation was also registered this month, courtesy of the Association's
president, Mace (Cooper-Maitland, *ca* 1968).

July saw the publication of a newspaper report related to the killing of a thylacine specifically identified as being a rogue specimen with an history of killing sheep. While charges against the species for its supposed sheep predation occur readily, and contemporary newspaper reports imply that killed specimens have probably been responsible for local sheep loss (for example, *Tasmanian News*, 10 July 1884), this is the first specific eye-witness report located for well over a decade that identifies a thylacine demonstrably involved in killing sheep:

> a large native tiger . . . was killed a few days ago . . . Mr. Archer's shepherd at
> Russel's Plains . . . found a sheep that had been just killed. He sent his dog away,
> and hearing it barking he hurried up and found the dog and a very large tiger
> having a severe fight. He shot the tiger, which . . . had killed four sheep previous
> to this. (*Launceston Examiner*, 6 July 1885; *Tasmanian*, 11 July 1885)

July also saw the donation of two further thylacines to the Tasmanian Museum by the treasurer of the Extermination Association, Turvey (Cooper-Maitland, *ca* 1968). The manner in which they were obtained, and Turvey's treatment of the shepherd concerned, drew forth an anonymous letter to the editor of the *Tasmanian Mail* (8 August 1885):

145

> SIR, – A shepherd in one of the Eastern back blocks recently captured two
> native tigers, and, as a matter of course, applied to a prominent member of a
> vermin destroying association for advice as to the best means of obtaining the
> reward offered for their capture. He was informed that nothing could be done
> until the next meeting of the society, but the member presented him with
> a small sum of money which, he wished it to be understood, was entirely
> independent of the reward he was entitled to from the board, promising, at the
> same time, to bring his case before the board at the next meeting. He did so,
> but in the form of an application for the sum he had advanced to the shepherd,
> purporting to be his own private subscription, which he stated at the meeting,
> he had paid on the board's account. He furthermore opposed a motion to make
> the amount up to what is usually given for the destruction of a tiger, stating that
> he considered the shepherd had been very well paid by what he had already
> received. The mortification and disgust of the unfortunate victim of this
> questionable proceeding know no bounds . . . It would be interesting to find out
> whether the Royal Society got the animals as cheaply as the gentleman who takes
> the credit of having supplied them to it. (Constant Reader, 1885)

As Turvey only received public acknowledgment for his donations to the museum, the last comment of 'Constant Reader' appears unfair. This purchase of two specimens for an unidentified 'small sum of money', rather than the £10 the shepherd would have received as a bounty, is most likely indicative of the sorry financial plight of the Association. Continued lack of support for the Association may have been one of the factors behind Gray's second petition on the thylacine to parliament in 1885.

On Friday 14 August 1885, 'Mr. Gray presented a petition from land-holders and sheepowners of Spring Bay, praying that some steps be taken to destroy the native tiger now infesting Crown lands in that district' (*Tasmanian*, 22 August 1885). The annual report of the Chief Inspector of Sheep for the year 1884/85 – which makes no reference to the thylacine when discussing stock problems – listed forty-one different sheep farmers in the Spring Bay area, holding 55 703 sheep as stock (Tabart, 1885). From this total population of the sheep farming fraternity twenty-six signed the petition. The House of Assembly agreed once again to print the petition.

The petition this time specifically refers to the Stock Protection Association and comments: 'contributions being voluntary, the funds have failed' (J. Gray, 1885). Note that in twelve months the Association is known to have paid out a full bounty on only six specimens, which along with two others, cost the Association £30 plus an additional 'small sum of money'. (Bounties paid on eagles also, at best, ran to only a small sum of money. Despite requests for eagle specimens as well, in the twelve months to August 1885, none of the eagles received by the museum, either as heads alone or as entire specimens, is able to be sourced to the Association.) Working on the basis of twenty-six contributing sheep-owners – holding, on average, from the annual report statistics, 1359 sheep – the total funds provided to the Stock Protection Association should have amounted to more than £63. That 'the funds . . . failed' with expenses of just over £30 suggests that, for all their enthusiasm in signing, only half of the twenty-six signatories to the petition were prepared to put their money where their mouths were, and contribute to the Association. Eight thylacine specimens captured in twelve months (with a nine-month gap in the middle) provide little support for the idea that thylacines were 'infesting Crown lands' or operating as a constant threat to the industry. Instead of a request for a state-funded bounty, once more the petitioners requested parliament to give up its own power and grant local government the legal power to intervene, and levy a tax on agricultural production. Again, it is not surprising that outside the rural-rump lobby the suggestion fell on stony ground.

Gray subsequently attempted to resuscitate the petition, requesting debate in parliament on 'the great ravages committed by tigers on sheep'. He withdrew his motion on the promise that something would be done about it: 'The MINISTER OF LANDS [N. J. Brown] admitted the seriousness of the matter, which would be dealt with by the Cabinet' (*Tasmanian*, 26 September 1885). Cabinet, however, did nothing. Thus N. J. Brown, through deliberate inactivity on the issue, first entered the political fray on the pro-thylacine side.

Stock protection associations, which brought together at the executive level the established, articulate and educated members of the farming community in an association supposedly speaking on behalf of all, appear to have become fashionable political entities in Tasmanian rural communities at this time. There is evidence for the existence of three other stock protection associations (sometimes paradoxically referred to as 'Thylacine Protection

Associations' – Guiler, 1973w, p. 131) purportedly directed against the thylacine, one of which, the 'Glamorgan Stock Protection Association', was established at an unspecified time in 1885 (Braddon, 28/5/1888). Unfortunately, no records of the Association's members or activities have been preserved (Guiler, 1985, p. 21). Its bounty payments were considerably lower than the £5 per head offered by Buckland–Spring Bay. All that is presently known of this Glamorgan association is that, in 1888, it increased the price of its bounty payment to £2 per adult and £1 per juvenile thylacine (Braddon, 28/5/1888). Guiler (1985, p. 21) mentions the existence of two other stock protection associations, a Midlands Stock Protection Association with H. R. Reynolds as secretary-treasurer; and the Oatlands–Ross Landowners Association with F. Burbury as treasurer. No contemporary records of these societies appear to have survived; their commencement dates, membership and activities are unknown. There is, however, no reason to imagine that, aside from their political influence within their respective communities, they were any more important or necessary than the Buckland and Spring Bay Association.

The year 1886 commenced with a *Launceston Examiner* comment upon the donation of a thylacine to the City Park Zoo, that 'the species is very rare' (18 January 1886). In June a live 'three-parts grown Tasmanian tiger' was brought to the office of the *Hobart Mercury* by Mr J. H. Cleary, prior to being placed on board the train to take it to the Launceston City Park Zoo (*Hobart Mercury*, 29 June 1886). 'Mr. Cleary stated that tigers were at present plentiful in the Bream Creek district, and were very destructive to sheep' (*Tasmanian Mail*, 3 July 1886). For a species that formerly was widely but thinly spread, reflecting as much as anything the social behaviour of the species, it might be expected that, in certain areas, it had become locally extinct: hence the *Launceston Examiner* comment. The corollary is that, in other areas, it would expect to be considered, at best, as continuing to be locally rare. Cleary's suggestion that it could suddenly be locally plentiful in a well-settled area of greatly altered farming habitat, with all its concomitant pressures on the social behaviour of individuals, would appear to be rather unlikely.

The annual report for 1886 of the Office of the Inspector of Sheep, (Tabart, 1886) provided a summary of the recent history of sheep farming in Tasmania, with reference to stock held, sales made and the problems facing the industry. Tabart noted the much reduced profitability and production of sheep farming over the previous eight years – 'the decrease in the number of sheep in Tasmania . . . is nearly a quarter of a million' (p. 4) – and suggested causes for this, primarily rabbits, helped along by drought, parasites and disease. Nowhere in the seven pages of the report is there any mention of the thylacine as representing a problem either to sheep farmers as a whole, or to some farmers at a local level. It is clear from the report that sheep farming had become both more difficult and less profitable over the previous eight years.

Gray's attempts in parliament to increase the power of local government in agricultural matters, through scapegoating the thylacine as a problem, and attempting to raise 'money for the benefit of the wool kings' had a lengthy historical precedent, but proved unsuccessful. It was not long, however, before other members of parliament tried a different tactic to achieve the same ends.

State politics: Politicians and the passage of the bounty

The problems claimed to be caused by thylacines on Crown land intruded upon the discussion in the House of Assembly on 14 September 1886 on the allocation of £500 for the destruction of rabbits, with W. Brown and Falkiner requesting 'a small sum for the destruction of tigers'. The Treasurer, Burgess, replied 'he would bring the matter under the notice of the Cabinet' (*Hobart Mercury*, 15 September 1886). There is no evidence that Cabinet concerned itself with the issue at all.

This debate on the Rabbit Bill that extended to cover thylacines produced an interesting comment in the press on the problems of misinformed popular mythology on the thylacine:

148

> It is quite time some other name was commonly adopted for the comparatively harmless marsupial generally spoken of as the 'tiger.' . . . It is not the ferocious brute the name implies . . . On two occasions I have met with recently arrived immigrants who objected to leave town to secure work in the country for fear they or their children might be devoured . . . In the House of Assembly last evening Mr. Falkiner is reported to have said that 'the tigers were in many places driving people off the Crown lands.' Imagine the effect of this and other similar references to the animal upon people in England contemplating emigration, and making enquiries previous, to determine what part of the world they would go. (*Hobart Mercury*, 16 September 1886; see also *Tasmanian Mail*, 18 September 1886)

Such critical analysis in 1886 parallels Meredith's concerns considered in the previous chapter and applies equally well to both popular and scientific constructions of the thylacine held more than a hundred years later.

As John Lyne now took over front position in parliamentary debate against the thylacine, and eventually achieved a desired government bounty against the species, a few background details are desirable.

From his earliest years Lyne had given evidence of his suitability for a career as a politician. He kept a diary and memoirs of his own and his family's activities. Concerning his migration to Tasmania he described how his father had obtained letters of introduction from none other than Earl Bathurst of the Colonial Office, and how the family left London on board the *Hugh Crawford* on 25 June 1826, bound for Van Diemen's Land via 'Tristan de Cunda' (Amos, 1963). Unfortunately, his father did not obtain letters of introduction from the Earl, merely from the Earl's son. Instead of leaving on 25 June 1826, the official records of this voyage of the *Hugh Crawford* indicate it left London on 15 July

1826 and the ship passed nowhere near Tristan de Cunha, an island far off the main shipping route to Australia (Amos, 1963). Whether John's diary was written at the time, as suggested by Nyman (1976), or merely constructed at a later date, is irrelevant. Either way it shows evidence of a personal aggrandisement, fertile imagination and disinterest in demonstrable facts that would later be to the forefront in his political career as he set new nadirs for the standard of parliamentary debate.

In his personal memoirs 'John . . . ridicules the idea of the prisoners receiving any corporal punishment' (Amos, 1963, p. 27). Amos proceeds to counter Lyne's gently idyllic view of convict life with a diary entry from Lyne himself in 1842 referring to 'two probation men' who received 'thirty six lashes each'. Further evidence, counter to John Lyne's view of the humane treatment of convicts and prisoners working on the Lyne holdings, may be found in a letter from Henry Tingley, a convict assigned to the Lynes who wrote of his experiences to his parents in 1835: 'all a man has got to mind is to keep a still tongue in his head and do his master's duty . . . if he don't . . . they would take you to the magistrates and get a hundred of lashes' (Nyman, 1990, pp. 134–5). At a direct, personal level, the *Tasmanian Mail* gave an account of John Lyne's punishment of a refractory 12-year-old servant boy, who: 'was still determined and obstinate, even in the presence of a member of Parliament, so there was nothing for it but a whipping. Mr. Lyne produced a whip, and proceed to administer . . . a sound castigation, as was proper' (*Tasmanian Mail*, 25 October 1884). 149

On the issue of corporal punishment, John Lyne behaved in much the same way as he was shortly to do on the issue of thylacines: his public statements and writings for posterity bear little resemblance to the reality of his personal experience and private behaviour.

Lyne's first significant political lesson occurred when he ran for election as one of six local councillors for the Glamorgan district municipality in 1860 (Nyman, 1976). He lost the election. Although rejected by his electorate he was undeterred. Possessed of considerable personal influence over a majority of the six elected councillors, the size of the council was increased, and to protect the community from the expense of further elections for such an unimportant issue as democracy, John Lyne, along with his brother William, were immediately appointed as new councillors, enabling both brothers to protect and promote the Lyne family's interests, while misrepresenting all those people in the electorate who had not voted for them.

John Lyne moved from local council to state politics with the retirement, due to ill health, of Charles Meredith in 1879. Meredith was the first elected member for Glamorgan in the Tasmanian Parliament and had continuously represented the district since 1856 (Reynolds, 1956). Meredith died in 1880, the year Lyne replaced him in the Tasmanian House of Assembly (Nyman, 1990). Lyne remained in state politics for only fourteen years, retiring from public life in 1893 'owing to increasing deafness' (Nyman, 1976, p. 37).

A war of attrition against the pro-thylacine block in parliament was now underway, with the attempt made at every quarter, against the wishes of the government, to introduce the thylacine into parliamentary debate. On 22 September 1886 Lyne gave notice of a question to be put before the Attorney-General on the following day: 'Mr. LYNE to ask the Hon. Attorney-General, if it is the intention of the Government, by new law or otherwise, to offer rewards for the destruction of the Native Tiger, *alias* the Tasmanian Dingo?' (Tasmanian Parliament, 1886). The government, however, could not find time to consider the question on the 23rd, so the question was held over to the following day. But the government could not find time to consider the question on the 24th either, so consideration was put off until the end of the next week. To ensure its consideration on the 30th, Lyne gave notice on the 28th and 29th of his continued intention to ask the question (Tasmanian Parliament, 1886). Finally, Lyne took the opportunity on 30 September 1886, during discussion on a closed season to prevent the killing of native game, to introduce a completely antithetical position, calling for: 'the Government to offer rewards for the destruction of the native tiger, *alias* the Tasmanian dingo . . . the ATTORNEY-GENERAL replied the Government had no intention of doing so' (*Tasmanian*, 2 October 1886).

150

Undeterred, on Thursday 7 October 1886 Lyne put forward a private member's Bill against the thylacine: 'to recommend the appropriation of a sum of £500 for the destruction of Tigers' (Tasmanian Parliament, 1886). Debate on the motion was recorded in detail in both the anti-thylacine, pro-rural-lobby newspaper supporting the opposition, the *Tasmanian*, and the pro-thylacine, anti-rural-lobby newspaper supportive of the government, the *Hobart Mercury*:

> Mr. LYNE . . . said he had been connected with a society for the destruction of tigers for some time, and that society had paid £3 per head for 15 tigers. He suggested that the Government should pay a moiety of the amount paid by societies . . . He had had as many as 14 sheep killed by these animals in one week on land rented from the Crown. (*Tasmanian*, 9 October 1886)

The *Hobart Mercury* noted that Lyne 'had to give up the run' on which fourteen sheep were killed in one week (*Hobart Mercury*, 8 October 1886).

As the parliamentary representative for the Glamorgan district, it is hard to imagine that Lyne could have been referring to being connected for 'some time' to a stock protection association outside his own farming community and political constituency. The Glamorgan Stock Protection Association was founded in 1885 (Braddon, 28/5/1888), but despite Lyne's claim of the Association paying £3 per head in 1886, the bounty being paid by the Glamorgan Stock Protection Association was considerably lower than this. The actual level is unrecorded, but it is known that the Glamorgan Stock Protection Association only increased its bounty payment to £2 per head in 1888, and wrote to the new Minister of Lands suggesting the government do the same (Braddon, 28/5/1888). This is merely the first indication of Lyne playing unfair and free with numbers in order to support his political cause. He made ever-increasing, extravagant claims in parliament of the numbers of sheep killed by thylacines on his own 'Gala' estate, and on the properties of other east coast farmers.

George Huston seconded Lyne's motion. Then N. J. Brown, the Minister of Lands, spoke; originally opposed to Gray's thylacine petitions, he switched sides and now supported the rural lobby. Finally some opposition was voiced: 'Mr. FENTON hoped the House would not agree to the motion, as he thought the sheep owners ought to kill the tigers themselves' (*Tasmanian*, 9 October 1886), but it was short-lived:

> Mr. PILLINGER said . . . there was a vast extent of land in this colony only fit for the pasture runs; and owing to the depredations of this animal, the tenants were in many cases compelled to give them up. He alluded to his own losses, and said that in many cases sheepowners were occupying land at a loss, owing to the damage sustained. (*Hobart Mercury*, 8 October 1886)

With respect to Pillinger's comments, and the forthcoming remarks of the Treasurer, certainly in 1886 many sheep-owners were occupying their land at a loss and some Crown land leases had been given up, but, as the Chief Inspector of Sheep's Report makes plain, this had everything to do with 'the great pastoral crisis' and 'the depreciation in the value of wool and stock' as a result of the rural depression and the glut of wool on world markets; along with the 'long-continued and disastrous drought', competition from rabbits, the persistent problem of sheep flukes and a new problem of intestinal worms, and nothing whatsoever to do with the thylacine (Tabart, 1886, pp. 4–5).

Braddon then mounted a challenge, directed at the intent to subsidise the political and economic activities of the unrepresentative stock protection associations. He: 'asked what Mr. Lyne would do if he got the committee? He knew there were many societies of not more than six persons who would be glad to receive presents of £500' (*Tasmanian*, 9 October 1886). His challenge went unanswered and the debate continued, with Sutton, Falkiner, Burgess, Lucas and Dooley speaking in favour of the motion.

In contrast, members Dumaresq, Bird and Mackenzie spoke against the motion, suggesting those affected should deal with the problem themselves. The position of the motion's opponents was summarised thus:

> Mr. YOUNG . . . thought that if any persons were able to take care of themselves it was the large sheepowners.
>
> Colonel St. HILL did not see that it would be a great calamity if the sheepowners were driven off the Crown lands . . . it would be much better to have them populated with settlers. (*Hobart Mercury*, 8 October 1886)

Finally Lyne replied, emphasising the problem and advocating the support of rural enterprise for the betterment of urban conditions. He:

> said it had been asked during the discussion whether this animal killed sheep. He could assure the House that it did, and kangaroos as well, and if the colony wished to keep its sheep and kangaroo it would have to get rid of the tigers . . .

> It was a selfish view to take of the question that some did not suffer, as they ought not to legislate on class principles. (*Hobart Mercury*, 8 October 1886)

Of course, in reality, there were 'class principles' very much at work here, in support of the squattocracy.

The House divided and the vote was then taken: 'Ayes, 10 . . . Noes, 10' (*Tasmanian*, 9 October 1886). The resolution of the tie was achieved in the following manner: 'The votes being equal, the SPEAKER [Dobson] said he would best fulfil his duty by voting with the ayes, according to the usual custom, so that the question might be further considered' (*Hobart Mercury*, 8 October 1886). Despite Alfred Dobson's appeal to 'the usual custom', in reality the Westminster tradition, true to its conservative values, is that the Speaker, in the case of tied votes, votes against new initiatives. While the actions of Dobson, in voting for Lyne's motion and thereby continuing its political life, may be interpreted as the actions of one committed to the principles of democratic discussion and parliamentary debate, an alternative explanation is possible. Dobson had been Leader of the Opposition in 1883 and 1884, prior to his acceptance in 1885 of the government's offer to make him Speaker of the House of Assembly (J. Reynolds, 1956). There was no question that his unstated political sympathies lay with the opposition members' tactics and concerns. There are no surprises that, when given the opportunity for effective political decision, he chose to ignore the Westminster tradition and support this private Bill of an opposition member.

In contrast to the prominence in reporting attached to the thylacine issue by the *Tasmanian* and the *Hobart Mercury*, the *Tasmanian Mail* discussed the debate in one short, cynical paragraph.

> THE NATIVE TIGER. – The tiger, or dingo, received a very bad character in the Assembly on Oct. 7; in fact, there appeared not to be one redeeming point in this animal. It was described as cowardly, as stealing down on the sheep in the night and wantonly killing many more than it could eat, as being worthless even for its skin. Many members declared that these ignoble brutes are fed on Crown lands, so that the question of their extermination was declared to be one for the Government. All sheepowners in the House agreed that 'something should be done,' . . . The House was divided in opinion . . . and finally committee was granted by the casting vote of the Speaker. (*Tasmanian Mail*, 9 October 1886)

While refreshing to read, it is unfortunate that this contemporary, cynical response to the claims of the sheep lobby has not been reflected in modern constructions of the thylacine's predatory behaviour.

It was a full month before committee was held on 4 November 1886. The debate extended well beyond the thylacine. A challenge to bring down the government on the thylacine issue was made, and horse-trading of votes went on, with the racists and white-supremacists in parliament trading their present support for a thylacine bounty for the rural lobby's future support of an anti-

Chinese immigration Bill. The debate is also noteworthy for demonstrating the accepted, contemporary use of the Aboriginal word 'legunta' as a common, colloquial name for the species; and for the record of the first published witticism involving the species. Two newspapers reported the debate in detail from their own editorial points of view: the anti-thylacine, anti-Chinese, anti-government, pro-rural position of the *Tasmanian* (13 November 1886) in contrast to the pro-thylacine, pro-government, anti-rural position of the *Hobart Mercury* (5 November 1886).

Portions of the debate were reported as follows:

> Mr. LYNE stated that sheepowners lost at least 50,000 sheep each year by the ravages of the tigers, which were bred on the Crown lands of the colony . . . and there was great difficulty in catching them. Societies had been formed with the object of getting them destroyed, and the amount he asked for was for the purpose of subsidising these societies, which gave a pound or thirty shillings per head for all that were brought to them, the country suffered a great deal by the pest directly, because lessees of Crown lands were obliged to throw up their leases. (*Tasmanian*, 13 November 1886)
>
> Mr. LYNE . . . said it might be taken that 100 sheep were destroyed per dingo, and the destruction of 500 dingos would preserve 50,000 sheep. He quoted some extracts in support of his contention that sheep were decreasing and dingos increasing. What he would propose was that £1 per head should be paid for every tiger. It was no joke. He reckoned 30,000 or 40,000 sheep were killed annually by dingos. They ran whole flocks down into gullies, and maimed more than they killed. The scientific name of the animal was *legunta*,[2] and it was one of the greatest pests the colony had. (*Hobart Mercury*, 5 November 1886)

153

On the positive side, Lyne's suggestion that the Glamorgan Stock Protection Association was only paying a bounty at £1 or £1.10.0 per head is a welcome return to reality, over his earlier claim it was paying out for thylacines at £3 per head (*Tasmanian*, 9 October 1886).[3]

Guiler has suggested that Lyne used the word 'country', not in a national sense, but in the sense of the rural areas of Glamorgan, his east coast electorate: 'J. Lyne of Swansea claimed in 1886 that 30,000 to 40,000 sheep were killed each year on the east coast . . . Later . . . Lyne raised his claim to no fewer than 50,000 head of sheep being killed annually by thylacines on the east coast' (Guiler, 1985, p. 19). Guiler's interpretation appears reasonable. Later, in parliamentary debate in 1887, Lyne was to estimate the total loss of sheep in Tasmania to the thylacine as some 700 000 head (*Hobart Mercury*, 27 August 1887). It appears reasonable to consider Lyne's suggested 30 000 to 40 000 to 50 000 sheep lost per year as just Glamorgan estimates.

The magnitude of Lyne's claim that thylacines were destroying up to 50 000 sheep per year in his east coast electorate is impressive. The Chief Inspector of Sheep's annual report suggests there were only 44 512 sheep in the Glamorgan district this year anyway (Tabart, 1886, p. 6). One either has to be impressed at the rapacity of the thylacine, that, at its peak, annually managed to kill not just the entire stock of sheep held in Glamorgan, but over 10 per cent of next year's stock as well; or alternatively, one has to be impressed at the capacity of a politician to make such a disingenuous claim. Contemporary criticism also took Lyne to task over his statistics (*Tasmanian Mail*, 3 September 1887). In addition, Lyne's claims that thylacines 'ran whole flocks down' and 'maimed more than they killed' are atypical thylacine behaviours, but very typical of predatory acts by feral dogs. It is possible that Lyne either had no ability, or no desire, to distinguish between the two.

The debate was then opened up, with two members speaking in favour of the address and two members speaking against. The 'for' position was led by the original petitioner Gray, who spoke about the Buckland and Spring Bay Extermination Association. He was followed by Davies who 'had been informed that sheepowners in the district he represented lost 20 per cent. of their sheep by the tiger' (*Hobart Mercury*, 5 November 1886), a claim almost as audacious as Lyne's.

Two members then spoke against the motion:

154

> Mr. FENTON . . . ridiculed the idea of £500 doing all the good that was expected of it. He did not believe that the Crown land runs had been given up on account of the ravages of the tiger, and attributed it to the low price of wool inducing sheepowners to give up the poorer runs they rented from the Crown. (*Hobart Mercury*, 5 November 1886)

Everyone, from the farmer to the politician to the Chief Inspector of Sheep, knew that the industry was in severe economic trouble. Outside of any environmental factors, the industry was depressed; there was a glut of wool and wool products available on world markets, leading to a severe drop in export demand for wool and a consequent price reduction. It was understandable that sheep farmers were giving up both sheep farming and their Crown leases on these grounds; it was less understandable to Fenton – who obviously had little experience of the attributive wiles of Australian farmers – as to why they should choose to blame the thylacine for the economic woes of the industry, rather than honestly appraise the result of an uncontrolled expansion of the industry and blame themselves.

Braddon, the member for West Devon, continued this line of argument:

> Mr. BRADDON said the animal must certainly be a wonderful one, as according to Mr. Lyne's statistics each tiger . . . must consume at least 100 sheep yearly. He did not know where Mr. Lyne got his statistics from, unless they came from the statistician [the Chief Inspector of Sheep], but it seemed to him that . . . (*Tasmanian*, 13 November 1886) . . . if the loss was 50,000 sheep a year it meant £20,000 at least, and he could not see why the sheepowners should not spend this £500 themselves to make such a gain. It would pay the sheepowners to spend £15,000 to save themselves from such a loss. He would be very willing to

assist the hon. member to quell this *leghunter*, but it was absurd to think it could be done by such a vote. It was wrong of the hon. member to drag in the Crown lands and heap the blame upon them. He should . . . take a straightforward course, and move a vote of want of confidence in the Government for not having caught the tigers. (*Hobart Mercury*, 5 November 1886)

The *Tasmanian*'s publication of Braddon's questioning the validity of the sheep farmers' statistics should not be constructed as support for his position. Instead, by exposing Braddon's desire for bureaucratic data-gathering and decision-making by the statistician (in this case the Chief Inspector of Sheep), the *Tasmanian* was pandering to the conservative, anti-intellectual, anti-government, anti-bureaucratic attitudes of its rural readership. In the next comment on the debate, the *Tasmanian* referred to a proponent of the Bill, N. J. Brown, as 'speaking from facts'. That is, you get the facts by going to the horse's mouth; ask the farmer, not any public servant statistician.[4]

The logic behind Braddon's argument is difficult to fault. If the exaggerated claims of stock loss by sheep farmers were to be accepted at face value, then they should be more than willing to find £15 000 amongst themselves annually, rather than sponge on the government for a paltry £500. His willingness 'to quell this *leghunter*' is a well-constructed, thinly veiled attack on Lyne's sexual proclivities, one for which Braddon would soon be admonished for 'lowering the tone' of parliamentary debate.[5] Braddon's challenge to take up the issue in terms of a vote of no confidence in the government was a political tactic aimed at negating Lyne's motion. Treating it as a private member's Bill, rather than an Opposition Bill, allowed individual government members to vote in terms of their 'conscience', or the dictates of perceived present or future political gain. If the thylacine bounty had been adopted as an Opposition Bill, thereby involving a potential vote of confidence in the government, parliament would have split upon traditional government–opposition lines, thereby removing the support of at least one government member, the Minister of Lands and Works, N. J. Brown, from Lyne's camp, and the Thylacine Bounty Bill would have been defeated.

Two further members spoke in favour of Lyne's motion. In an attempt to win more government member support for this private Bill, both Sutton and Pillinger stressed the benefits of more mutton at lower prices. Offended at the innuendo against Lyne, Pillinger castigated Braddon (the West Devon representative) for 'lowering the tone' of parliamentary debate and attempting to make humorous this serious issue: 'He hoped hon. members would realise the fact that the question was one affecting the interests of towns as well as the country, and that the question would not be made a funny one, as the hon. member for West Devon had made it' (*Hobart Mercury*, 5 November 1886).

An opponent of the Bill then spoke, and, in two short sentences, summed up the problems involved in accepting the thylacine as a consistent predator on sheep and the need for the government to concern itself on the statistics provided for the rural lobby: 'Mr. DUMARESQ could not support the vote, as he did not see that the tigers could be any more numerous now than they were formerly. Nor could he understand why they were not destroyed, if they did so much damage, by the sheepowners themselves' (*Hobart Mercury*, 5 November 1886). Given the century's dramatic increase in Tasmania's human population, associated with significant habitat change, Dumaresq asked how reasonable was it to suggest that, in response, the thylacine population had dramatically increased in size? Only a developing commensal relationship with human society could justify this idea, and beyond the dubious claims of a few rural-rump politicians and one economically challenged company there is no evidence to suggest the thylacine discovered a sudden empathy and adaptability towards urban, suburban or rural existence.

In response to Dumaresq, Hawkes, the representative from Ringarooma, suggested that habitat change in Tasmania was even greater than city and rural incursion would suggest; that thylacines were forced to flee their destroyed native environment for the less hostile and more accommodating rural environment:

> Mr. HAWKES said as the miners of his district had driven these animals out into the hon. member's district by appropriating their natural food, it was only a matter of justice that he should support the hon. member; but he might shortly have to ask for a vote to exterminate another animal his constituents were suffering from – the yellow agony – and he would then ask for the hon. member's support.
> Mr. BARRETT: What, the Chinamen? (*Hobart Mercury*, 5 November 1886)

If the bush had been so successfully blitzkrieged by the miners (and trappers and timber cutters) then thylacines, as habitat-less and home-range-less refugees, would have disappeared within a generation, along with most other native species. The designation of the Chinese migrants as less than human – 'another animal', 'the yellow agony' – does not become softened in time. The purchase of votes for the thylacine bounty from the racists and bigots in parliament was crucial to the narrow success of the Bill, and, on balance, neither altered the morality, nor shifted the standard of intellectual debate, associated with viewing the thylacine as a major threat to the sheep industry.

The Speaker of the House, Dobson, then attempted to amend Lyne's motion. While prepared to accept the principle of government-funded bounty payments for agricultural pests, he expressed the gravest concern that 'So many extraordinary statements had been made as to the habits of this animal and its terrible ravages that he thought it would be worth while to refer it to a Select Committee to see what the facts were' (*Hobart Mercury*, 5 November 1886). Such amendment was anathema to Lyne and the anti-thylacine lobby. The last thing they wanted was a serious, objective examination of the issue. Unsurprisingly, the amendment was not accepted.

If Dobson had had the courage of his previously expressed convictions as Speaker of the House, and behaved 'according to the usual custom'

(8/10/1886) having lost the amendment he would have opposed Lyne's motion when it came to the vote, and the bounty would have failed to have been passed. The committee was being chaired by Henry Lette, representing the city-based electorate of North Launceston. In the event of a tied vote, his political sympathies would have lain with the government members opposing the Bill. Instead, when it came to the vote, Dobson, as the former Leader of the Opposition, appeared unable to vote against his political allies, and refused to cast a vote on the issue in committee, effectively ensuring the bounty was passed, by twelve votes to eleven, on the strength of his missing vote (*Hobart Mercury*, 5 November 1886).

The passing of the bounty was the final straw for the *Tasmanian News*, which produced a stinging editorial attack on 'The Wool-Kings and the Public Revenue':

> Year after year this pampered industry [has] . . . held the leading place in
> the favourable attentions of the Legislature . . . In regard to the compulsory
> poisoning clause of the Rabbit Act, the House of Assembly has placed itself
> in a most humiliating position. After its rejection by the House, a few large
> landholders met, and, urging their views on the accommodating Attorney-
> General, they got the rejected clause reinstated. Thus the wool-kings govern
> the House, and they get whatever they ask . . . During this session they have
> obtained whatever they desired for the protection of their industry. They had
> £500 voted to them for the slaughter of rabbits on the Crown Lands. They
> had another £500 granted for the killing of tigers. They have had a Rabbit
> Destruction Bill passed, which, according to the estimate of Mr. Pillinger, will
> cost the public revenue £12,000 a year . . . So far as the financial position of the
> large sheep-owners is concerned there was no reason why one sixpence should
> come out of the revenue for the destruction of rabbits or anything else.
> (*Tasmanian News*, 9 November 1886)

157

A largely unsupportive government now found itself encumbered by an unwanted Bill and an unwanted expense. It behaved in typical fashion, ignoring the existence of the resolution and omitting it from budget estimates and proposed expenditures. Despite the supposed ravages of the thylacine against the sheep industry and the urgency requested in government response, it was ten months before members of the anti-thylacine lobby discovered they had been rolled once again on the issue.

During this hiatus, the captures of two thylacines were recorded in the press. Another supposed capture by the French brothers was noted, together with reference to the sheep problem and the inactivity of the government on the issue:

> On the 19th inst., whilst the brothers French . . . were out on their run looking
> after and setting snares for tigers, they came upon one, and succeeded in
> rushing it into a snare set close by and caught him . . . it is a very nice specimen

of the Tasmanian tiger, and is nearly fully grown. Already a gentleman has made an offer of £6 for it, but which has been refused by the Messrs. French. These settlers . . . keep on catching and killing every year, and consequently their good work must be a great benefit to the sheep farmers of the district of Spring Bay . . . The loss of sheep during the last 18 months has, however, been great . . . and no doubt is caused by the ravages made by tigers. Settlers in the district are anxious to know something about the £500 their member moved for at the last session of Parliament, and hope that he will keep the thing alive at the coming session, and thus aid sheep farmers to eradicate the pest from amongst their flocks. (*Tasmanian Mail*, 30 April 1887)

One likely 'gentleman' who was constantly seeking live thylacines was William McGowan, curator of the Launceston City Park Zoo. If this specimen existed – and reasonable doubt attends most French brothers' claims – £6 was reasonable and generous.[6] However, this specimen was not sold to Launceston Zoo, and, despite stated desires and attempts made to obtain specimens, neither Melbourne nor Adelaide Zoos were able to purchase thylacines in 1887. This supposed specimen was either traded elsewhere to an unknown source, died in French captivity – without a Buckland–Spring Bay bounty claim or donation to the museum – or never existed at all. Once again, when positive proof should be readily available for a French brothers' thylacine capture, the specimen apparently disappears into thin air.

By the time the bounty commenced in April 1888, the French brothers claimed to have caught around seventy-five thylacines in the previous six-and-a-half years (East Coaster, 1929; Sharland, 1971w, 1971x). Despite repeated requests for specimens from the Tasmanian Museum and the French brothers' intimating to the press that they would supply the museum with specimens, two, donated in June 1885, are the only individuals from the claimed French captures that made it into the museum collection. The French brothers, supposedly killing on average eleven thylacines a year on their property, were no doubt looking forward to the government bounty. Unfortunately, tragedy struck the family. As soon as the bounty scheme commenced their thylacines disappeared! In the twenty-one years of its operation the French brothers were only able to submit two thylacines for bounty payment (payments approved on 16 March 1894 and 10 June 1901)!

The French family were doubly distressed however, for as soon as the bounty ceased in 1909, according to family history, the thylacines apparently returned. The brothers claimed to have continued killing and capturing live specimens on their property until at least 1922 (Morgan, 9/4/1928). Despite newspaper advertisements from the Tasmanian Museum during this time offering £5 for dead thylacine specimens, none of the claimed French brothers' kills made after the cessation of the bounty were ever offered to the museum. In an unpublished letter, William Ferrar, now owner of the Frenchs' 'Tiger Hill' property, recalls Abraham French's account of the brothers' last live capture. One brother supposedly saw a thylacine in the bush and grabbed it by the tail, and the other brother took a possum snare from his pocket and noosed the animal and they dragged it home: 'The tiger was kept for five months, till a traveller called, and offered the boys five pounds for it. In the year 1914 five pounds was a lot of money, so the Tasmanian Tiger was sold' (Ferrar, 24/8/1981).

Nothing more is known of this supposed specimen or unidentified traveller, which, by itself, is not unusual. What is unusual, however, is that the French brothers chose to sell it for a paltry £5. Commencing in early May and continuing until August 1914, the Tasmanian Museum, desperate for specimens, ran a series of advertisements in at least three different newspapers, detailing desired purchases. 'Tigers' headed the list of wanted animals, and interested individuals were invited to contact the museum. Enquiry resulted in a mailed circular from the museum, which commenced with the specific request for ' "Tigers" alive in good condition, £12 to £15', also offering £5 for dead tigers (and free rail carriage), but preferentially requesting that: 'All animals, if possible, to be sent in alive' (Tasmanian Museum, 1914).

Why the French brothers chose to sell their last live thylacine to a stranger for only £5 when they could have sold it to the museum for three times that amount is, by itself, unusual. In the 1880s they were aware of the museum's advertised requests for thylacine specimens and indeed offered to donate them, but only two were ever donated out of their claimed total of seventy-five killed thylacines; this, by itself, is also unusual. Apparently, when the bounty commenced all their thylacines disappeared, but when the bounty was discontinued they all returned once again. This must be considered an unusual account of thylacine migratory behaviour. Why the kills they claim to have made after the bounty finished were not translated into £5 per specimen from the Tasmanian Museum is also, by itself, unusual. Each of these occurrences, by themselves, present difficulties of explanation. To accept one, or even two of these unlikely events as having actually taken place might be permissible. Although such significant modern figures as Guiler and Sharland have been able to swallow the French brothers' claims, I find such a leap of faith beyond my poor capacity. While acknowledging how important the contemporary publication of their claims may have been to reinforcing popular perceptions of the thylacine as a sheep-killer, I consider that the French brothers' claims of phenomenal numbers of thylacines killed during the years 1883 to 1922 fail to meet any of the available criteria that exist for assessing their validity. I judge their claims about the thylacine to be the most exploitative and unlikely ever made about the species – outside, of course, parliamentary debate.

The elections in 1886 eventually led to a change in government. The coalition government of the so-called 'continuous ministry' that had been in power since 1879 could no longer hold itself together (J. Reynolds, 1956, p. 163). In the absence of any formal political parties, individual members realigned their political allegiances in March 1887 under Premier Fysh (in the Legislative Council) with Braddon as Government Leader in the House of Assembly (*Hobart Mercury*, 29 March 1887). Lyne remained an Opposition member.

The anti-thylacine lobby finally got its act together again on 26 August 1887 and Lyne, seconded by W. Brown, reintroduced the issue into parliament (*Tasmanian News*, 27 August 1887). Braddon, as Government Leader in the lower house and the new Minister of Lands and Works, oiled his way out of the non-budgeting issue. He congratulated Lyne on his success with the motion the previous year: 'It was his policy, his whole policy and nothing but his policy, and it was a proud moment when he got it through'. But while feigning sympathy with the problem, 'There was no doubt something would have to be done'; he suggested that a bounty 'would probably effect very little' and that some undefined alternative means should be found to keep Crown tenants happy (*Hobart Mercury*, 27 August 1887).

The government, through Dodds, again tried to defuse the issue, attempting to set it aside pending further data-gathering and discussion, and received support for this stalling tactic from Dooley (*Hobart Mercury*, 27 August 1887), but then Mackenzie, who had opposed Lyne's original motion in 1886, agreed to support Lyne's new motion and grant formal debate in committee on a Bounty Bill. Young presented the key issues of the pro-thylacine case. He: 'thought that whenever sheep were killed or lost in places where tigers were known to resort, it was always attributed to these animals. In his opinion the sheepowners should do a great deal more towards getting rid of the pest than they do at present' (*Hobart Mercury*, 27 August 1887).

Finally, in his inimical style of exaggeration and blather, typical of a true-hearted politician from a parliamentary minority, Lyne summed up his proposal:

> Mr. LYNE said his patience had been sorely tried to sit and hear the remarks of hon. Members on the motion. He could state from his own experience what devastation the tigers caused. He rented 2,000 acres of Crown lands upon which he had not been able to put one single sheep for the past five months in consequence of the ravages of the tigers. In his district the sheepowners paid £3 per head for all tigers that were killed there, and each of them had to keep a huntsman for the express purpose of killing these animals. On the Malahide estate Mr. Rigney had been paying 25s. a head for them during the last five years, and had paid for 50 of them. The fact remained that there were 700,000 sheep less in the colony than there should be, and the tigers had a great deal to do in causing this loss to the country. He had followed the matter up very consistently, and he hoped the House would grant him his committee.
> (*Hobart Mercury*, 27 August 1887)

The motion was put before the members and agreed on the voices (*Formby North-West Post*, 30 August 1887).

The passing of Lyne's thylacine motion did not proceed without negative comment in the press. The following letter was published in the *Tasmanian News*:

160

SIR, – I see the honorable member for Glemorgan [*sic*] wants £500 of public money to destroy hyenas (so-called tigers.) There are comparatively few sheepowners troubled with those vermin, and it is no affair for Parliament to be troubled with, for it would be a great injustice to the whole country in paying for the benefit of a few wealthy sheepowners. (Farmer, 1887)

Cynical comment was also made upon the debate in the weekly *Tasmanian Mail*:

Tiger Lyne, as the honorable member for Glamorgan is now very generally called, is on the war-path again on the look-out for 'manslayers' or rather 'sheepkillers,' and if all he tells us is true the jungles of India do not furnish anything like the terrors that our own east coast does in the matter of wild beasts of the most ferocious kind. According to 'Tiger Lyne,' these dreadful animals may be seen in their hundreds stealthily sneaking along, seeking whom they may devour, and it is estimated that in less than two years they will have swallowed up every sheep and bullock in Glamorgan. (*Tasmanian Mail*, 3 September 1887)

Such contemporary cynicism serves to introduce an attempt to critically evaluate Lyne's stories and statistics.

The significance of Lyne's last general claim 'that there were 700,000 sheep less in the colony than there should be and the tigers had a great deal to do in causing this loss to the country' (*Hobart Mercury*, 27 August 1887) needs to be viewed in terms of the total number of sheep in Tasmania. The annual report of the Chief Inspector of Sheep for 1887/88 records the presence of 1 171 064 sheep in the colony (Tabart, 1888). Lyne's claim that there would be an additional 700 000 sheep were it not, primarily, for the thylacine, suggests that thylacines were largely responsible for the predatory loss of 60 per cent of the sheep flocks of Tasmania. One would imagine that such activity could hardly fail to be noticed at some time by one of the Chief Inspectors of Sheep. However, no evidence exists in the official annual records of the sheep industry from 1884 to 1891 that gives any support to the claims aired in parliament by Lyne and the other rural-rump politicians. Despite their cynical reception in parliament, the closeness at all times of the debate and vote, the absence of supporting statistics, and the often cynical reporting of the issue in the contemporary press, these outrageous claims experienced largely uncritical acceptance in the late twentieth century, and have wielded inordinate power over modern constructions of the thylacine's behaviour.

Lyne claimed that the east coast farmer Rigney had paid for the capture of fifty thylacines at £1.5.0 a head over the previous five years. Given the implied commonness of the species, and the claimed capture of, on average, ten specimens a year, it was most unfortunate that when a representative from the Calcutta Museum called on Rigney, desiring thylacine skeletal material, Rigney was only able to produce Tasmanian devil bones (*Launceston Examiner*, 24 May 1887;

Tasmanian, 4 June 1887). Equally, if the claimed capture rate was genuine, with the introduction of the government bounty Rigney should have been able to convert his thylacine problem into a windfall of £10 per annum. It is not only in an inability to provide museum specimens wherein Rigney's experiences match those of the fabulous French brothers. In similar manner to the French brothers' claims of consistent captures of thylacines, immediately the bounty started Rigney's supposed thylacine problem disappeared. In the twenty-one years of the bounty's operation Rigney was unable to submit even one thylacine for bounty payment! Lyne was either an innocent, misinformed, unassuming politician, or he lived up to his name as a deliberate purveyor of misinformation to parliament.

Finally, it is time to turn to Lyne's personal claims that, as a farmer in the Swansea district, 'He could state from his own experience what devastation the tigers caused' (*Hobart Mercury*, 27 August 1887); that 'He had had as many as 14 sheep killed by these animals in one week' (*Tasmanian*, 9 October 1886); and that thylacines 'ran whole flocks down into gullies, and maimed more than they killed' (*Hobart Mercury*, 5 November 1886). Outside of claims aired in parliament about other east coast farmers with supposed thylacine problems, it appears that Lyne's personal experience of the 'ravages of tigers' (*Hobart Mercury*, 27 August 1887) was distinctly at odds with the preserved published comments of his Swansea district neighbours.

Louisa Anne Meredith, farming at Cambria near Swansea, originally accepted the popular mythology that thylacines were a problem to sheep (1852), but noted that in farming on the east coast in the 1880s, 'we do not hear of them very often now' (Meredith, 1880, p. 66). Lyne's and Meredith's positions on the thylacine were diametrically opposed.

The suggestion that Lyne's was the unreliable and untenable position, that, at best, he was guilty of the grossest exaggeration, if not a deliberate pedlar of mistruth, is demanded by the published comments of stock problems given by the sheep farmers Alfred, James and Lewis Amos at 'Glen Gala' and 'Cranbrook' in 1884. The Lyne estate 'Gala, and the Amos' properties adjoined one another; the Amos were Lyne's neighbours. Both were the subject of description in one of a long-running series of travels through Tasmania published in the *Tasmanian Mail* in 1883 and 1884, by a 'Special Correspondent'. The pasture land on the Lyne property, 'Gala', received universal acclaim; in contrast, the Amos' pasture land was more uneven, well developed close to the homesteads, but wilder and less developed in the back runs in the hills. Sheep in the back runs of the Amos' properties were likely to experience far greater levels of predation than those experienced by sheep on the longer-running, more carefully managed back runs in the hills of the Lyne property. Alfred Amos admitted to significant stock predation problems in the hills and identified the cause: 'Wild dogs are a nuisance on the back runs, frequently destroying sheep . . . Mr. Amos dreads these runaway curs more than tigers . . . tigers are comparatively scarce' (Special Correspondent, 1884).

Lyne's neighbours supported Meredith's (1880) position on the rarity of thylacines in the district and their lack of impact on the sheep farming industry.

Instead, Amos expressed 'dread' at the attacks of feral dogs. Lyne suggested that his thylacines supposedly killed up to fourteen sheep a week, ran whole flocks down and maimed more than they killed. As discussed earlier, these behaviours are characteristic of attacks by wild dogs, not thylacines. The identification of a wild dog problem (and scarcity of thylacines) by his neighbouring Swansea farmers suggests that, despite the self-aggrandisement involved in the glorious appellation of 'Tiger Lyne', Lyne was either cognitively incapable of discriminating between feral dogs and Tasmanian tigers, or politically disinclined to do so. Neither alternative does anything to support the validity of Lyne's position and posturing on the thylacine.

Retrospective comment from old-time Swansea district farmers made to Guiler in the 1950s has continued to isolate Lyne. An aged descendant of the original settlers, R. Amos, still living at Cranbrook, identified a different stock problem, but expressed a similar position on the status of thylacines: 'Lost 150–200 sheep per annum in the 1880s but all were stolen. No thylacines about' (R. Amos, cited in Guiler, 1985, p. 121). Sheep-stealing was also identified as a major problem at Kelvedon, another east coast property near Swansea, where thylacines were held responsible for only 6 per cent of stock lost: 'losses in the 1880s were about 200 sheep per annum', of these 'about 12 per annum to tigers' (T. Cotton, cited in Guiler, 1985, p. 120). F. Shaw, another Swansea landholder, also suggested that at Red Banks there were 'not many tigers' and that, all told, they 'didn't lose many sheep on account of good shepherds'. Loss of stock to thylacines was also estimated to be at about the 6 per cent level. Balancing this, however, he suggested 'we didn't use the back country on account of tigers' and that 'about six [tigers] a year were claimed as a bounty' (Shaw, cited in Guiler, 1985, pp. 20, 121). Guiler has analysed the twenty-one years of bounty data for the locality of capture and determined that of the more than 2000 claims presented for bounty payment, only ninety-one came from this east coast region. Guiler politely comments: 'Shaw's claim of six thylacines per annum presented for bounty does seem excessively high' (1985, p. 121) – particularly so when my research on the bounty ledgers shows that at no time did anyone called Shaw ever receive a bounty payment. With Shaw's relatively modest account labelled as 'excessive' one is at a loss to find suitable words to describe Lyne's claims about the thylacine.

It is apparent that 'the dogmatism of unenlightenment' – a phrase coined by Montgomery (1912) to describe the persistence of the belief amongst nineteenth-century farmers and bushmen that marsupial young are born and fully develop on the teat – is equally applicable to those same individuals persisting with the misological position of blaming the thylacine for the rural depression, feral dog predation, sheep-stealing and increasing rabbit population affecting the profitability of sheep farming.

Did the numbers of thylacines submitted for bounty payment by members of the Amos and Lyne families reflect the different claims of ravages and densities of thylacines on the neighbouring properties? The Amoses, who claimed they were fortunate enough not to experience thylacine problems,

163

were, however, able to submit two thylacines for bounty payment (payments approved 5 July 1901 and 31 July 1901). At face value, the Lynes were not only unfortunate in apparently being besieged by thylacines, they were doubly unfortunate in being apparently unable to catch them effectively, for the Lynes were also only able to submit two thylacines for bounty payment during the twenty-one years of its existence (payments approved 2 August 1894 and 28 August 1898). Alternatively, as Lyne's verbal claims in parliament were never matched by any hard data in terms of specimens produced, it is possible to suggest that perhaps the same low density of thylacines was to be found on both the Lyne and Amos properties. As Guiler noted in his consideration of the comparative distribution of bounty payments, with respect to the ninety-one identified above, that 'there were relatively small numbers produced for bounty payments in the regions on the east coast where it was alleged that thylacine predation took place' (1985, p. 71).

Beside the extravagant, unsubstantiated claims by a few sheep-owners and their parliamentary representatives in the 1880s, claiming a massive number of thylacines in existence, other sources of contemporary comment in Tasmania (covered in the previous chapter) express concern at the rarity of the species and its likely extinction, and note the difficulties experienced by zoological garden and museum curators in obtaining specimens. Fortunately at least one independent source exists that provides insight into thylacine population density at the time: the deaths of thylacines recorded over fifty years in the station diaries of the Van Diemen's Land Company property at Woolnorth, from 1874 onwards, as analysed by Guiler (1961w, tables ii and iv, pp. 208–9; 1985, tables 2.1 and 6.2, pp. 22, 98–101).

Provided no behavioural change or new innovation is introduced into the relationship between predator and prey (in this case human and thylacine respectively), interaction and successful predation between them should occur at a rate reflecting the population density of both species. Rather than any evidence of increase in thylacine numbers in simple response to the presence of sheep, or even stability in thylacine population numbers, reflecting a balance between a new, introduced prey species accompanied by significant habitat alteration, the data on captured and killed thylacines at Woolnorth suggest a species in decline. Of the 154 thylacines recorded as killed on Woolnorth during the fifty years from 1874 to 1923, just under half of them, some sixty-six specimens, were killed in the first ten years. In a typical illustration of species decline, in the second ten-year period, from 1884 to 1893, only twenty-nine thylacines were killed. Such classic illustration of decline did not continue. Significant behavioural change occurred in the thylacine population towards the end of the next ten-year period. At the turn of the century, the debilitating epidemic disease that affected thylacines and increased their ease of capture spread across the island. In two years, 1901 and 1902, twenty-eight thylacines were destroyed on Woolnorth. In total, fifty-five thylacine specimens were killed during this third ten-year period, from 1894 to 1903. These figures represent the effective end of the species on the extensive Woolnorth property. During the next ten years, 1904 to 1913, only one thylacine was killed at Woolnorth (in 1906). In the next decade, 1914 to 1923, three thylacines were killed (all in 1914), and these were the last thylacine specimens ever to be caught on the extensive Woolnorth property.

This Woolnorth data, at the moment, represent the best available data showing the population dynamics of thylacines in Tasmania in the late nineteenth century and early twentieth century. Instead of the thylacine population increasing in size, or holding steady at levels likely to cause consistent, significant damage to the sheep industry – as claimed by Lyne and the rural-rump lobby in parliament – the Woolnorth data, starting ten years before the first government debate on the thylacine in 1884, show a species in dramatic decline, increasing in rarity and rapidly heading towards extinction.

'Thursday next' (that is, 1 September 1887), as specified in the parliamentary resolution, came and went without discussion of Lyne's motion. The formal process of ensuring that parliament discussed the resolution was put into place when Lyne gave 'Notice of Motion' on Wednesday 14 September 1887 of his intention to call for his 'Committee to Address the Governor in Council re £500 Grant for Destruction of Tigers' (Tasmanian Parliament, 1887). Committee was achieved the next day and considered as the last major piece of business late in the evening of the 15th. By now, much of the sting had gone out of the debate, the essential points had been made, positions declared, and the motion was likely to be passed.

For the last time in parliament, Young raised the issue of the thylacine as the scapegoat, the excuse for poor farming practice. He: 'thought there should be more information before the committee before it agreed to the resolution ... It was very easy for a shepherd to say tigers had taken the sheep when a number were short ... if the loss of sheep was so large it warranted a special inquiry' (*Hobart Mercury*, 16 September 1887). Braddon made a final attempt to undermine the proposal by arguing once again, on the basis of figures previously introduced in debate, that if the thylacine caused a £25 000 loss to sheep-owners, then the farmers should be more than willing to pay for its extermination themselves. He received some support in this idea, but the debate was heading in only one direction.

The committee's decision was largely a foregone conclusion. It was past 11 at night, in speeches there had been three significant defections from the No vote that had narrowly lost the previous November. With the lateness of the hour and the outcome obvious, the resolution was agreed to on the voices and no formal vote was held (Tasmanian Parliament, 1887).

Accepting the inevitable, the government now set about finding administrative processes and money to fund the bounty for 1887/88 at a typically leisurely pace, but with sufficient activity to convince Lyne and his supporters that there was nothing to be gained by demanding further debate to familiarise and commit parliament to a bounty scheme that had already been twice voted on and accepted. The bounty finally commenced in 1888, but with a marked lack of support from the Chief Inspector of Sheep (Tabart, 1888).

On 4 February 1888, Braddon, as Minister of Lands and Works, circulated his brilliant idea to make rural communities pay for the bounty by using municipal funds as the source of rewards. Presentation of thylacine specimens was to be made to municipal wardens who would authorise payment from local council coffers. The circular produced outrage and the obvious political backlash. It was countermanded by a further departmental circular:

> The Colonial Auditor having objected to the expenditure of Municipal Funds
> in rewards for Native Tigers destroyed, this Office Circular of the 4th instant
> is hereby superseded by the following instruction:- Out of the sum of £500
> appropriated for rewards for destruction of tigers during 1887–1888, there shall
> be paid out of the Consolidated Revenue £1 for every full grown Native Tiger
> destroyed, and 10/- for every partly grown Native Tiger destroyed. These
> rewards will be paid by the Hon. the Treasurer on the certificate of a Municipal
> Warden or 'Police Magistrate' of a Police District that the animal has been
> destroyed, and so disposed of that a second reward cannot be claimed for it.
> (Braddon, 29/2/1888)

166 The first bounty claimant was Roger Marshall, of Glen Alyse, Bichenou, who made claim for a kill on the day of the bounty's introduction, 4 February 1888, before his local municipal warden. His claim, caught up in the administrative and political cross-fire over the source of funding and the necessity for signed declaration of destruction, took 217 days to be processed. The first bounty approved for payment was made on 28 April 1888, to James Harding of Ross.

The positions held by Lyne and espoused by the rural-rump lobby in parliament and a select few sheep-owners in the community, that thylacines were consistent and significant predators on sheep and that they existed in large, stable or significantly increasing numbers, have not proved tenable when compared to other contemporary newspaper reports and zoo and museum records. They have had, however, infamous effect. At the time their much publicised claims helped support a growing popular perception of the thylacine as a significant killer of sheep, a perception that was, too unquestioningly, adopted into most twentieth-century constructions of the animal. A critical analysis of sheep predation records from the nineteenth century shows a consistent picture of thylacines as infrequent, incidental predators on sheep – of minor importance compared with the effects of other factors such as sheep-stealing, wild dog predation, competition with rabbits, parasites and disease. The reality is that at no stage did the species represent any significant economic threat to sheep farming, and, as a result of the European invasion and habitat alteration, the thylacine population was in significant decline well before the parliamentary brouhaha and establishment of the government bounty.

At no stage did bounty payments ever approach the £500 budgeted level. The specimens caught and submitted for bounty payment in 1888 were so much under the expected level as described in the parliamentary debates that the initially budgeted figure of £500 for 1889 in the estimates of 5 July 1888 was reduced to only £250 for the year in the revised estimate of 18 September 1888, and then further reduced in the next revised estimate ten days later only allowing for an expense of £150 per annum for the 'Destruction of Native Tigers' (Tasmanian Parliament, 1888). The maximum paid out in any one year was £162.10.0 in 1901 (for 157 adults and 11 juveniles). During 1901 the Lands and Surveys Department paid a bounty on a thylacine, on average, once every two days. When the bounty was finally terminated in 1909 – after payment had been made for some 2207 specimen claims – the Department was paying a bounty on a thylacine, on average, once every five months. Instead of meeting the original budgeted estimate and costing an expected £10 500 for the twenty-one years in which the bounty operated, the decimation of the species to the point of no return was obtained through the expenditure of a mere £2132.10.0. It remains a challenge for all modern-day economic rationalists to identify another situation of such significant biodiversity reduction, where so much wanton environmental destruction has been obtained for so little reason and so little money.

Notes

1 Government in Tasmania has a long history of seriously considering destructive, inaccurate and insensitive petitions from the economically powerful, and later converting them into bounty schemes. These date back to April 1828, when a petition from the 'gentlemen settlers' of Campbell Town – supported by petitions from Longford and Bothwell residents – requested the Governor to provide a solution to the 'Aboriginal problem' (S. Morgan, 1992, p. 149). These resulted in the introduction of a bounty scheme, offering a £5 reward for the capture of an adult Aborigine and £2 for each child (H. Reynolds, 1995, pp. 113–14, 131).

2 Lyne's designation of the east coast Aboriginal word 'legunta' (Milligan, 1859) as the species' scientific name discloses his perfect ignorance of scientific knowledge of the species. Given the well-documented history of the Lyne family's violence against the indigenous east coast peoples – Amos (1963) and Nyman (1976, 1990) provide specific details of incidents in which John Lyne was personally involved in shooting Aborigines – it is unlikely that Lyne would knowingly use an Aboriginal word, or give the appearance of being sympathetic towards the knowledge and existence of Tasmania's Aborigines at any time; let alone in a debate that was to link the anti-thylacine lobby with the anti-Chinese lobby amongst the members.

3 Note, however, that the *Hobart Mercury* suggests that Sutton provided this detail.

4 It is unfortunate that, in terms of the statistics they used in the thylacine debate, the rural-rump lobby had apparently enquired at the other end of the horse!

5 John Lyne's interest in members of the opposite sex, apparently obvious to his parliamentary colleagues familiar with his Hobart-based, away-from-home behaviour, remained with him all his life. He married for the last time in 1891 at the age of 82 (Nyman, 1976).

6 In 1886 McGowan sold three different thylacines to Melbourne Zoo for £5.5.0, £7.10.0 and £8.10.0.

168   If the fate of the thylacine was problematical before the bounty's introduction, extinction was inevitable by the time the bounty was dismantled in 1909. Concern about the rarity and potential extinction of the thylacine, following the introduction of the bounty in 1888, continued to be expressed, for example: Beddard (1891); *Launceston Examiner* (9 March 1892; 22 March 1899); Lydekker (1895); Sterland (1892); *Tasmanian Year Book*, (1899/1900); and E. P. Wright (1892). As the bounty continued into the twentieth century there was a pressing need to halt the killing of thylacines. However, the wholesale slaughter of Tasmania's fauna that took place in the nineteenth century (Stancombe, 1968, p. 8) continued unabated. The visiting anthropologist and anatomist, Hermann Klaatsch, noted that Tasmania's animals 'were being totally exterminated, with no understanding and with no pity, the hard-hearted and ignorant colonists shooting everything on sight' (Klaatsch, 1905, cited in Kolar, 1965, p. 63).

Growing conservationist concerns

Reflecting this destruction, museum and zoological gardens specimens became ever more difficult and expensive to obtain. H. H. Scott, director of the Queen Victoria Museum, Launceston, described the problem to Oldfield Thomas at the British Museum:

> The Government years ago put a price upon their heads . . . this has thinned them down, & good specimens are rare and not easy to obtain . . . If a man kills an animal in the back blocks he has only to chop its head off & claim 20/- reward, this does not spoil the skin for certain markets, & if he cares to bother

with it he can get another 20/- that way, which is far more easy than lugging the animal for some miles to a railway station in the hope of sending it to a museum, & asking more. (H. H. Scott, 3/8/1901)

Scott expressed himself similarly in correspondence with Spencer at the National Museum of Victoria (25/4/1904); and also with Hoyle at Manchester Museum: 'the restricted "*Thylacinus*" are now rare, & . . . worth £6-0-0 at the trap with all risks' (25/4/1904, see also 13/9/1904). Hall, curator of the Tasmanian Museum, Hobart, was so 'anxious' to get any thylacine material that he wrote to the Commissioner of Police, seeking help from the police force to obtain any thylacines, 'even poor specimens' (14/12/1910).

Similar problems confronted zoological gardens. In London it was noted by the Secretary of the Zoological Society that the thylacine has 'now become extremely scarce and seldom seen in captivity' (Sclater, 1901), the reason for this being that 'the thylacine . . . became an object of persecution to the Tasmanian shepherds, whose fierce hostility has now brought it to the verge of extinction' (Sclater, 1904). William McGowan, as superintendent of the Launceston City Park Zoo, had had his concern for the species publicly recorded in the *Launceston Examiner* (9 March 1892), and he continued to privately express his concern for the species. He expressed his belief 'that Tasmanian wolves . . . were almost extinct' in a letter to Melbourne Zoo in November 1907 (RMZAS, 1910, pp. 134–5) and he wrote to the National Museum of Victoria that 'they are becoming very scarce' (31/8/1911) and about his experience of 'the increased difficulty in procuring these animals' (10/7/1912). Mary Roberts, the original owner and curator of Hobart Zoo, spoke publicly about 'the rare and handsome Tasmanian tigers' as a 'species now almost extinct' (*Hobart Daily Post*, 30 October 1909). At Melbourne Zoo, W. H. D. Le Souëf recognised that thylacines 'are now getting scarce, as every man's hand is against them' (1907, p. 176) and that personally, he had already started the countdown: 'These animals will probably become extinct before many years' (monthly report, 1/6/1914).

During the first twenty years of the twentieth century broad environmental concerns surfaced in Tasmanian public debate. The close of the nineteenth century saw the founding in 1895 of the Tasmanian Game Protection and Acclimatisation Society, a potent pressure group that represented the 'sportsmen's' interests of conservative politicians and members of the landed gentry (Tasmanian Animals and Birds' Protection Board, 1934, p. 159). Its political power was evident in that the Patron of the Society was the Governor, Viscount Aarmanston, and the President of the Society was Tasmania's Premier Edward Braddon. However, it was not long before the Tasmanian Field Naturalists' Club was established in September 1904. Traditionally, for the time, it supported both an appreciation of native flora and fauna, alongside support for the continued importation and acclimatisation of foreign game species (Reid, 1907). In time, its conservation concerns became focussed on the necessity to prevent the indiscriminate killing of native birds and animals, and the desire to reserve national parks in different areas of the state (Hall, 1925; *Hobart Mercury*,

169

11 August 1908; Wall, 1955). In these political activities the Club was frequently joined by members of the Royal Society of Tasmania.

At the same time the chorus of concern voiced by scientists about the specific fate of the thylacine continued to grow. In 1914 the Professor of Biology at the University of Tasmania, T. T. Flynn, a frequent contributor to the Naturalists' Club publications and meetings, 'advocated the establishment of some safe retreat, ... where ... [thylacines] should be allowed to live' (Flynn, 1914, p. 53). However, nothing came of this desire. The recognition that it requires both political and direct action to change government attitudes, rather than just the constant repetition of scientifically demonstrable data in professional and occasionally public spheres, grew too slowly to save the thylacine.

A recognition of the parlous state of the species was not restricted to professional scientists, it was recognised by the community at large. The capture or killing of a thylacine became significant events within local communities, exciting interest and attracting visitors from far and wide. For example, a half-grown male thylacine was captured live from a light springer snare at Meunna, in north-west Tasmania in the winter of 1916. It was taken back home to Lapoinya, and all the neighbours came from miles around to see it (Doherty, 16/12/1972, 1977, 2/9/1981):

> By nine that night, the bush telegraph had summoned from far and near, those others who had or had not seen a thylacine alive. For miles the roads were blocked – phaetons, gigs, traps, buggies, jinkers and spring drays, hacks and bicycles and almost as wonderous [sic] as the tiger, there was a motor car. (Doherty, 1977)

Sometimes the mere presence of a thylacine in the vicinity was enough. Col Bailey has published details of an interview with 'Bill', an old-timer from Tasmania's north-west, who never saw a thylacine, alive or dead. However, 'Bill recalls the time when, as a youngster [in the 1920s], his father got all the children out of their warm beds at 10 o'clock one cold, still night to listen to a hyena calling in the distance, along the Inglis River' (*Derwent Valley Gazette*, 26 June 1996).

The rural lobby, having worked to scapegoat the thylacine throughout the nineteenth century, and desperate to preserve their holy trinity of excuses – drought, flood or Tasmanian tiger – was determined to fight against any protection of the thylacine and mounted a traditional campaign. With popular perceptions about the rarity of the species rapidly gravitating towards the scientist's position, there was a necessity to construct and publish an argument against both popular and professional perceptions of the rarity of the species, and ensure that administrative structures and processes were such as to mitigate against any change from the status quo.

The first of these aims was met in the rural lobby's treatment of the thylacine's vocal ability. Comment upon thylacine vocalisations commenced with the initial scientific description of the species (Harris, 1808), continued throughout

170

the nineteenth century and was used as the basis for the previously presented categorisation of the different types and contexts of thylacine vocal expression. Notwithstanding the published evidence to the contrary, in the early twentieth century the rural lobby opposed to protecting the thylacine seized upon an hypothesised lack of thylacine vocal ability in order to argue against the rarity of the species, to split the popular perceptions of its rarity along a city versus country dimension, and then to attempt to convince country people that the natural position for those in the rural community was one of complete opposition to any protection for the thylacine. This position was best expressed in the regular column by H. W. Stewart published in the conservative weekly newspaper, the *Bulletin* (11 November 1919), that succinctly encompassed the first political aim of the anti-conservation campaign:

> From the very early days there have been arguments between bushmen and scientists whether there are two species of Tasmanian tiger. The scientists say one, the bushmen say two, defining them as bulldogs and greyhounds. The specimens in any museum seem to bear out the bushies' view. There are thick-set big-headed ones, and there are others more the size and build of a large collie. The tiger is still more numerous in Tasmania than would appear from a casual inquiry. Its habit of hunting at night, quite silently, and taking the kill home to its den, where it lies close in the daytime, helps it to elude observation.
>
> Not long ago a large specimen was shot on Tullachgorum, near Avoca . . . The den was half-full of bones, amongst them those of a half-grown calf, it is hardly likely that the tiger could pull down a strong calf, but that it dragged even a weakling to its cave gives a good idea of the brute's strength . . . tigers and devils will clean up the carcases of 20 big kangaroos in a night.
> (H. W. Stewart, 1919)

171

There was nothing subtle about this argument. Its blatant expression requires little additional analysis. It commenced by implying that scientists could not be relied upon to even classify the thylacine correctly, suggesting that on the basis of their own data there were two different species, as country people suggested, rather than just the one, as maintained by scientists.[1] The fact that any 'bushie' who had any practical knowledge of the animal at all knew there were cranial and body size differences that were representative of the age and sex of different specimens was not mentioned – obviously Stewart lacked any first-hand knowledge himself and was just pushing a political wheelbarrow. It was, however, telling to be able to imply that city-based scientists were wrong in comparison to country people, and so constrained by their error as to be unable to recognise the truth, even when it was apparent in the museum data that scientists displayed themselves. The argument also suggested that there were many more thylacines around than any casual, city-based enquiry would discover. No professional, city-based scientists wandering around the bush in the middle of the day were going to see a thylacine, and, in the unlikely event of these scientists wandering around the bush in the middle of the night, they would be unaware of the subtle signs of the thylacine's presence: as thylacines do not possess a voice, they do not make any noise and therefore cunningly give the impression that they are not there at all! But the argument suggested that thylacines were not only present, but that their blood-thirstiness, that drove them to kill up to twenty kangaroos a night, suggested thylacines not only had the potential to continue to damage the sheep industry, but they had also the potential to affect beef and cattle production as well.

Some twentieth-century memoirs have maintained this politically generated belief in the muteness of thylacines. In an unpublished letter S. Mitchell (25/8/1981) remarked 'they are silent hunters [I] never heard them bark'. Frank Darby, who claimed to be a former keeper at Hobart Zoo, suggested in an interview with the naturalist Graham Pizzey (1968) that the last Hobart zoo thylacine 'never made a sound – his bearing and silence were uncanny'. Unquestionably, Pizzey's interview with Darby – containing such intimate details of the last thylacine as its name being 'Benjamin' – has proved the most widely replicated oral history account on the thylacine. Michael Sharland noted the belief of a few bushmen with supposed knowledge of the species, who denied thylacines had any vocal ability: 'some say it cannot even growl' (1962, p. 3). On the basis of his own observations, he expressed personal doubt whether the thylacine had any voice at all (1961, 1971x) and suggested that records of thylacine cries heard in the bush were probably made by owls (Sharland, 1971x). From such twentieth-century opinions has the point of view arisen that, in contrast to the nineteenth-century position, thylacines were usually or normally mute (for example, Dixon, 1989; Griffith, 1971; Rounsevell, 1983: S. Smith 1981) with a few exceptions recognised, usually the vocalisation records of Gould (1851) and Harris (1808).

The cumulative nature of science encourages an assumption that the more recent the observation, the more accurate it is likely to be. As previously mentioned, while this assumption often has a distinct practical utility in scientific endeavour, it is a confounding variable when applied to the knowledge of a recently extinct species. The selective weighting that has been granted to the few published twentieth-century reminiscences about the muteness of the thylacine, over nineteenth-century observations on the vocal abilities of the species, is a case in point. An attempt was undertaken in the 1980s to provide a definitive summary of the knowledge of each Australian mammalian species through a federally funded project that resulted in the published volumes of the *Fauna of Australia: Mammalia.* Joan Dixon was invited to write the chapter on the thylacine, and included a detailed section in the text, based on nineteenth-century sources, covering different thylacine vocalisations in different situations. However a member of the editorial board requested the section's removal on the grounds that it was highly contentious whether the thylacine had any vocal abilities at all (J. M. Dixon, personal communication). As a result of this editorial intervention, all that was approved for inclusion in this definitive summary of the species was a one-sentence quote from Harris (1808) on the short guttural cry, and the single sentence: 'Usually they are mute, but when anxious or excited they make a series of husky, coughing barks' (Dixon, 1989, p. 549). That a purely political manoeuvre on the part of anti-conservationists at the beginning of the twentieth century – one that was totally divorced from reality and obviously incorrect to anyone with a knowledge of the animal or the published literature at the time – could nevertheless constrain scientific publication at the end of the twentieth century demonstrates, once again, the fallacy behind the common assumption that scientific data improve in objectivity, accuracy and approximation to the truth over time. Until scientists as research workers or editors are educated to consider the social, cultural and historical contexts in which data are constructed and knowledge generated, they will continue to make unfortunate errors in analysis and interpretation, and unwise alliances for industrial or government support.

Instances of unwise alliances, in a non-financial context, may be found in the history of the politics of protection for the species, in which the last of the aims of the anti-thylacine lobby – to ensure that administrative structures and processes are such as to mitigate against any change from the status quo – appears very much to the fore. Scientists and naturalists in Tasmania ultimately failed to preserve the thylacine from extinction because they were prepared to play by the rules, through genteel and proper social and political activity, attempting to put into place legal measures preventing the continued killing of the species.

173

Bureaucratic obfuscation

The Tasmanian Scenery Preservation Board was set up in 1915 for the preservation of land with scenic or historic interest in Tasmania, and its activities resulted in the establishment of the first scenery reserves, game reserves and the National Park (Hall, 1925). In 1920 a separate Board was set up to administer the National Park, on which the Royal Society of Tasmania, Field Naturalists' Club (with C. E. Lord as its nominee), and the University of Tasmania were represented. This Board set out to protect the flora and fauna 'to the best of its ability' (Hall, 1925, p. 218) but its effects were limited, due to its legal field of operation being restricted to the existing National Park.[2] The state's administrative structure for preserving Tasmania's biodiversity was thus spread between two different bodies, the Scenery Preservation Board and the National Park Board, and this division of responsibility was inevitably accompanied by a divergence in opinion and recommended policy on the use of limited government monies: either to create new historical sites or national parks, or to develop the potential of existing national parks. (There was a pressing need for fencing the park, and getting the four timber mills actually located and operating within the national park moved out.) This lack of unity allowed a considerable influence to be exerted on the government on conservation issues by the older and powerfully connected Tasmanian Game Protection and Acclimatisation Society. Working against the Society from a broader-based but inferior political position was the Tasmanian Field Naturalists' Club, and the more narrowly constructed Royal Society of Tasmania, both of which constantly kept up, via meetings, conferences, political deputations and in publication, calls for the immediate conservation and protection of native fauna.

In 1923 the state government was prompted by Australian federal government action to establish 'The Tasmanian Advisory Committee re Native Fauna' with the responsibility to inform the federal Minister for Trade and Customs of endangered species requiring federal protection through the prohibition of exportation (*Launceston Examiner*, 22 June 1923; Robson, 1928). When the Committee took up a brief for the protection of the thylacine, gathering data and publishing its conclusions, it was immediately and effectively countered by rural-rump politicians and members of the conservative establishment who prevented, for as long as possible, any positive action being taken to preserve the thylacine, even as irrefutable evidence of the destruction of the species mounted.

Added to the specific warnings of the rarity and potential extinction of the thylacine and the calls for its protection, expressed during the first twenty years of the twentieth century, were the general concerns for conservation raised in the 1920s (Gregory, 1921). Throughout Australia, alongside broader community interest in conservation expressed in the public profile and political value given to the establishment of national parks and reserves, were public warnings and debate over specific faunal problems. With six million animals slaughtered for their pelts from 1919 to 1921, the Australasian Association for the Advancement of Science began to express concerns over the fur trade (Powell, 1988), in particular, the fate of the Australian mainland animal with the highest public profile, the koala or native bear, *Phascolarctos cinereus*.

Despite the suggestion by Lucas and W. H. D. Le Souëf that koala fur 'fortunately is not valued in the market' (1909, p. 92), they nevertheless noted that 57 933 koala pelts had been sold just in the Sydney markets in the previous year (p. 6). Within half a dozen years, through the actions of trappers, the koala was completely exterminated in South Australia. The other mainland states proceeded merrily on their way. 'In the two years 1920 and 1921 ... the huge total of 205,679 koalas were killed for the fur market' (Wood Jones, 1923, p. 186). The destruction continued exponentially, almost reaching passenger pigeon proportions: 'in 1924 the colossal total of over two million [koala pelts] were exported from the eastern States' (Troughton, 1941, p. 137; see also *Hobart Mercury*, 23 July 1928). This slaughter was initially accomplished without much public protest because the data were well disguised, 'since in the fur trade Koala pelts pass under the name of "Wombat"' (Wood Jones, 1923, p. 186). Public recognition of the destruction resulted in belated political activity, first of all in the now koala-free state of South Australia: 'in 1923, the South Australians made a conscience-stricken, and successful, attempt to establish Victorian animals on Kangaroo Island' (A. J. Marshall, 1966). Then, with the koala in a precarious situation in all states except Queensland – where it had been placed on the protected species list in 1919 – the Queensland state government declared a

174

one-month open season on koalas in August 1927. Despite an enormous Australia-wide outcry the killing went ahead. The Queensland government received licence fees from 10 000 registered trappers, and in that one month 584 738 koala pelts were obtained (A. J. Marshall, 1966).

In August 1928 a group of naturalists from the Queensland Field Natural-ists' Club set out to determine the status of the koala in Queensland, through letters to city, municipal, town and shire councils and the different Dingo Boards in Queensland. Sixty-nine out of eighty-one local bodies that still had koalas in their districts recorded them as very scarce or practically exterminated and supported their protection. In fact, all but one of the eighty-seven local bodies that provided data on the distribution of the koala in 1928 favoured some form of protection; seventy-nine of these local bodies suggested total permanent protection (A. J. Marshall, 1966). This data, presented to the Queensland Minister of Agriculture, had immediate effect. The koala was returned to the protected species list in Queensland, and has remained there ever since. No koalas have been commercially killed in Queensland, or in the whole of Aus-tralia for that matter, since that time.

Tasmania, lacking koalas, was not a part of this mainland debate, and the mainland states, now lacking thylacines, paid little attention to their status in Tasmania. Attempts to achieve thylacine protection proceeded in isolation, largely unsupported and unrecognised by active conservationists in the other Australian states, a situation which definitely weakened the impact of the move-ment in its conflict with rural industry, agriculture and economics.

Throughout the 1920s, awareness of the scarcity and likely extinction of the thylacine continued to grow steadily. A large degree of credit for this must go to C. E. Lord, Director of the Tasmanian Museum, Hobart. Together with his fellow museum director, H. H. Scott, from the Queen Victoria Museum, Launceston, he published a handbook of Tasmania's animals in 1924. More than just *A Synopsis of the Vertebrate Animals of Tasmania*, it was also a call for the public recognition of the need for regulations to protect Tasmanian wildlife – including the thylacine – as a national asset. This point of view was acknowl-edged in the *Launceston Daily Telegraph* (12 September 1924), where the thylacine, described as 'unique', was used to introduce the concept of required legislation to protect Tasmanian wildlife. Three years later, in a paper, presented to the Royal Society of Tasmania, 'Existing Tasmanian marsupials', Lord began with a call for 'the need for a better system of conservation' (1927, p. 17). He devoted more text in the paper to the thylacine than to any other species, calling for the creation of a completely protected reserve for the species and encour-aging the attempts of the curator of Hobart Zoo, Arthur Reid, to breed the species in captivity.

As well as being Director of the museum, C. E. Lord was also secretary of the Tasmanian Advisory Committee re Native Fauna. His cousin, Col. J. E. C. Lord, Commissioner of Police and Chief Inspector of Fisheries and Game, was chairperson of the Committee. As secretary, on 27 March 1928 C. E. Lord wrote to his cousin, referring to the decision made at the last meeting of the Committee, requesting that consideration be given to 'placing the Tasmanian Marsupial Wolf (*Thylacinus cynocephalus*), on the totally protected list'. The reasons given were the rarity of the species and the inability of Tasmanian institutions to afford the competitive prices offered for specimens by mainland and overseas institutions. It was pointed out that total protection would still allow permits to be granted for the live capture or killing of thylacines for Tasmanian zoos and museums: 'In regard to prices, Beaumaris Zoo will give £20 for a live *Thylacinus cynocephalus* and the Tasmanian Museum will give £10 for a dead one, provided same is in good condition'.

Wearing the different political hats of Commissioner of Police and Chief Inspector of Fisheries and Game, and aware that an anti-conservation position reflected the sentiments of the government, J. E. C. Lord was disinclined to act too hastily on the issue. Instead, he decided, as a temporary delaying tactic, that there was a need to possess an accurate assessment of the status of the species. In an attempt to obtain such data, he enclosed a copy of C. E. Lord's letter (with a typed memorandum to the superintendents of all four Tasmanian police districts, initially asking: 'Can any useful information bearing upon the subject [the thylacine] be obtained from members of the Force who have been stationed in parts of the State near to where these animals were to be found [?]. It is considered that protection is advisable') (J. E. C. Lord, 3/4/1928). However, while the last sentence reflected the personal position of both Lords on the subject, on reflection, J. E. C. Lord recognised its overly direct political position and, not wishing to antagonise non-conservationists, when the letters were signed and sent out, J. E. C. Lord altered the last sentence of each letter by hand to read: 'Is it considered that protection is advisable?' This, of course, gave the memorandum a significantly different slant.

In 1928 two attempts were made, in different Australian states, to gather data on the status of two endangered species: koalas in Queensland and thylacines in Tasmania. While the data obtained from both attempts were similar, the treatment of this data, their evaluation and the ultimate response thereto, could not have been more dissimilar. The Queensland approach was instigated, organised, gathered and summarised by naturalists and scientists, and had an immediate and lasting effect: total permanent protection for the koala. The Tasmanian approach, however, was to deliberately ignore the abilities and expertise of the states' naturalists and scientists. As they appeared to be unanimously calling for the protection of the thylacine before any hard data had been obtained, they were obviously biased and untrustworthy. The Tasmanian Fauna Committee turned instead to what might at best be described as a naïve, supposedly neutral data-gathering source: the police. However, by their very nature, Police Departments are conservative and designed to protect the status quo; they are not designed to be sympathetic to ideas suggesting the overthrow of accepted community beliefs or the desires of prominent establishment figures, represented in rural communities by the landed gentry and squattocracy. While individual members of the police force responded accurately to the request for information, the summarised responses sent to Col. Lord reflected the political reality.

The replies of all four superintendents of Tasmanian Police Department regions have been preserved in the archives of the Tasmanian Animals and Birds' Protection Board, together with many of the original responses from individual police officers. All told, there are responses originating from staff at thirty-eight different police stations throughout Tasmania. These responses are summarised in Table 7.1.

Not one response refers to immediate knowledge of the presence of the thylacine in 1928 (see Table 7.1). Close encounters in the past are designated at best as 'some years ago'; the most recent being two records from 1922. While Eyles' comment for Circular Head in 1922 of 'a few species in the . . . District' may be taken at face value, Morgan's comment for Buckland in 1922 is less impressive – not just because of the uncertainty of whether a specimen was seen or killed, suggesting Morgan was always some distance away from talking with anyone who had first-hand knowledge of the supposed incident but more importantly, through his reference to the source of this information as 'a family named French', members of whom, as previously considered, were notoriously untrustworthy when it came to claiming thylacine kills.

Notwithstanding an absence of evidence for the continued existence of thylacines, the political nature of potentially protecting the thylacine was referred to by a number of police. Trooper Harrower of Chudleigh made no comment whatsoever about the present status of the thylacine, merely noting that 'having made enquiries from old residents of this district, farmers and others . . . they all state that they do not consider it advisable to protect the Tasmanian Marsupial Wolf' (26/4/1928). Sergeant Collins of Deloraine had no contemporary evidence for the thylacine either, but agreed with Harrower: 'This seems to be the opinion of large stock owners' (28/4/1928). While Inspector Eyles at Longford stated 'it is many years since the Tasmanian Wolf has been seen', nevertheless he commenced his response by writing: 'Enquiries made from Stock Owners in the Division show that they are against the protection of the Tasmanian Wolf' (2/5/1928). Superintendent Gunner at Launceston summarising the responses of police in the North-east Region, after stating that 'the Tasmanian Wolf has not been seen in this District for very many years', concluded with an emphatic single sentence paragraph: 'Sheep owners who have been interviewed are against the animal being protected' (8/5/1928). Caught between responding accurately to the request of his Police Commissioner, and responding to the conservative pressures from significant members of his rural community, Superintendent Lonergan of the Southern Region forwarded to Lord a single masterly sentence, logically dissonant but politically correct: 'I think these animals are now very scarce, and I do not think their protection is desirable' (17/4/1928).

While these responses were being gathered, both Lords and the curator of Hobart Zoo, Arthur Reid, attended a two-day 'Commonwealth Fauna Conference' held in Melbourne on 17–18 April. As a result of a motion passed late in the afternoon of the first day of the conference, a sub-committee, including C. E. Lord as the Tasmanian representative, was set up to work that evening to

Table 7.1 Memoranda responses of police to the status of the thylacine in 1928, by region and police station

Police Station	Author	Comments
North-east region		
Avoca	Goss, A.	'unable to give any information concerning these Animals.'
Beaconsfield	Herbert, F.	'No Marsupial Wolves in this district, and no information can be obtained about them.'
Branxholme	Goss, R. C.	'the animal is not known to me.'
Chudleigh	Harrower, G. D.	(returned without comment on present status).
Deloraine	Collins, A. F.	(returned without comment on present status).
Exeter	Sleeth, J. H.	'there are no Tasmanian Marsupial Wolf in any part of my District.'
Fingal	Beresford, C. W.	'I know nothing about these animals, and I have never heard of one being captured in any district which I have served in.'
Georgetown	Berryman, C.	'there are no Tasmanian Marsupial Wolves in my district . . . They used to be fairly plentiful years ago . . . although none have been seen for years.'
Gladstone	Donohoe, J.	'it is very many years ago since one was seen.'
Gladstone	Medwin, E. H.	'the animal is not known to me.'
Launceston	Gunner, E.*	'the Tasmanian Wolf has not been seen in this District for very many years.'
Longford	Eyles, D. N.	'it is many years since the Tasmanian wolf has been seen – even in the bush country in this Division.'
Moorinna	Friboth, A. W.	'no knowledge of the animal at all.'
Patersonia	Gillespie, R.	'I have never been in a part where the Tasmanian Marsupial Wolf is to be found, in any numbers.'
Ringarooma	Goss, N. A.	'the animal is unknown to me.'
St Helens	Blair, E. R.	'I have never heard of the above animal being seen in this District.'
St Leonards	Gunton, W.	'it is very scarce in almost every part of the State.'
St Marys	Marshall, A. R.	(Memo 'noted', signed and returned with no additional comments, implying an absence of 'any useful information'.)
Scottsdale	Donahue, J.	'Marsupial Wolfe [*sic*] . . . is almost unknown in this part of the Coast.'
Metropolitan		
Alonnah	Morgan, W. E.	'they have never been known to have been here.'
Buckland	Morgan, W. S.	'in 1922 . . . there was either one seen or killed some where about that time by a family named French.'
Glenorchy	Byers.	'unable to supply any information regarding the Marsupial Wolf.'
Glenorchy	Hill, H.	'cannot supply any information.'
Glenorchy	Newman, C. J.	'unable to supply any information regarding the Marsupial Wolf.'

Table 7.1 (continued)

Police Station	Author	Comments
Hobart	Oakes, N. G.*	(Memo signed and circulated, unsurprisingly without comment upon the thylacine's existence in the City of Hobart.)
Moonah	Brooks, J.,	'not heard anything of this animal for many years.'
Moonah	Broomhall, J. A.	'unable to supply information in reference to this animal.'
Swansea	Bourk, N. F.	'from 1919 to 1925 not in any instance did I hear the Tasmanian Tiger have either been seen, nor caught . . .'
Swansea	Morrison, C. C.	'heard nothing of this animal.'
Tasman's Peninsula	Bourk, N. F.	'did not hear of this animal either as having been caught or seen.'
Tasman's Peninsula	Morrison, C. C.	'if it was ever in these districts it must have been many years ago.'
Southern		
Campbelltown	Goyen, J. L.	'"Native Tigers" are very rare indeed in this District: for quite a number of years none have been heard of.'
Great Lake	Nibbs, L. R.	'a few years ago . . . I had occasion to visit the Lake St. Clair . . . I saw Two Native Tigers . . . mother and . . . cub . . . near the China wall . . . some weeks later I saw another near the lake and . . . I saw another . . . [near] the mouth of the Derwent.'
New Norfolk	Lonergan, J. L.*	'I think these animals are now very scarce.'
Oatlands	Cooper, J. N.	'skin buyers . . . state that they [have] not bought any skins of Native Tigers for a number of years nor have they heard of any being caught.'
North-west		
Burnie	Duran, J. J.*	'I have had enquiries made throughout this District . . . all members of the Force give as their opinion that the time has arrived when these animals should be protected.'
Circular Head	Eyles, D. N.	'six years ago there were a few of these animals in the Mt. Balfour District . . . but they were getting very scarce.'
Preolenna	Bourk, N. F.	'from my own personal knowledge this animal was to be found at Preolenna . . . about 18 years ago, but even then it was very rarely seen or caught.'

*Superintendent of region.

construct a list of mammals and birds to be totally prohibited exports; the list to be submitted for discussion at the opening session of the conference the next day. The task was to share between state representatives the distribution and population estimates of the mammal and bird species within each state, then to decide which ones were 'in danger of extinction' (Robson, 1928) and to prepare a list of these species for submission to the conference with a recommendation that they be prohibited exports.

This task was, frankly, an impossible one. The sub-committee worked on into the early hours of the morning, reviewing the status of at least 689 bird species, but at the opening of the second and last day of the conference, unsurprisingly, had only 'compiled such a list in regard to birds, mammals had not yet been dealt with' (Robson, 1928). Nor were they dealt with. The delegates spent the rest of the conference, debating and adding species to the bird list and discussing minimum cage-size requirements for animal export. What was a productive conference for the protection of Australian avian species was, through lack of foresight, time and organisation, totally unproductive for the other half of its agenda, the intention to deal with the protection of Australia's mammalian species. There can be little question that, as all three Tasmanian representatives at the conference in 1928 were personally and professionally committed to the immediate total protection of the thylacine, had the sub-committee had time to deal with mammals, or, had a separate sub-committee been set up to deal with mammals, the thylacine would have been listed as in danger of extinction, and its export would have immediately been totally prohibited by the federal government. The opportunity for an Australian federal government regulation protecting the thylacine was lost through time constraints, conference mis-management and a lack of foresight by scientists. While at that time the existence of a federal law protecting the thylacine would not have legally prevented anyone from killing them in Tasmania, it would nevertheless have severely damaged the credibility granted to the anti-thylacine lobby within Tasmania and, in all likelihood, would have speeded up the process of legally protecting the species.

In May 1928 J. E. C. Lord passed the file of responses from the police to C. E. Lord, who tabled it at the next Native Fauna Advisory Committee meeting. On 3/7/1928 he returned the files to J. E. C. Lord, reporting as he did so that: 'The matter was considered at the last meeting of the Committee and the Committee decided to request that if possible the Thylacine be placed upon the totally protected list'. A week later J. E. C. Lord made a statement to the press, acquainting them of the decision of the Advisory Committee re Native Fauna. It received significant notice in the press, presented as an item in the 'Notes of the Day' section on the editorial page of the *Hobart Mercury* (5 July 1928). Such prominent publication produced little response. Only one negative 'letter to the editor' on the issue was published in the next three weeks (R. Stevenson, 1928), and the archives of the Tasmanian Animals and Birds' Protection Board preserve only the disgruntled complaints and correspondence of one Thomas Edwards (17/7/1928 and 2/8/1928) writing directly to the Committee over the newspaper report.

By now the movement for the protection of the thylacine had gathered sufficient data and such momentum that the only thing that Premier John McPhee's government, representing significant rural and hunting lobby interests, could do to stop it effectively, and pander to their rural constituency, was to carry out an administrative shift of responsibilities.

Cabinet, therefore, offered the carrot of state recognition and power if the role of the federally orientated Advisory Committee re Native Fauna was taken over by a newly created Tasmanian Animals and Birds' Protection Board. The Advisory Committee, with its federal orientation and independence, was encouraged to dissolve itself and devolve its conservation responsibilities to a new, official state government organisation, with its members selected by Cabinet and appointed by the Governor. This was supposed to increase its local power and formal position over state issues of conservation, at the same time as it would be able to respond officially and more effectively to federal issues on conservation. It appeared to be too good an opportunity to refuse, and the issue of protecting the thylacine was placed on hold as the Advisory Committee was dismantled and the new Board set up. Unfortunately for the thylacine, the new Tasmanian Animals and Birds' Protection Board (later renamed the Tasmanian Fauna Board) was numerically controlled by a majority of newly appointed members selected by Cabinet, representing vested interests supporting the timber industry, the hunting and snaring of native game, and the uncontrolled expansion of agricultural enterprise.

C. E. Lord lost no time after the new Board's establishment. At its second meeting on 14 May 1929, with reference to the decision of the former Advisory Committee, he requested the immediate placement of the thylacine in the 'wholly protected' schedule. However, the Board, reflecting the conservative attitudes held by the majority of its members, 'gave consideration to the subject but decided to defer the question until a future meeting' (Tasmanian Animals and Birds' Protection Board, 1934, p. 16). The same tactic was used when he raised the issue again at the next meeting in June: 'further consideration of the question of protection of the Tasmanian wolf was deferred until a future meeting' (Tasmanian Animals and Birds' Protection Board, 1934, p. 24). By the time it was eventually discussed by the Board at the meeting of 20 August 1929, C. E. Lord had lost control of the debate. Instead of discussing a motion calling for the thylacine's placement on the wholly protected list, the motion before the Board, proposed by A. K. McGaw and B. R. Reynolds, was 'that the animal should remain in the schedule of Animals unprotected' (Tasmanian Animals and Birds' Protection Board, 1934, p. 38). When put to the Board this motion was passed.

Partial and impartial protection

Having lost the main debate, C. E. Lord tried a flanking manoeuvre and raised the issue of prohibiting the export of thylacines. The difficulty Tasmanian

181

museums and zoological gardens experienced in competing with mainland and overseas institutions was already a part of the thylacine protection debate (C. E. Lord, 27/3/1928, 23/7/1928), and on an issue relating to improving the status of Tasmanian scientific institutions, individual members of the Board would likely be united. Furthermore, by raising the issue of prohibiting its export, the new Board would be raising federal issues as well and would therefore be publicly seen to be supposedly continuing with the traditional area of the disestablished Advisory Committee re Native Fauna. The Board considered the question and agreed in principle to 'investigate' the possibility of preventing exportation (*Hobart Mercury*, 21 August 1928; Tasmanian Animals and Birds' Protection Board, 1934, p. 38) and then rapidly moved on to further business. Consequently, at the September meeting C. E. Lord had to start the issue once again from scratch, whereupon a legal objection to his proposal was raised: as the animal was totally unprotected, would it be legally possible to prohibit its export? No decision was therefore made by the Board but C. E. Lord was authorised 'to confer with the Board's solicitor and report' (Tasmanian Animals and Birds' Protection Board, 1934, p. 47).

182   C. E. Lord reported to the November meeting the legal opinion that the thylacine, being on the wholly unprotected schedule, could not be restricted in its exportation. As the animal was 'not sufficiently plentiful to justify unrestricted exportation' he gave as a notice of motion that 'the Native Tiger be transferred to the "partly protected" schedule with a short close season' (Tasmanian Animals and Birds' Protection Board, 1934, pp. 65–6). This time round C. E. Lord played his politics well, and with the timely and supportive intervention on two occasions by J. E. C. Lord in the chair, the motion, involving protection of the species for the shortest time possible, one month, was discussed at the meeting in February 1930 and eventually passed (*Hobart Mercury*, 12 February 1930; Tasmanian Animals and Birds' Protection Board, 1934, pp. 74–5).

Events, for once, followed rapidly and positively, aided by some organised support from Australian mainland scientists (something that had never been attempted in the days of the Native Fauna Advisory Committee). At the April 1930 meeting correspondence was tabled from the Linnean Society of New South Wales requesting protection for the thylacine and, in line with this, the Board's solicitor was formally directed to draw up 'a regulation providing for a close season for the native tiger from 1st December on each year to the 31st day of December in the same year both days inclusive' (Tasmanian Animals and Birds' Protection Board, 1936, pp. 80–1, 84).

No records have been preserved in the minutes that suggest why December was chosen as the minimum one-month closed period. One data source available to interested Board members was the old Lands and Surveys Department account books on the government bounty scheme. Given that the peak bounty payments for the capture of young occurred mid-year, and an assumption that independent young were minimally at least six months of age before their likely capture, it is certainly possible to interpret December as the peak month for breeding from the bounty record data (Guiler, 1961x, Guiler and Godard,

1998), and these ledgers may well have been the source for the suggestion that 'this month was believed to be the breeding season' (Guiler, 1985, p. 30).[3]

By the time of the Board's meeting on 6 May 1930 the wording for the Governor's proclamation of the thylacine as 'partly protected' had been agreed to and regulations fixing the close season for December and prohibiting the thylacine's export were adopted and forwarded to the Attorney-General (Tasmanian Animals and Birds' Protection Board, 1934, pp. 92–3, 100–1). Both were included in the complete list of 'Regulations and Amendments to the Regulation, made by the Animals and Birds Protection Board' published in the *Hobart Gazette* (23 May 1930).

In January 1932 Mr Tom Clarke of Quorn Hall, Campbell Town, as a private individual, offered C. E. Lord '£100 towards the expense of enclosing an area of land in which the Tasmanian Thylacine might be kept so as to endeavour to save it from extinction' (Tasmanian Animals and Birds' Protection Board, 1934, pp. 274, 279). It was hoped that breeding would take place therein and C. E. Lord was authorised to investigate the possibility. In June 1932 C. E. Lord raised the stakes again, the Board approving his suggestion for the appointment of a committee to deal with 'the proposal for the greater protection of the Thylacine' related to the offered donation and 'the thylacines preservation and protection . . . in the interests of science'. A. L. Butler and B. R. Reynolds joined C. E. Lord on the Committee (Tasmanian Animals and Birds' Protection Board, 1934, p. 290). In October 1932 the committee suggested 'the practicability of placing some Thylacine on De Witt Island' and the Board encouraged further investigation of this suggestion (Tasmanian Animals and Birds' Protection Board, 1934, p. 297). By early 1933 complete protection of the thylacine and the establishment of a protected reserve in a national park into which thylacines could be liberated and encouraged to breed appeared imminent, but then a twist of fate took over.

Unexpectedly, C. E. Lord died suddenly in July 1933. In the published obituaries it was noted that 'Mr Lord had on foot proposals for the preservation of some specimens of the thylacine (native tiger)' (*Hobart Mercury*, 19 July 1933). The conservationists on the Tasmanian Animals and Birds' Protection Board considered it might be an opportune time to unite the warring factions over the thylacine as a fitting memorial to C. E. Lord:

> It seemed an opportunity for the Game Protection and Acclimatisation Society to co-operate with the Fauna and National Park Boards to endeavour to safeguard, in satisfactory conditions, some specimens of native tiger . . . Such a work would be a lasting monument to Clive Lord, who was giving attention to the matter when he died. (*Hobart Mercury*, 19 July 1933)

It would, indeed, have been a highly appropriate and fitting memorial, but the desire to co-operate in any memorial to Lord was not found in the hearts and minds of the members of the Game Protection and Acclimatisation Society, whose anti-conservation arguments, outside the original Fauna Advisory

183

Committee and inside the Fauna Board, had been so politically effective in preventing protection for the thylacine over the past ten years. The Board was forced to act alone. At the September 1933 meeting, Butler and Burbury (and Reynolds as well, according the *Hobart Mercury*, 20 September 1933) were appointed to a Committee of the Board to continue C. E. Lord's work for the preservation of the thylacine (Tasmanian Animals and Birds' Protection Board, 1936, p. 337).

Finally, in its eighth year of existence, at the meeting of 2 June 1936, the Fauna Board agreed to complete protection. A draft proclamation was completed and included with a letter from J. E. C. Lord to the Attorney-General, requesting the Governor's approval of the transfer of the thylacine to the 'wholly protected' schedule (25 June 1936). Jumping the gun, without waiting for the Governor's signature, J. E. C. Lord immediately had printed for distribution through police stations, a four-page pamphlet on *Animals and Birds, Tasmania: Close Seasons, Prohibitions, Restrictions* dated 1 July 1936, which indicated in the opening section that the 'Tasmanian Marsupial, Wolf, or Native Tiger' was now 'wholly protected'. Legally, the thylacine did not actually become wholly protected until 10 July 1936, the date when the Governor signed the proclamation (*Government Gazette*, 14 July 1936).

184

Whether anybody wandered around to Hobart Zoo and told the last thylacine the good news is unknown. Unequivocally, the species was totally protected for the last fifty-nine days of its existence.

Notes

1 One scientist thought otherwise. Krefft (1868w, 1868x, 1871) made an attempt to establish two different species (the 'dog-headed thylacine' *T. cyanocephalus* and the 'short-headed thylacine' *T. breviceps*) on the basis of cranial sexual dimorphism. The idea was exposed as sex-based and unnecessary by Allport (1868) and Thomas (1888), and, equally, received scant acknowledgment in the popular press (*Launceston Examiner*, 25 March 1869).

2 It was even more limited with respect to the thylacine, as the five-day flora and fauna survey of the National Park in January 1923 'saw no trace' of thylacines. An immediate call was made by Professor David (1923) for 'a special sanctuary for these animals' to be established.

3 Note, however, that combining the bounty data with police, magistrate and local council records from the locality where each individual bounty claim was originally made, more accurately dates the time of capture for young thylacines, and suggests instead, a late winter/early spring peak breeding season.

Chapter 8

The Last Tasmanian Tiger

Just as contingency affects extinction in the wild, so too is contingency evident   185
in the extirpation of the last known thylacine specimen at Hobart Zoo. Being
the only living thylacine ever to have been fully protected made no difference.
It could easily have lived much longer in captivity, but it did not, and the reason
for its failure to survive lies in a series of insensitive and offensive administrative
decisions made by a bureaucratic management structure with no representation
from keeper or curatorial staff, no expertise in animal care or zoo management
and, ultimately, on economic grounds, no interest in the zoo's continuation. To
place this behaviour in context a brief history of Hobart Zoo under the Hobart
City Council's control is required.

The Hobart Zoo had its origins in the private collection of animals of Mary
Grant Roberts. Known as the 'Beaumaris Zoo' it was first opened to the public
in Hobart in 1895 (E. A. Bell, 1965; Evans and Jones, 1996; Guiler, 1986). On
Mary Roberts' death in 1921 the collection was first placed in the care of the
Trustees of the Tasmanian Museum and Art Gallery, before finally being passed
into the hands of a somewhat reluctant Hobart City Council, whose Reserves
Department relocated it to the Domain and reopened it in 1923, under the
curatorship of Arthur Robert Reid (appointed on 18 April 1922), a founding
member of the Tasmanian Field Naturalists' Club, an active campaigner for the
preservation of Tasmanian wildlife, and noted breeder of both native and
introduced game bird species (Reid, 1907). He was initially paid at a salary
commensurate with that of the Superintendent of Reserves, Mat (T. M.)
Lipscombe (Reserves Committee, 30/9/1924). Reid's position involved responsibility for the restocking of the zoo, the care of the animals, and establishing
and organising the duties of the animal keepers. However, other aspects of the
functioning of the zoo were not relinquished to the curator; the cleaning of the
zoo grounds and paths, repair of the zoo cages and the planting of the zoo
gardens remained under the control of the Superintendent of Reserves.

In the absence of any recognised form of training in the curatorship of zoological gardens, in both European and also Australian zoos the position frequently passed from father to son (Mullan and Marvin, 1987; Strahan, 1991). In Tasmania, the curatorship of the Launceston City Park Zoo passed from William McGowan snr to William McGowan jnr, and on the mainland, the directorship of the Adelaide Zoological Gardens was successfully handed down through three successive generations of the Minchin family (Rix, 1978), thus preventing access to the Le Souëf family who, between them – father and four sons – at different times controlled all the other major Australian mainland zoos.

Unfortunately for the thylacine, Arthur Reid's only offspring interested in the zoo was his daughter, Alison (5 June 1905 to 20 June 1997). Trained in taxidermy by her father, she mounted hunting trophies for 'sportsmen', and departed pets for their grieving owners (interview 27/2/1994). Her skills in this area were soon recognised by the Tasmanian Museum, where she was employed as a taxidermist from May 1922 on a salary of £104 p.a., raised to £120 p.a. in April 1923 and £150 in August 1925 (Museum Trustees, 1928, pp. 117, 125; 1945, pp. 61, 65, 74). Outside her working hours at the museum, Alison assisted her father at the zoo 'rearing the pheasants and rearing quail and things like that, and bringing up the baby animals' (interview 27/2/1992) – amongst them the two lion cubs Sandy and Susie and the leopard cub Mike, all bred in the zoo (see Plate 8.1). In addition she was paid for weekend work selling tickets at the turnstile (interview 24/6/1996).

In 1927 a full-time, seven-days-a-week position of gatekeeper/turnstile attendant was created for Alison in the zoo. Nominally it was to involve the supervision of the weekday turnstile operator, Miss K. Scanlon; and the continued covering of the turnstile on the weekends, together with secretarial and actuarial duties for the zoo from an office in the gatehouse. Such a position allowed ample time for the seven-days-a-week responsibility for the animal nursery established within the Reid family home. Alison was the only genuine applicant for the position (Miss Scanlon also applied, but did so on the basis of a preparedness to work part-time only), and seemed assured of appointment, but the new Superintendent of Reserves, another member of the Lipscombe family, L. J. Lipscombe, encouraged a soldier friend, R. Manson, to apply for the position, after the closing date for applications had passed, and then proceeded to roll his preferred candidate through the Reserves Committee (10/1/1928). As a limbless soldier Manson was not a great success as a secretary, he was neither inclined towards nor physically capable of working in the animal nursery, which involved the rearing of the large carnivore cubs, and he was reluctant to work seven days a week.

Disappointed, but not downcast, in response Alison decided for professional as well as personal reasons to travel to Europe. In February she sought, and was granted unpaid leave from the museum from a sympathetic and supportive Clive Lord, but the Trustees over-ruled Lord's decision and demanded her immediate return to work. In consequence, Alison resigned (Museum

Plate 8.1 Alison Reid with lion club 'Sandy', outside the animal nursery in the curator's cottage, 1927. (Author)

Trustees, 1928, pp. 162, 163). Carrying letters of introduction to the British Museum and the Zoological Society of London (interview, 25/6/1992), with her savings of £70 – and a weekly remittance of £2.10.0 from her father – she set off, staying with friends and relatives in London and Scotland, along the way visiting Regent's Park Zoo while its last thylacine, a former inhabitant of Hobart Zoo, was on display (interviews, 27/2/1992, 25/6/1992).

On her return to Tasmania at the end of 1928 she continued to assist her father in looking after the zoo and, once again, took over the rearing of the young animals in the nursery. With Manson refusing to work on the weekends, she even returned to paid employment at the zoo, working on the turnstile on Saturdays and Sundays. Armed with new taxidermic knowledge from abroad, she was re-employed at the Tasmanian Museum in January 1929 on a salary of £150 p.a., increased to £200 p.a. in July 1929 (Museum Trustees, 1937, p. 20; 1945, pp. 88, 90).

As the effects in Tasmania of the world-wide economic depression increased, a reduction in the expenses of the zoo was achieved by sacking zoo employees, including the positions of the turnstile operators, Alison Reid and Miss Scanlon. The turnstile entrance was closed and access to the zoo made instead through the tea rooms, with tea room staff given the responsibility of ticketing visitors to the zoo (Reserves Committee, 4/8/1930, 25/1/1932), under the 'supervision' of Manson.

The museum's initial response to the Depression and the decreasing size of its government grant was to preserve the employment of staff, but reduce their salaries. Alison's salary was dropped to £175 p.a. in October 1930, and briefly restored to £200 in May 1931, before being dropped to £180 p.a. in July 1931 (Museum Trustees, 1937, p. 40; 1945, pp. 96, 98, 99).[1] As the effects of the economic depression continued, by majority vote the Trustees of the museum decided in August 1932 'That Miss Reid's appointment be terminated at the end of September' (Museum Trustees, 1937, p. 100). Clive Lord and some of the Trustees were concerned about her treatment, given that Alison had commenced employment at the Museum in 1922, but their attempts at the next two Trustees' meetings to undo or soften the decision were unsuccessful (Museum Trustees, 1937, pp. 102, 103). From October 1932 Alison effectively became her father's full-time, unpaid assistant, a role that steadily increased in importance to the functioning of the zoo as time moved on.

One evening, towards the end of 1930, Arthur Reid heard a commotion in the zoo after-hours and upon entering the gardens he was viciously attacked by A. E. Fischer of Battery Point, who was attempting to steal some birds (Reserves Committee, 21/11/1930). One of the kicks and punches he received destroyed his left eye. The damaged eye and an associated growth were removed in December 1931 (Reserves Committee, 11/1/1932), but a carcinomatous growth remained active in the eye socket. He continued to work as curator for another three years, but gradually the effects of the cancer became more and more obvious. During 1934 and 1935 Arthur became less and less able to work. When he became physically tired in the afternoon, Alison would take over entirely as de facto curator, managing the zoo's income and expenditures, ensuring the late afternoon feeding of the carnivores took place and checking all the animals to make sure they had access to their sleeping quarters for protection during the night, before closing and locking up the zoo in the evening (Reid, interview 24/6/1996). The records of the Reserves Committee make it quite clear that the members knew what was going on, frankly acknowledging that, in an unpaid capacity, Miss Reid 'took charge of the zoo . . . when her father was injured' (Reserves Committee, 30/10/1935).

Arthur's illness was seen by the then Superintendent of Reserves, Bruce Lipscombe, and the members of the Reserves Committee, as an heaven-sent opportunity to reduce the expenses of running the zoo. The Town Clerk, William Brain, after strong prompting from Lipscombe (12/9/1935), presented Arthur with a proposal 'That he should accept an appointment as a part time Officer only', with a reduction in annual salary from £370.10.0 (£7.2.6 per week)

188

to only £100 (£1.19.6 per week). Arthur refused and 'made the following alternative suggestion, that his daughter who is aged 31 years, should take charge of the Zoo, at a salary of £3.10.0 per week, equal to £182 per annum'.[2] Arthur was prepared to resign and continue to work 'without charge to help his daughter in every possible way' should she be appointed the new curator (Brain, 29/10/1935).

Brain expressed himself as 'rather diffident of recommending' the transference of the curatorship from father to daughter (30/10/1935). The Reserves Committee concurred (30/10/1935) and asked Brain to approach Reid on the matter once again. One of the reasons Brain suggested to Alison for his expressed diffidence was the desire for the Reserves Committee to be seen to appoint openly and fairly, and not to be tainted by any charge of nepotism (Reid, interview 24/6/1996). However, given that the new Superintendent of Reserves, Bruce Lipscombe, was now the third consecutive member of the Lipscombe family to have been appointed by the Reserves Committee to the senior public position of Superintendent of Reserves, this explanation appeared more than a little rich. Obviously this crude and fantastic suggestion was given merely as a smoke-screen to hide a deeper reason for denying the curatorial position to Alison.

Arthur's physical condition deteriorated rapidly. He became too ill to work in November and died on 13 December 1935 (*Hobart Mercury*, 14 December 1935). Bruce Lipscombe officially took over the zoo until a replacement curator could be appointed (Reserves Committee, 27/11/1935, 22/1/1936).

Despite the fact that the zoo had been founded by a woman, Mary Roberts, and that, as the Reserves Committee acknowledged, Alison Reid had been effectively acting as curator for the previous twelve months, the Reserves Department would not countenance the idea of appointing a female, even in an acting capacity, to the vacant curatorial post. The Reserves Committee recognised 'that after the Curator's death there appeared to be no one in the Council's employ with any knowledge of the care of animals needing attention' (22/1/1936). Note that Alison had ceased to be a Council employee after the abolition of the turnstile operator positions in 1932. Nevertheless, her expertise was sorely required. Bruce Lipscombe obtained this by allowing Alison and her mother to remain in the curator's cottage rent free, so long as Alison was prepared to continue to work as unpaid curator of the zoo:

> The Superintendent of Reserves . . . had arranged for Mrs. Reid and her
> daughter to remain in the cottage . . . No rent was to be paid for the cottage, but
> in consideration of this, Miss Reid (who has considerable knowledge of fauna)
> was to give her services to the Superintendent free of charge. (Reserves
> Committee, 22/1/1936)

With a supposedly secure roof over their heads, but the absence of any salary, Alison and her mother were 'getting close to poverty . . . we lived like slaves' (Reid, interview 24/6/1996). Mrs Reid applied for a gratuity respecting the

189

length of Arthur's service to the Committee and his death resulting from being assaulted while performing his duties (Reserves Committee, 22/1/1936). After some delay while legal advice was sought about the Committee's responsibility for the effects of the assault and eventual death of the curator, the Council's Finance Committee finally approved a small gratuity payment (Guiler, 1986; Reid, interview 27/2/1992).

Alison, with the interest of the animals at heart, continued to work as unpaid curator, but with one major difference in professional responsibility. As soon as her father was forced to stop working in November 1935, Bruce Lipscombe 'took the keys from me' (Reid, interview 27/2/1992). Alison now had to wait in the morning for a representative from the Reserves Department to arrive and open the zoo before she could gain access to the animals, and she was unable to check all of the zoo exhibits and specimens at the end of the day, being forced to leave the zoo before the gates were locked in the afternoon. More importantly, she no longer had access to the zoo grounds out-of-hours, to deal with emergencies or the effects of irresponsible care. This callous insensitivity to both Alison and the zoo animals reflects back upon the values and priorities possessed by the Superintendent of Reserves and the members of the Reserves Committee.

190

Thylacines at Hobart Zoo were usually kept in a large rectangular pen opposite the kangaroo and deer paddocks, just down the hill from the curator's cottage and the rear entrance gate to the zoo. Approximately 8 m by 4 m, the pen was partially shaded by a large deciduous tree, and contained two wooden doors, the larger one for keeper access to the cage, the smaller one for thylacine access to the protected sleeping quarters or den, a covered and enclosed retreat space of approximately 3 m by 1 m.

The thylacine, along with the placental carnivores, figures prominently in the Reserves Committee records – notes on the purchase, health and death of various thylacines are found therein. Unfortunately, the Reserves Committee minutes have not been preserved between 21/6/1932 and 22/5/1934, and it was during this two-year interval that the last known thylacine was purchased by the zoo.[3] With reference to the last thylacine in the zoo Alison recalled: 'That one came from Tyenna, I think. That would be … about a year before my father died', that is, about 1934 (interview 25/6/1992). The naturalist Michael Sharland had no 'exact knowledge of where the last Thylacine to be kept in the old Hobart Zoo was trapped. The only record I have is that given me by the then Zoo manager, A. R. Reid, that this particular animal came from the Florentine Valley' (Sharland, 17/12/1972).

In this letter Sharland also suggests 'you may be interested in an interview with Elias Churchill', whom Sharland accepted as being the capturer of the last specimen.

Elias Churchill was working as a timber cutter and snarer around Adamsfield and in the Florentine Valley, with various companions, in the 1920s and 1930s. He snared eight adult thylacine specimens between 1924 and 1933, two were taken alive, the others were either strangled in the snare or so badly damaged by the snare they were destroyed on the spot (*Hobart Mercury*, 1964; Sharland, 1957x, 1980). His first live capture was a female with three pouch young which were initially exhibited at country shows and carnivals by Walter Mullins, before being sold to Hobart Zoo (Reserves Committee, 19/2/1924, Sharland, 1957). Details of Churchill's second live capture in 1933 are available from interviews with Churchill (*Hobart Mercury*, 19 September 1964; Sharland, 1957x).

Two people remember the thylacine in Churchill's care. In an unpublished letter to the Director of the Tasmanian Museum, Algie Chaplin (31/8/1954) suggested that 'twenty years ago one was captured in the fields by Churchill'. In another unpublished letter the 80-year-old V. Stanfield recalled: 'The last Tiger to be caught in Tasmania was away up behind Fitzgerald Tyenna way by a Mr. Churchill' (31/8/1981). Once its sale to Hobart Zoo had been arranged, the thylacine was trussed up, placed on the back of a pack horse and taken to Tyenna (Sharland, 1957) where it was placed in a cage to be railed to Hobart. E. L. Gossage remembers how, as a young girl, she saw the 'tiger that was being sent to the Hobart Zoo. It had been caught out beyond Maydena by E. Churchill . . . and was being transported by train to Hobart, and at National Park station we were allowed to take a "peep" at it in its cage in the Guard's Van' (Gossage, 24/8/1981). Soon after its arrival at the zoo in 1933 it was filmed for 62 seconds by David Fleay, who received a bite on his hindquarters during the filming, thus becoming the last person known to have been attacked by a thylacine (Fleay, 1956, 1979). The film, together with other still photographs, clearly identifies the last specimen as a mature but relatively young adult female, Plate 8.2.

There is a general comment about thylacines being on display in the Hobart Zoo at this time, in a book published by the Tasmanian Government Tourist Bureau (1934), but only one published Tasmanian reference has been discovered that specifically refers to this specimen while it was still alive. On a lecture tour of Tasmania, the Japanese Christian philosopher Dr Toyohiko Kagawa, accompanied by his secretary K. Ogawa, visited Hobart Zoo on 6 April 1935. Ogawa remarked that 'The Tasmanian tiger was also of great interest to us' (*Launceston Examiner*, 22 April 1936).

In the morning if the thylacines were out, they were encouraged to return to their den and the door was closed, constraining the animals while the cage was cleaned and the water dish filled. Then the den door was opened and the thylacines were 'encouraged' to leave, whereupon the door was closed again to prevent their return; keeping them on public view all day with no retreat space available. In the late afternoon it was the responsibility of the keepers, after the feeding of the carnivores which commenced at 3 p.m., to reopen the door to the den and allow access to the protected retreat area during the night (Reid, interview 27/2/1992).

Occasionally in her father's active time in charge of the zoo, the zoo staff would leave in the afternoon without opening the door to the thylacine sleeping quarters. Through the use of the coughing bark the thylacines would attract the attention of Arthur and Alison Reid to this problem, and the Reids would enter the zoo and correct it (27/2/1992). When Alison's repeated attempts to obtain

Plate 8.2 The last living Tasmanian tiger, Hobart Zoo, 1933. (Archives Office of Tasmania)

another set of keys from Lipscombe, for such evening and emergency use, were rejected, she went over Lipscombe's head and appealed directly to the Town Clerk, William Brain, for help. Brain refused, implying that it was only through the incompetent management of the staff by her father that there was any need for 'the habit of inspecting the Zoo each night, in order to see that the animals were alright and properly locked away in their cages'. Now that Bruce Lipscombe had 'been appointed to take charge [of the zoo] during the absence of Mr Reid' he would ensure that the zoo staff lived up to their responsibilities – 'it will be the duty of the attendants out there to see that everything at the Zoo is satisfactory'. Thus, under the new management, there would be no need for Alison 'to make her nightly inspection' or reason for providing her with an alternative set of keys (Brain, letter, 29/11/1935).

The Hobart winter of 1936 was particularly severe. So too was Tasmania's economic depression. The zoo had lost experienced staff, through death, retirement and retrenchment (Reserves Committee, 17/7/1935, 28/8/1935, 13/11/1935, 22/1/1936). They were not replaced. Instead, the zoo was filled with 'sussos', sustenance workers, who, by experience, were totally unqualified, and by inclination totally uninterested in the positions to which they were arbitrarily appointed to work for the dole. Encouraging the replacement of experienced, professional staff with sustenance workers (Reserves Committee, 8/10/1935) was a major act of bureaucratic and administrative vandalism perpetrated by the Reserves Committee. The so-called staff present: 'in the zoo were mostly unemployed. They got work there you know, as sort of giving them a job. As somebody said . . . "There was so many unemployed, looking at them, there was more unemployed than animals"' (Reid, interview 25/6/1992).

In the absence of an officially appointed curator, the zoological gardens could be seen, by anyone who chose to do so, as effectively rudderless. As Alison remarked:

> Women couldn't do anything, you see in those days, that was the attitude . . .
> I could have looked after it [the zoo] but they got this idea in their head that
> a woman couldn't do it . . . If I'd been a boy it'd be like they would have
> continued it, but [there was] no equality in those days whatever.
> (Reid, interview 27/2/1992)

193

She had neither the gender nor the legal or hierarchical status to negotiate or enforce work objectives or duties upon the sustenance workers, appointed to work under the supervision of Lipscombe. In 1935, five individuals, Brett, Fleming, Newman and both Reids, were responsible for the daily care, cleaning and feeding of the animals in seventy-eight enclosures (A. R. Reid, letter 23/8/1935). In 1936 the daily care for the animals was physically beyond the capacities of just Brett, Fleming and Alison, particularly within the narrow time-constraints imposed by denying out-of-hours access. Alison could only make suggestions as to the daily work requirements of the sustenance workers, and hope the animals, and her instructions regarding them, would be respected and acted upon in recognition of her knowledge and experience in the running of the zoo. In the main, she suggested and hoped in vain. The sustenance workers 'had to be watched all the time', unsurprisingly, 'they never had . . . any sense of responsibility to the animals, they were just there to do as they were told [by Lipscombe] and go home as quick as they could' (Reid, interview 27/2/1992).

'It was pretty awful at the zoo after my father died . . . you'd see everything getting neglected and killed off' (Reid, interview 27/2/1992). With control legally in the hands of a laissez-faire and largely absent Bruce Lipscombe, content to turn the zoo into a ghetto for the unemployed, cages were not cleaned and animals were unfed or left to eat the previous day's rotten food before being fed again (Reid, interview 27/2/1992). Because of the expense involved, Alison's requests to Lipscombe and then Brain for veterinary treatment for the animals were refused (Brain, letters 19/3/1936, 22/4/1936; Reserves Committee, 1/4/1936, 23/4/1936, 20/5/1936). 'The animals were dying off from being neglected . . . it was pathetic the way those animals were treated' (Reid, interview 25/6/1992). Due to indifferent feeding and exposure (being locked out of its covered sleeping quarters at night and left on open display 24 hours a day), the once magnificent black panther died of pneumonia (Reserves Committee, 19/2/1936).

Only once did Alison's protests to the Town Clerk receive any form of positive recognition. The last thylacine, caged at the back of the zoo, away from the zoo office, storeroom, workroom and public entrance (but in front of the curator's cottage) was consistently neglected and left exposed both night and day in the open, wire-topped cage, with no access to its sheltered den. Alison used this consistent neglect in a last appeal to Brain in early May 1936 for a set of keys for emergency use. While Brain refused her request for the keys, he nevertheless accepted Alison's criticisms and concerns over the treatment of the thylacine and discussed it with Lipscombe on the morning of 13 May 1936. Finding Lipscombe unresponsive to such criticism (that obviously emanated from Miss Reid) Brain felt the necessity to place his request to Lipscombe about the thylacine's treatment formally, in writing: 'having it out from 10 a.m. to 5 p.m. is too long'. In order to ensure it was given access to its sheltered den at night, Brain gave instruction that it should be the first animal to be put away in the afternoon. The thylacine display should be 'arranged from say 11 a.m. to 4 p.m. (at the most)' (Brain, 13/5/1936). Lipscombe did not take kindly to this suggestion and no change was evident to Alison in the thylacine's treatment. He dismissed the letter's content, having already chosen to dismiss Alison.

Such dismissal, since Alison was not officially employed by the Reserves Committee, consisted merely of demanding the Reids pay rent for the zoo cottage. Bruce Lipscombe first threatened Alison with this disciplinary measure, before eventually putting it into practice. In the succinct words of the 91-year-old Alison: 'Bruce Lipscombe . . . became a perfect bastard' (Reid, interview 24/6/1996; see also the *Sunday Tasmanian*, 3 April 1988). Less than a month after Mrs Reid received the written offer of the curator's cottage rent free in return for Alison's professional services to the zoo (22/1/1936), Lipscombe set out to change the tenancy agreement (Reserves Committee, 19/2/1936), charge rent for the cottage and place one of the existing zoo hands therein (Reserves Committee, 22/4/1936). The Reids were given advance warning, that it had been 'decided . . . that a leading hand should be appointed at the Zoo, and he should have the house at present occupied by you'. For the time being 'nothing definite will be done in the matter . . . but I am informing you of the position so that you will have plenty of time to make arrangements for another cottage' (Brain, 17/3/1936). With the Depression still raging, unemployed and unable to pay the rent or continue living in the zoo cottage, in June 1936 Alison and her mother were forced to leave the zoo – which father and daughter had looked after for over fourteen years – and seek shelter with relatives (Evans and Jones, 1996; Reid, interview 24/6/1996).

194

With genuine distress in her voice, Alison recalled to me the last weeks of her life at the zoo in 1936. Powerless, keyless and shortly to be dismissed from the zoo and turned out of her home, she listened at night to the distress calls of the zoo's remaining carnivores: the last thylacine, a Bengal tiger and a pair of lions, all too frequently locked outside in the open to face the cold, rain and snow of the Hobart winter, with no covering, protection or access to their sheltered dens. Her continued attempts to recover a set of keys for emergency use from Lipscombe and Brain were refused. Neither the marsupial nor the placental tiger survived this maltreatment.

September, and the new spring season arrived, with the deciduous tree covering the thylacine's cage still leafless from the winter. Locked out of her den, the thylacine remained at the mercy of the weather. In the last ten days of her life, minimum daily ambient temperatures varied from freezing 32°F (0°C) to 46°F (8°C), and maxima from 47°F (8°C) to 63°F (17°C); while on the floor of the open cage, thermal conductivity ranged from a terrestrial minimum of 26°F (–3°C) to a solar radiation maximum of 107°F (42°C), [*Hobart Mercury*, 31 August 1936 to 9 September 1936]. Without access to her den, the thylacine was unshaded from the sun by day, and shelterless from the cold by night.

Thus, unprotected and exposed, the last known thylacine whimpered away during the night of 7 September 1936 (Reserves Committee, 16/9/1936), as much a victim of sexual as species chauvinism.[4]

195

Notes

1 On both these occasions of salary reduction Alison was singled out by the Trustees for special, discriminatory treatment. In October 1930 all staff at the Museum had their salaries reduced by 5 per cent – except for Alison, who suffered a 12.5 per cent reduction. In July 1931 Alison had to endure a 10 per cent reduction in salary, while her co-workers experienced at a maximum a 5 per cent reduction, and, at a minimum, no reduction in salary at all (Museum Trustees, 1937, p. 70; 1945, pp. 97, 98).

2 A woman being paid only half the wage of a man carrying out exactly the same duties was not unusual in Australia in the 1930s. The practice of equal pay for equal work was only established in Australia with the election of Gough Whitlam's Labor Government in December 1972.

3 Guiler (1986, p. 154) anomalously records the purchase of a female thylacine on 21 July 1935. But this is merely unfortunate transcription or typographical error, as the details given for this specimen relate to a purchase on 21 July 1925.

4 For the record, the Bengal tiger died on 24 July 1936 (*Hobart Mercury*, 25 July 1936). The pair of lions survived a further year, but once the decision to close the zoo on 25 November 1937 had been made, at seven and ten years of age no purchaser could be found for them, so they were shot (Evans and Jones, 1996).

Post-Extinction Blues

196　　In an 1871 publication Krefft, at the Australian Museum, Sydney, counselled the staff of the Tasmanian Museum and members of the Royal Society of Tasmania in the following manner: 'Let us therefore advise our friends to gather their specimens in time, or it may come to pass when the last Thylacine dies the scientific men across Bass's Straits will contest as fiercely for its body as they did for that of the last aboriginal man not long ago' (Krefft, 1871, p. 7).[1] Krefft effectively predicted the fate of the species but not the behaviour of the scientists. When Bruce Lipscombe reported to the Reserves Committee in September 1936 that 'the Tasmanian Tiger died on Monday evening last, 7th instant', he also noted that 'the body had been forwarded to the Museum' (Reserves Committee, 16/9/1936). Regrettably, the museum appeared totally uninterested in the specimen and no attempt was made to preserve it.[2]

In the early 1930s the Fauna Board was in a position to investigate occasional reports of thylacines made by police in response to memoranda requesting lists of local species of game. On investigation, all of these claims were found to relate to historical rather than contemporary events. At the Fauna Board's meeting of 9 February 1937 it was 'decided to . . . ascertain in what places . . . the "Native Tiger" was likely still to be found . . . through the "Press" and the Police' (Tasmanian Animals and Birds' Protection Board, 1948, p. 72). A special committee of J. E. C. Lord, Joseph Pearson and A. L. Butler was set up to direct this. These three individuals hold a special place in the narrative history of the thylacine, for they decided to publish the conclusion that the thylacine was extinct, to see if this would generate any reliable information on the continued existence of the species. Thus the Fauna Board suggested, both in print and in the news broadcast statewide by Tasmanian radio stations of the Australian Broadcasting Commission (Marthick, 10/2/1937), that: 'members of the Animals and Birds Protection Board are concerned lest the Tasmanian

Thylacinus (native tiger) should have become extinct . . . it is feared that the animal . . . may have ceased to exist . . . We have no reliable evidence of the present existence anywhere of these animals' (*Hobart Mercury*, 10 February 1937). Although the newspaper report and radio broadcasts resulted in a generous correspondence, the investigation of these 1937 claims came to nothing. Unsurprisingly, all claims of sightings investigated since then have also come to nothing. J. E. C. Lord, Pearson and Butler deserve acknowledgment as the first people to publish notice of the species' extinction – a mere 156 days after the extinction event occurred.

The belief in Benjamin

As the older generation of professional scientists and naturalists with first-hand knowledge of the species proceeded to follow the thylacine into oblivion, it was recognised that there was also a large, but ageing, population of non-professional individuals in Tasmania with personal experience of the thylacine and significant attempts were made to record this information.

L. D. Crawford, zoologist at the Queen Victoria Museum, Launceston, was 197 the first person to consistently record oral accounts of the species. He set about interviewing old-timers in the early 1950s, not just to obtain contemporary information as to the likely whereabouts of the species, but to question and record people's memories of the species as a whole. Much original information on the behaviour of the thylacine was provided by people who recalled not just the last fragmentary days of the species in the 1920s and 1930s, but whose memories stretched back into the nineteenth century, before the bounty returns reached their peak. This Launceston tradition resurfaced again in the late 1960s and early 1970s with the three-man Thylacine Expeditionary Research Team of Jeremy Griffith, James Malley and Bob Brown. While their primary aim was to locate contemporary thylacines, they also recorded many pre-1936 accounts of the species from trappers and old-timers.

Tasmanian naturalists and scientists unfamiliar with the species were able to indulge themselves by interviewing trappers and old-timers whenever found, but opportunities were far more restricted for their mainland counterparts. The well-known mainland naturalist, Graham Pizzey, had a weekly nature column that was syndicated in Melbourne and Sydney newspapers and his books, containing both text and his memorable photography of Australian animals, were popular 'coffee-table' best-sellers (for example, Pizzey, 1966). When one Victorian resident, Frank Darby, appeared, claiming to be the keeper of the last thylacine at Hobart Zoo, it appeared too good to be true. So eager were mainland scientists to chase away the post-extinction blues that they failed to test the credentials of the individual concerned and the validity of his information, choosing instead to welcome additional data on the behaviour of the species into the scientific canon. The opportunity to publish new data on the species led Pizzey to suspend his disbelief during the interview and publishing process, and his status as the best-known mainland Australian naturalist of his time encouraged a suspension of disbelief in virtually all the professional as well as lay readers of his report.

The newspaper article derived from this interview was published in May 1968:

> This week I talked with Mr Frank Darby, of Belmont, Geelong, who had the care of this animal, Benjamin and is thus one of the few people alive who can speak with personal knowledge of an undoubted live Thylacine.
>
> He said he used to feed Benjy a live rabbit night and morning, and the speed with which the creature despatched the prey was impressive. The whole rabbit was eaten, crushed by immensely powerful jaws – almost crocodile-like. Benjy was extremely quick in his movements.
>
> 'Sweeping the 50-foot enclosure,' said Mr Darby, 'Benjy would be down the far end one moment, the next he'd be right behind you. Quick as lightning.'
>
> He was tame, could be patted, but was morose, and showed no affection.
> He stood a good deal like a pointer, and used his nose a lot. He never made any sound – his bearing and silence were uncanny. (Pizzey, 1968)

198

Of all the information contained in this report that has entered the public and professional domain, none has been stronger than the imagery associated with the idea that the last known thylacine specimen was affectionately known as Benjamin, or Benjy. The data from this Darby interview have been widely repeated by others writing on the species. Popularly, the idea of 'Benjamin' has almost become a cottage industry in itself.

A ten-minute colour film, designed for use in Australian senior primary and junior secondary school classrooms, emphasises the importance of conserving and protecting Australia's wildlife through examining the fate of the thylacine. This film, *Tiga* (Clutterbuck, 1989), is 'dedicated to Benjamin – the last Tasmanian Tiger in captivity'. The teacher's guide that accompanies the film covers hand-outs for students, classroom activities and assessment tasks on the issue of conservation, and all of these persistently use the word 'Benjamin' as an identifier for the last known thylacine specimen (Scholes, 1990). The idea that Benjamin was the name of the last thylacine in existence is an encouraged and growing concept in the minds of the present generation of Australian school-children.

At this stage it is worth introducing the idea that if the last thylacine specimen (as illustrated in Plate 8.2 and Plate 9.1) was genuinely called 'Benjamin', then a dreadful accident must have happened to him in his youth. Male mammals frequently display important sexual characteristics in the vicinity of their hind legs. Although male thylacines possessed a pouch in which the testes were often carried, they were never permanently hidden from view (Plate 9.2. See also the photograph of an erect male thylacine with pendulous scrotum at Berlin Zoo in 1905 [Raethel, 1992, p. 61]). There is no evidence of the presence of testes in any of the surviving film or still photographs of the last thylacine

Plate 9.1 Rear view of the last living Tasmanian tiger, Hobart Zoo, 1933. (Queen Victoria Museum)

199

specimen. Instead, these records suggest, in parallel with the size and shape of the head and body, that the last known thylacine specimen was a mature, but still relatively young, adult female.

So, Darby got the sex and name of the last known thylacine wrong. Does it matter? The idea of Benjamin has obviously struck a chord in both lay and professional readers alike. Much has been made of it, including an environmentally and politically desirable educational unit on conservation issues for use in Australian schools, that is based around Benjamin's existence. Should this be upset? The answer to both these questions depends on a consideration of two additional factors: first, the quality of the other information on the thylacine provided by Darby, and secondly, the possible reasons for the unquestioning adoption of the Benjamin story.

Leaving aside the issue of nomenclature and sex of the last captive specimen, there are at least three other aspects of the Darby interview that should have engendered some caution in the minds of readers and later replicators of this interview. First, Darby expressed the anti-conservationist position on the muteness of thylacines, an argument created only in the second decade of the twentieth century and designed to suit a particular political need – to help prevent the preservation of the species (see chapter 7). The extensive vocal capacities of thylacines at Hobart Zoo were well known to the curators. Secondly, rather than a 50-foot enclosure, film and photographs of the cage suggest the size of the enclosure was closer to only 25 feet (7.6 m). Thirdly, Hobart Zoo was an internationally recognised zoo of the highest category, and registered as such under the Commonwealth Department of Health Quarantine Act on 20 July

Plate 9.2 Male thylacine on display, Hobart Zoo, *ca* 1916. (photo: Joan Dixon)

1927. Ignoring the cruelty, morality and ethics of the Act, no zoo in the 1920s that fed live prey to its carnivores could possibly have obtained or continued to hold a Grade A rating. The implication that her father, or that she herself, allowed or connived at the feeding of live prey to their carnivores was always deeply offensive to Alison Reid, and is an aspersion she wished to see corrected in the literature (27/2/1992, 25/6/1992).

Hence much of what Darby told Pizzey about 'Benjamin' appears factually inaccurate and plainly incorrect. Both the last curator, Alison Reid (interview, 27/2/1992), and the former 'publicity officer' for the zoo, the naturalist Michael Sharland (17/12/1972), have denied the existence of 'Benjamin' as the pet name for the last thylacine specimen. What Darby told Pizzey about himself as the former keeper of thylacines in Hobart Zoo is also open to question. Alison Reid emphatically affirms that no one called Darby was ever employed in any capacity at the zoo during the years her father or she herself was in control (25/6/1992). In support of her statement, I have found no records of any salary payment to a Frank Darby in either the accounts tabled at Reserves Committee meetings, or in the cash book of the Tasmanian Museum. Nor have I been able to find any mention of Darby in the existing diaries and letters of Mary Roberts who founded the zoo in 1895.[3]

I have not been able to find any evidence that Frank Darby was ever employed at the zoo at any time during the forty-three years of its existence. I am unable to explain why Darby chose to present himself to Pizzey as the person 'who had care of' the last thylacine in Hobart Zoo, but none of the information he provided about the animal, in terms of its sex, name, size of cage, feeding régime and vocal ability, has withstood investigation of its validity. It appears that Benjamin's only existence was in the mind of his maker, Frank Darby.

Interestingly, there was enough internal inconsistency in the original newspaper report to question Darby's claims. Why then was this not done? Certainly, at one level, we are dealing with data that were accepted and published by Australia's most well-known naturalist. However, it is likely that other characteristics of the Benjamin story have also led to its replication and acceptance. Since it has proven such a powerful story in the literature, popularly accepted and intricately tied in with conservation values, this raises the question again, does the correction of this error really matter?

201

I can only conclude that there are aspects of this story of the last thylacine specimen as a male called 'Benjamin' that appeal to professional and lay readers alike. I can see no reason from an examination of the historical development of Australian zoology and comparative psychology why the name 'Benjamin' should be special. I am forced to suggest that it is simply the maleness of the animal that is important. The 'noble' masculine tradition of bravely pursuing a lost cause and following it to its logical end: the last, lone, male survivor on an isolated outpost, losing the battle of life and dying hopelessly but grandly as the last representative of its species. Perhaps this picture finds a resonance in the minds of male scientists, who make up the bulk of practitioners within the sciences of zoology and comparative psychology, as well as a majority of those who refer to themselves as naturalists. If the reason for the ready acceptance of the Benjamin story lies in its resonance with patriarchal scientific assumptions, then the question 'Does it matter?' needs to be answered with a resounding 'yes'!

Causal factors in extinction

The extinction event of any species is going to be the result of a set of inter-related factors. The most well-known examples of recent vertebrate extinctions are all associated with a deliberate human assault upon the species concerned. The quagga, *Equus quagga*, became extinct on 12 August 1883 (Willoughby, 1966), the passenger pigeon, *Ectopistes migratorius*, became extinct on 1 September 1914, the Carolina parakeet, *Conuropsis carolinensis*, became extinct on 21 February 1918 (Bridges, 1974), the toolache wallaby, *Macropus greyi*, became extinct on 30 June 1939 (Robinson and Young, 1983) and the dusky seaside sparrow, *Ammodramus maritimus nigrescens,* became extinct on 16 June 1987 (Walters, 1992). Species in which the last representatives did not die in captivity, but were slaughtered for amusement or the table, departed at less precisely known times, but are still recognisable by their absence; amongst these are the great auk, the dodo, the Antarctic wolf, Stellar's sea cow, and several species of insular giant tortoises. The direct cause of the extinction for all these species is unequivocally human behaviour.

The list of causal factors involved in the extinction of the thylacine includes, (1) the destruction of the original Tasmanian environment and its replacement with an agricultural community; (2) the concomitant destruction of the thylacines' native prey species through environmental alteration, or for food or the fur trade; (3) human predation through the direct killing of thylacines for protection of agricultural industry or for 'sport'; and (4) non-human predation involving competition from introduced carnivores and the effects of introduced disease micro-organisms. It becomes of relatively minor academic interest to ask what additional factors, unrelated to human activities, need to be included in the list?

202

Two aspects of this model require amplification here. First, human predation for the protection of agricultural industry extended deep into the Tasmanian wilderness, well beyond sheep farming. It was not unusual for trappers and snarers after kangaroo, wallaby and possum pelts to lay strychnine-laced carcasses before setting snares, and along snare lines once snaring was underway. This indirect human predation was undertaken to reduce the numbers of thylacines and devils in the vicinity of snaring, to increase the population of desired game species and to prevent snared specimens falling easy prey to the marsupi-carnivores. Secondly, while reference to the epidemic disease has been made previously, some specific details are required on the effects of this introduced disease micro-organism. This, however, prompts a small aside.

Some tension is evident in the published literature and oral history records of the behaviour of snared thylacines. Some snarers recalled that, when caught, thylacines did not fight against the snare, made no attempt to bite it, and appeared just to give up, to the point where, with little sign of struggle, they were commonly found dead in the snare. Others, however, have recalled how desperately thylacines fought to escape; attacking the snare, the sapling to which it was attached, and any stray human that got within reach. For once, such differences in opinion on the behaviour of the species prove easy to resolve.

An epidemic disease passed through the thylacine population at the end of the nineteenth century. A preliminary analysis of the bounty data, just for northern Tasmania, suggests it took about six years for the disease to spread from the east coast thylacine population to the west coast population. Prior to the spread of the disease it was unusual to find thylacines dead in the snare, and prospective bounty claimants were officially warned of this potential hazard: 'Tigers do not choke themselves with the snares, – it is a very rare thing to find one dead' (Braddon, 28/5/1888). But after the appearance of the disease in the

1890s, it was another matter. In the wild it was anecdotally described as 'distemper' or 'mange'; and distressed individuals, exhibiting significant hair loss or scabs over the head or body, were easily killed, and when snared, frequently made little attempt to free themselves, and often died as a result of the additional trauma of capture.

Some authors have assumed the disease was fatal, and have considered it a significant factor in the thylacine's demise. My research on captive thylacines, however, suggests the disease was episodic and debilitating, without necessarily being fatal. Mild symptoms involved the appearance of only one or two small skin lesions. In the most extreme cases, severe symptoms involved blood (and hair) loss through numerous lesions on the body, limbs and tail, accompanied by diarrhoea, potentially associated with internal bleeding and, unsurprisingly, loss of appetite. These symptoms could persist for up to four days. On the occasions when a captive animal survived a first bout of illness, it recurred at two- to three-monthly intervals. Given the stress of captivity to wild-caught specimens, much higher rates of mortality to the disease would be expected in captive over wild populations. Allowing for this, however, the loss of captive specimens could be horrific. As individual thylacines succumbed to the disease at Melbourne Zoo, they were immediately replaced with more specimens from Tasmania, which, if they did not bring the disease with them, soon picked it up from the infected stock already present in the zoo. By early 1903 Melbourne had lost to the disease sixteen out of the seventeen thylacines it had on display in the preceding two years.

Admittedly, Melbourne Zoo is the most extreme case. During this period, as the disease passed through the wild population in Tasmania for the first time, wild-caught thylacines also entered other zoological gardens. There is little evidence of the disease affecting stock in Adelaide, Berlin, Cologne, Hobart, London, Sydney or Washington Zoos at this time. The disease was certainly present and caused the deaths of some specimens on display in Launceston and New York, but at nothing like the levels experienced at Melbourne Zoo. Captive thylacine records suggest that some thylacines exposed to the disease never picked it up, while others experienced its effects only mildly, or were naturally immune. One captive thylacine that exhibited the most severe symptoms of infection, and was given up for dead by the curatorial staff, nevertheless survived three bouts of the disease over a six-month period, then lived for a further five years in captivity without ever showing signs of the disease again. In the wild it is reasonable to expect far less terminal effects from the disease, and a greater capacity for wild specimens to survive its debilitating and episodic characteristics and develop immunity – particularly so if, at the time, the individual thylacine remained a member of a small family group.

While the origins of the epidemic disease are unknown, it would appear far more likely to have arrived as an invasive micro-organism from the wealth of foreign species, deliberately and accidentally introduced into Australia by Europeans, rather than a chance introduction from a migratory bird or bat. In so far as the episodic disease was, in a minor way, a contributing factor to the thylacine's demise, its primary cause most likely relates to the behaviour and actions of humans.

While the evidence suggesting a primary hominid responsibility for the extinction of the thylacine appears obvious and unquestionable it is nevertheless surprising to examine the literature and discover how many authors – in particular, Australian scientists and naturalists – attempt other explanations. Most commonly they blame extinction on the thylacine itself, using hypothesised arguments about the nature and abilities of the species, and that of marsupials in general.

The construction of 'scientific innocence'

The sense of loss associated with extinction, and the acceptance of responsibility, guilt and blame for allowing it to happen, are not easily borne. So long as human activity can be considered largely responsible for the extinction of the thylacine, the obvious question, 'which humans were responsible?' is bound to produce some nifty footwork by those close to the action. An inference of 'scientific innocence' has been constructed by Australian scientists – after the event – that denies the ineffectiveness of scientists as extinction took place. Rather than admitting their failure and examining the reasons why, an argument denying any responsibility of scientists for the extinction of the species has been vigorously pursued in the recent scientific literature. The inference of 'scientific innocence' has been obtained through a selective use of historical records, based on a five-part argument that: (1) the thylacine's extinction was the fault of the species itself; (2) there were too few scientists around in nineteenth-century Tasmania for them to know what was happening to the species; (3) as the animal was dull, boring and uninteresting it did not attract the interest of the few contemporary scientists anyway; (4) unfortunately, by the time scientists became aware of its plight it was too late to save the species; and (5) as thylacines did not breed in captivity, scientists could not have saved the species from extinction, even if they had been aware of the problem.

This construction of 'scientific innocence', however, may be challenged by a deconstruction of imperialistic attitudes, first towards past and present evolutionary arguments for the 'primitiveness' of Australian marsupials, and secondly towards the colonial naturalists and scientists of the time. Granted the contingency associated with the extinction of the species, the rest of this chapter will focus upon deconstructing this five-part argument that distances scientists from accepting responsibility for the extinction process.

1 'INNOCENCE' AS A RESULT OF MARSUPIAL INFERIORITY

The argument absolving humans from primary or sole responsibility in exterminating the thylacine has been presented by many authors: 'No other species has declined so rapidly, probably too rapidly for hunting to be the sole cause, and we have to look for other factors which may have been partly responsible' (Guiler 1985, p. 27). These other factors are bound up with the idea of the thylacine as a 'living fossil' (Sharland, 1924), 'one of the most ancient animals left in the world' (Vaughan, 1914, p. 128). This was a product of the construction that 'Australia was sealed off . . . from the great evolutionary advances that took place elsewhere on earth, and primitive creatures like the thylacine could survive' (Pizzey, 1968). Various aspects of thylacine physiology and anatomy have been identified as indicative of a supposed close relationship to reptiles (Lydekker, 1915, p. 216) and dinosaurs (Raethel, 1992).

More recently, L. R. Green has asserted that: 'Never was one of nature's creatures so unadaptable and so ill-fitted for survival in a changing world . . . so limited is the flexibility of its closely articulated spine that it hardly qualifies as a vertebrate in the accepted sense . . . [they are] animals that evolution passed by' (ca 1975). There is a fairly strong agenda present here, when, even tongue-in-cheek, the sub-normal thylacine's sub-phyletic status is questionable. The nineteenth-century Tasmanian description of the thylacine as 'the monarch of the forest' parallels perceptions of the African lion as 'the king of beasts', but contrasts starkly with twentieth-century scientific descriptions of the thylacine as barely a vertebrate. 205

Scientific comment on the conformation of the thylacine's head has been just as extreme. Owen said of the thylacine that 'its head is of disproportionate magnitude' (1841, p. 258), and its proportionately larger size in comparison to a dog was also noted by Professor T. T. Flynn (1914). Sharland used this characteristic in his damning description of thylacine physiology:

> The long and semi-flexible tail, long hind limbs with the knee seemingly too low, the short forelimbs, together with a head that looks out of proportion with the body, all give the impression of the [Tasmanian] Tiger's lines being wrong and suggest it is badly formed and ungainly and therefore very primitive. (Sharland, 1962, p. 3)

At times the constraining power of placental chauvinism upon scientific thinking almost beggars belief! Were the animal photographed in Plate 9.3 assumed to be a placental carnivore, the large head in proportion to the body would be interpreted as an obvious sign of an animal of high intelligence. However, as the photographed animal is known to be a marsupial, then the large head is a sign of a primitive and ungainly physiology. Truly, if you are born a marsupial, you just can't win!

The roots of this imprecation lie deep in the nineteenth century: 'marsupial quadrupeds are all characterized by a low degree of intelligence' (Owen, 1834, p. 360) and, of course, 'The lowest marsupial is the Tasmanian wolf' (Packard, 1889, p. 575). More recently it has been suggested that 'thylacines are not very adaptable' (Griffith, 1972, p. 75). It was on this basis that the thylacine's demise was viewed by Frank Baker, Superintendent of the National Zoological

Plate 9.3 Seated thylacine, Hobart Zoo, *ca* 1923. (Queen Victoria Museum)

Park in Washington DC, as a natural consequence of imperialism and colonial practice: 'Australian animals, having no enemies to combat and an abundance of food, have come to be the stupidest animals in the world, which accounts for their rapid extermination when the country was settled by the British . . . the thylacine belongs to a race of natural born idiots' (*Washington Post*, 8 March 1903).

Not only were they primitive and stupid, they did not have any feelings either. Alison Reid, the late curator of Hobart Zoo, recalled an operation performed in 1927, by the medical surgeon Wilfred Giblin, to amputate a thylacine's foot. Giblin had previously operated, under anaesthetic, upon a Russian bear's injured foot. But with the thylacine:

> There was one come in, its foot had been caught in a snare and . . . they had to amputate it. And Dr Giblin came out, and he wouldn't give it an anaesthetic, poor brute . . . he just tied it down [and] just chopped its foot off . . . there was an enamel dish there, [the thylacine] bit straight through it. (Reid interviews, 27/2/1992, 25/6/1992)

While taking a piece out of a metal basin is an impressive bite, it is also an impressive expression of distress.

These comments merely scratch the surface. In addition to those references already quoted, many other comments have been made about the thylacine, based upon an assumption of placental superiority. Griffith's and Sharland's conclusions are representative: 'What the picture adds up to is a primitive try at making a mammal – an evolutionary prototype that really has no business to be living on at all' (Griffith, 1971, p. 2); or, the thylacine is 'an unproportioned experiment of nature quite unfitted to take its place in competition with the more highly-developed forms of animal life in the world today' (Sharland, 1924). In short, the animal was said to be primitive, unfeeling, unadaptable and stupid – an evolutionary experiment in every way inferior to placental mammals – and really only had itself to blame for becoming extinct.

The concept of nineteenth-century Australia as a 'palaeontological penal colony' (Desmond, 1982, p. 104) has its roots in scientific knowledge construction associated with the imperial expansion of the time. From a palaeontological perspective it was easy to view nineteenth-century Australia as a faunal backwater, a land of living fossils, both invertebrate (for example, *Trigonia*, *Nautilus* and the malacostracan *Anaspides tasmaniae*) and vertebrate (the marsupials and monotremes), supposedly a bastion of Mesozoic mammals. Fossil marsupials were known from the Mesozoic rocks of Europe (Owen, 1841, 1846) but in Europe marsupials had become extinct. It was a palpable intellectual and cultural construction to paint a scenario that hypothesised direct competition between placental and marsupial mammals that, from an imperialistic perspective, was bound to be won by the higher European mammals, the Placentalia. Remnant marsupial populations survived as supposed fragments in South American and South-East Asian faunal assemblages, ultimately destined for extinction as these weaker mammals were replaced by invading placentals from the north (H. H. Scott, 1903). As biogeographical explanations for the distribution of species at this time were based on assumptions of island hopping, or sunken land bridge or sunken continent hopping, Australia, the most isolated and least hopped into (but very much hopped upon) continent, was still awaiting the placental invasion. The process of colonisation could readily be seen as an expected and natural role for the seafaring British nation, and likely to have but one conclusion: the eventual replacement and extinction of out-moded species by superior placental types.

Recent fossil evidence and the acceptance of continental drift have destroyed the basis for assuming placental superiority. The oldest known marsupial fossils in the world are North American, dated at 90 to 95 million years of age. From North America marsupials rapidly spread to Europe, with extreme regional outliers being recorded from North Africa and west central Asia (Rich, 1991). It is still possible to suggest that, on these continents, they came into direct competition with placental mammals, the result of which was a 'victory' for the placentals with the complete extermination of Eurasian, North American and African marsupials. However, it needs to be recognised that the 'victory' took a considerable period, in excess of 70 million years, with the last known European and North American marsupials becoming extinct only 17 million years ago.

In dealing with different possible worlds resulting from the contingencies of history in the evolution and extinction of individual species, groups of species, or entire biological systems, S. J. Gould (1989, p. 297) reflects on the rarity of instances where global distribution and isolation have 'replayed life's tape'. Gould presents an illustration in the establishment of animal types in the dominant carnivore niche in the Eocene to Pliocene, whereby mammalian predators prevailed over the giant predacious birds in Europe and North America, while in South America, the giant predacious birds prevailed over the mammals.

While it is hard to argue, beyond the species level, that these 'replayings of life's tape' with different end results were common events, similar interactions resulting in different conclusions may not be as rare as might initially be considered. Scientists need to change their perspectives on such interactions, replacing overt or unacknowledged assumptions about superiority and progress in evolution with a recognition of the contingency inherent in any sensible account of historical development. Bearing Gould's analysis in mind and returning to the specific context of the mammalian fossil record, what is now established is that the interaction between placental and marsupial mammals has been played out on different continents, with different outcomes.

Marsupial and placental mammals first appear in the South American fossil record 70 million years ago. The relative proportions of the two groups have fluctuated over time (Archer, 1993) but they have continued to exist together to the present day. Despite frequent claims of insignificant South American marsupial numbers, there are some ninety-five recognised species living in South America today (L. G. Marshall, 1984); in Australia, about 140 species are recognised (Strahan, 1983). Marsupials represent an important and significant element in the contemporary South American faunal assemblage. Seventy million years of interaction between placentals and marsupials have produced a coexistent compromise between these two different mammal types, in a situation that permits of no conclusive evidence for either placental superiority or marsupial superiority.

As the Gondwanaland supercontinent broke up, Australia remained in effective contact with South America, via Antarctica, until the early Tertiary. Monotremes, only extant today in Australia and New Guinea, are first recorded in Australia 120 million years ago with the fossil *Steropodon galmani* (Archer, Flannery, Ritchie and Molnar, 1985). Monotremes managed to cross over into South America, via Antarctica, and have been recorded in Patagonian fossil deposits dated at 63 million years of age (Pascual, Archer, Ortiz Jaureguizar, Prado, Godthelp and Hand, 1992). Marsupials travelled in the opposite direction and arrived in Australia from South America, via Antarctica, between 45 and 70 million years ago (Archer, 1993), until direct contact between Antarctica and South America ceased. The final separation of Australia and Antarctica occurred between 35 and 45 million years ago. There are no dominant marsupials in the Antarctic fauna today, but at the risk of stating the obvious, this can hardly be taken as a sign of marsupial inferiority, there are no indigenous terrestrial placentals either. Until recently, all known Antarctic terrestrial mammalian fossils were marsupials (Rich, 1991), however, Woodburne and Case (1996) have now uncovered placental and marsupial fossils from Seymour Island, which show strong affinities to Patagonian fossils from around 50 million years ago.

At present, the oldest known Australian marsupial fossils come from Murgon, in south-east Queensland, and are dated at 55 million years of age. Shortly after this time, Australia became separated from both South America and Antarctica. This left Australia effectively isolated for the next 30 million years, during which time many of the continent's indigenous families of marsupials evolved. The whole scenario of marsupial isolation, freedom from competition, and therefore potential for inferiority would appear to be supported by the Australia fossil mammalian record – almost, but not quite. In the oldest known marsupial deposits at Murgon there is also the tooth of a condylarth-like placental mammal, *Tingamarra porterorum* (Godthelp, Archer, Ciselli, Hand and Glikeson, 1992). It now appears likely that both placental as well as marsupial terrestrial mammals arrived in Australia from South America 55 million years ago. Once again, placental and marsupial mammals came into direct competition, but this 'replaying of life's tape', however, resulted in 'victory' for the marsupials with the complete extermination of Australia's early placental mammals.

The different results of the interactions between marsupial and placental mammals on different continents provide additional illustrations of the effects of historical contingency upon evolutionary processes. In Europe, North America and Africa, placentals prevailed over marsupials. In Australia marsupials prevailed over placentals, and Antarctica (with a predominance of marsupial forms over placentals in its known terrestrial fossil fauna) appears to have been headed that way as well. In South America and South-East Asia both marsupials and placentals coexist significantly in the mammalian biotas. On the basis of fossil evidence and the known results of previous interactions, scientists are not in a position to label either reproductive type as superior or inferior. What we have here are different outcomes of the same interaction between different orders of mammals that reflect contingencies in the different environments in which the interactions took place. The placental chauvinism expressed in arguments about marsupials in general, and the thylacine in particular, as being primitive and unadaptable, is a scientific construction created by a cultural preconception, rather than a careful analysis of the fossil record.[4]

The related argument, touched upon previously, that marsupials are stupid and unintelligent, has its basis in the same placental chauvinism which assumes that placental superiority in intelligence accompanies their supposed superiority in reproduction. The argument is worth examining in some detail for its further illustration of how scientific preconceptions can distort data and, via the 'great man' hypothesis, enter the established scientific canon.

The first published statement that may be construed as suggesting a lack of mental ability in the thylacine occurs in the original description of the species

209

by Harris (1808, p. 175): 'It from time to time uttered a short guttural cry, and appeared exceedingly inactive and stupid'. Owen repeated this description and incorporated it into his argument that, because marsupials have only an imperfectly developed reproductive system, they would be expected to have only an imperfectly developed brain (1834, pp. 359–60; 1841, p. 259). Harris, however, provided the context for his observation. The thylacine 'remained alive but a few hours, having received some internal hurt in procuring it' (1808, p. 175). Harris made it perfectly clear that his behavioural descriptions applied to the last few hours of life of a mortally wounded specimen which, in a manner likewise known from placental mammals dying of internal injury, occasionally 'uttered a short guttural cry, and appeared exceedingly inactive and stupid'.

While Owen was the most influential biologist of his time to argue for a general marsupial intellectual inferiority, he was not the only one to use Harris' 'inactive and stupid' descriptor to typify the thylacine as a species. Others also misread and misinterpreted Harris and drew on his 'exceedingly inactive' phrase to construct the common nineteenth-century description of the thylacine as 'sluggish'. Scott wrote in 1829 (also 1830, p. 53): 'it has a sluggish appearance', and this pulmonatory description was repeated by Melville (1833), Martin (1839), Bunce (1857) and Ireland (1865). The imputation of stupidity by Harris and Owen was also taken up by Martin Duncan (1884), and by Vogt and Specht (1887, p. 203): 'it is fierce, but stupid'. The popular reflection of this concept may be found in an article on the thylacine in the Australian monthly *Agricultural Gazette* (June 1897), which described the thylacine as 'this stupid, cowardly animal'. At the opposite end to Owen on the intellectual spectrum of attachment to evolution, Lydekker (1894, p. 153) also described the thylacine's diurnal behaviour as 'dull and inactive . . . moving with a slow pace'.

These perspectives of the thylacine in the nineteenth century were also reflected in the thylacine's iconography. The earliest illustrations of the thylacine all presented it in typically dog-like poses, depicting it as an upright, noble animal with a pronounced stop on the head, walking erect on all four legs or seated on its hocks (for example, Cuvier, 1827; Desmarest, 1822; Harris, 1808; T. Scott, 1823; Swainson, 1834). The new depiction of the thylacine as low, mean and skulking, untrustworthy, sluggish and stupid was soon reflected in illustration. W. H. Lizars provided an early illustration of this in Waterhouse (1841) with a cowering, relatively flat-headed thylacine, Plate 9.4.

An additional discourse exists to the concept of stupidity as used in the nineteenth and early twentieth century, not readily recognisable within a post-Darwinian framework. Statements about an animal's stupidity or intelligence also carried a message about the perceived ability of an animal to be domesticated, to accept a divinely ordained subjugation to human will and desires, or, at a minimum, to recognise human power and to flee from its presence. Ritvo (1987) points out that domesticated herbivores (sheep and cattle) were by this definition intelligent, and undomesticated carnivores (placental tigers and wolves) were by this definition stupid and depraved.[5] It is not surprising to find in the literature on captive thylacines statements that link a suggested stupidity

Plate 9.4 Lizars' thylacine engraving. (from Waterhouse, 1841)

in the species with instances of untameability in captivity. The previously quoted article in the *Washington Post* (8 March 1903), suggesting that thylacines were 'natural born idiots' was entitled 'Leaders in Stupidity', and justified such label- ling through accounts of the largely undisciplined, undemonstrative and undomesticated behaviour of the zoo's captive thylacines to most humans with whom they came into contact. Such is the nature of cultural constraint upon scientific thinking that the relationship between stupidity and untameability is assumed to be causative and based on the supposed mental ability. In an article on thylacines at New York Zoo entitled 'Too stupid to tame', the thylacine is described as 'one of the most stupidest of animals; its lack of intelligence is the cause of its untameableness' (M, 1916). As Ritvo indicates, the relationship has significant cultural meaning and aetiology when viewed in the opposite direc- tion: historically, an 'untameableness' and lack of domestication in a wild species were readily labelled as 'stupidity'. The connotations attached to the concept of 'stupidity' in the nineteenth and early twentieth centuries do not always easily translate into an early twenty-first-century scientific context, and modern authors who persist in labelling the thylacine as stupid do so at con- siderable distance from the reality of its original expression.

Only one published specific denial of Harris' 'exceedingly inactive and stupid' claim has been located from the nineteenth century, that of Breton (1846): 'It is said to be stupid and indolent; but this is a mistake'. Others, however, have denied it by anecdote or example.

Various experienced trappers and snarers of the thylacine have strongly expressed this alternative viewpoint: 'My word, they are not stupid' (Cooper, 14/10/1970), and thylacines were 'certainly not stupid' (Unsigned, questionnaire, 1970). Earlier, Troughton (1941, p. 51) suggested that the thylacine possessed 'a cunning and adroitness beyond that of the average dog'. One of the two distinct uses of the word 'curiosity' when applied to the thylacine also supports this viewpoint, that is, curiosity in the sense of a general personality characteristic of the species, of a readiness to observe and learn. The thylacine in the wild has been described by various people as curious: 'the two I have seen and not caught appeared to be as curious and surprised to see us, as we were to see them, and did not race away in a hurry' (Willoughby, 4/10/1970); 'cunning' (D. A. Bell, 1/9/1981; O'Shea, 1981) and 'inquisitive' (Cartledge, 17/10/1970; Cooper, 14/10/1970; and in *Tasmanian Westerner*, 7 October 1982).

Finally, a few general comments on supposed marsupial dim-wittedness are required. The comparative analysis of animal intelligence is never easy, more particularly so if one of the species under consideration happens to be extinct. Marsupials are not infrequently used as specimens in psychology laboratories in South America and, less commonly, in North America (VandeBerg, 1990), and a detailed overview of marsupial learning has been published (in Spanish) by Papini (1986). Despite an enthusiastic start in the 1930s with the work of George Naylor on Australian marsupial learning (1935), nothing like the South American experimentation is carried out in Australia today (Wynne and McLean, 1999). Macphail (1982) reviewed the published literature on marsupial learning and suggested that 'marsupials should not in general be considered poor learners' (p. 271). He concluded that, with respect 'to the intelligence of marsupials and monotremes, there are in fact no data to suggest that their intellect is in any way different from that of any other non-human animals' (p. 271). The frequent labelling of marsupials as stupid is just another illustration of the power of placental chauvinism, and represents not a limitation in the mind of the marsupial but a limitation in the mind of the placental observer.

Placental chauvinism is not innate. Its origins lie in the acceptance of a progressive model of evolution. For all the modern damning of the falsely overlaid concept of progress upon Darwinian thinking, the attempt to present evolution in a progressive, predictable light as a natural unfolding, conforming to and supporting basic British religious, racist and sexist beliefs was not a product of non-Darwinians attempting a shabby compromise between science, society and religion but a product of the selling of evolution itself by some of the most influential figures closest to Darwin. Spencer, supported by Huxley, intruded the construct of progress on to 'descent with modification' – against Darwin's express wishes and concerns – argued for this 'evolutionary' model within the intellectual arena, and remains largely responsible for the acceptance of progressivism in evolutionary thinking (S. J. Gould, 1996) and thus of chauvinism in placental beliefs. It was T. H. Huxley, after all, who expanded and encapsulated, within the classification of mammals, basic assumptions about progress overlain upon the three different mammalian reproductive types, when he established the labels of proto-, meta- and eu-theria (Huxley, 1880).[6]

The idea of marsupial equality, vigorously pursued by some scientists (such as Michael Archer), in the late twentieth century, and now centred in a return to original, non-progressive models of evolution, is not new but has a long and respectable nineteenth-century history – unfortunately written by all the 'wrong' people. Desmond (1982) warns of the danger of accepting scientific propaganda at face value since, by and large, history perpetuates the partisan viewpoint of the victor: 'The early evolutionists doubled as historians and did what any nationalistic group does to forge its self-identity – they wrote the kind of heroic history encouraged by Victorian positivism, portraying the forces of light triumphing over religious obscurantism' (Desmond, 1982, p. 56). Not all the ideas of those who lost the evolutionary debate deserved to die with them.

Marsupial equality with placentals was an argument newly constructed and vigorously supported by those who believed in separate creations. It was a necessary requirement that the same levels of excellence and perfection in biological design should be readable in all the separate acts of creation, whether separated in distance or in time. That distant Australia's marsupials were the equally advanced ecological parallels of placentals was just further evidence of an externally directed programming in creation.

The problem with animals perceived as perfectly adapted to their environments, and perfect ecological parallels between different species in different environments, is their potential not just to support arguments for the constructive power of natural selection, but to support creationist viewpoints of life developed by an external, omnipotent and omniscient creator (S. J. Gould, 1991). Darwin and his supporters were very much aware that it was the imperfections and shoddy compromises apparent in biological design that best illustrated evolutionary processes at work. Thus, either marsupials were accepted as ecological parallels of placentals and largely ignored as of little use in evolutionary argument, or else deliberate attempts were made to identify sources of primitiveness within marsupial physiology and behaviour that, while damaging their status as mammals, supported evolutionary models of their development by natural selection in supposed isolation from competition with superior, more perfectly constructed placentals. Darwin typified the first approach, Huxley the latter, and while evolutionists as a group failed to find consensus in their treatment of marsupials, a new-found perfection was discovered in marsupial physiology and behaviour by the creationists.

On viewing Australian fauna in New South Wales in January 1836, Darwin noted in his diary: 'An unbeliever in every thing beyond his reason might exclaim, "Surely two distinct Creators must have been at work"' (Darwin, 1836, quoted in de Beer, 1963, p. 107). In the section on analogical resemblances in *The Origin of Species*, Darwin included a paragraph on the similarities of

213

dentition between the thylacine and dog (1859, p. 584). For Darwin, the distant biological relationship between marsupi-carnivores and placental carnivores did not involve a value judgement on their respective levels of development or adaptation. Although, after summarising his argument on analogical variation, Darwin recognised that a scientist could, by arbitrarily raising or lowering the status of different animal groups, extend the analogous argument over many more species and use this ordering in different classificatory schemes (p. 585) – which is what T. H. Huxley set out to do.

It took some time for the more foolhardy Huxley, bent upon an arbitrary process of sinking the value of marsupials to meet a progressive model of evolution, to decide on which analogous characteristics in a developmental sequence were most appropriate to his progressive classificatory model (1880). His inconsistency in this, together with his championing of placental superiority over marsupials, was welcomed by the creationists and used, with particular emphasis on the thylacine, as examples of the illogical and inconsistent thinking of evolutionists. Their view was that marsupial and placental equality, particularly evident in the thylacine's parallel with dogs, represented revealed evidence of pre-ordained design in the creation of species.

214

The most prominent use of the thylacine in creationist arguments was made by St George Mivart, who, for a long time, kept a foot in both camps before his disillusionment and eventual excommunication from the evolutionary church (Desmond, 1982). In a set of arguments against natural selection as the sole constructive power behind speciation, Mivart raised the issue of analogical structures and remarked upon the improbability of their arising independently (1869, p. 41). He argued for the existence of a creative constructive power in speciation, and illustrated this by discussing the status of marsupials in general and the parallels between the thylacine and the dog in particular (1869, pp. 48–50), concluding that 'similarity of structure is produced by other causes than merely "natural selection"' (1869, p. 51). He persevered with the same argument in his book *On the Genesis of Species*, mentioning Huxley's fluid interpretation of marsupial classification and backhandedly thanking Huxley for pointing out the difficulty for natural selection of the similarities between thylacines and dogs (1871, pp. 76–8).

Marsupial equality quickly became the standard fare of the creationists, as, in time, placental chauvinism became the bread and butter of progressive evolutionists. These differing perspectives of the animal were, again, reflected in iconography. Anti- or at least non-Darwinians (whether from creationist or Lamarkian viewpoints) projected the animal as noble and grand (see J. Gould, 1851), representing the thylacine as the 'monarch of the forest' (*Launceston Daily Telegraph*, 12 November 1884), often appropriately placed on top of the mountains, as 'lord of all', looking down upon the rest of marsupial creation (W, 1855; and Figuier, 1870). As progressive evolutionists became the dominating intellectual force they ignored these interpretative illustrations of the thylacine. In 1894 Lydekker, noted evolutionist and friend of Huxley, returned to Lizars' 50-year-old illustration in Waterhouse (1841) of a primitive, skulking thylacine (Plate 9.4), to serve as the illustration accompanying his own work.

This section has attempted to discuss the first identified argument in the construction of 'scientific innocence' in the face of extinction, by denying there was anything intrinsic to the species *Thylacinus cynocephalus* that predisposed it to extinction prior to the human invasion of its environment. The imputations of primitiveness, unadaptability and stupidity have their origins in the fantastic constructions of placental chauvinism, powered by an imperialistic rhetoric and progressive models of evolution. Modern attempts currently underway, including the work of Archer (1993) and S. J. Gould (1989, 1996) that deconstruct progressive models of evolution, will, as they become accepted, not only elevate the status of marsupials but also the scholarship of the all-too-easily-forgotten biologists who argued on the losing side of the evolutionary debate.

2 'INNOCENCE' AS A RESULT OF THE RARITY OF SCIENTISTS

The second part of the argument for reduced responsibility for scientists in the extinction process suggests that there were too few naturalists and scientists present in nineteenth-century Tasmania for them to concentrate upon, or be able to share, knowledge about the species.

This claim is based on the argument that 'we did not know what was going on'; that there were insufficient numbers of scientists around in nineteenth-century Tasmania for them to be aware of any biodiversity problem. The acceptance of this argument requires the denigration and dismissal of most of the nineteenth-century records of scientific endeavour encouraged in Tasmania.

Authors of books on the thylacine have commented in the following manner: 'Little is known of the habits of this species except by hearsay and this is often conflicting' (Guiler, 1960); and 'The thylacine was mercilessly hunted to the point of extinction before any extensive study of its characteristics was made' (Beresford, 1985). Not unsurprisingly, the colonists involved in the destruction of the thylacine have had their attitudes and recorded opinions on the species categorised as biased, subjective and non-specialist (Salvadori, 1978). While it is one thing to castigate opinions recorded about the thylacine by colonists and graziers, those with vested political and economic interests in the animals, it is another matter to denigrate nineteenth-century sources of knowledge on the thylacine. With the arrogance of youth and assumptions that the accuracy of scientific knowledge increases and improves over time (even for extinct species), twentieth-century scientists and naturalists have been known to judge and dismiss their nineteenth-century counterparts intemperately (Frauca, 1963, p. 24). Joines, from a position suggestive of professional strength, argued: 'Unfortunately for the Tasmanian tiger and for science, Harris's brief description . . . was destined to constitute the first and virtually only scientific interest in the species until well into the twentieth century' (Joines, 1983, p. 5).

There is nothing new about warnings in the scientific literature on the thylacine about knowledge construction by untrained (or trained, but contextually absent) observers. They can be traced back to R. C. Gunn's comments in 1838 criticising the too hastily drawn deductions by European scientists over the thylacine's laterally flattened tail and presumed semi-aquatic behaviour. The point that needs re-emphasis here is that, by 1838, Gunn was prepared to argue that there was a body of competent and knowledgeable naturalists and scientists generating and sharing information about Tasmania's flora and fauna that was both valid and reliable, in context and in content. There is something deeply offensive and worrying, in terms of what it says about the lack of historical perspective in Australian zoology, when the same criticisms of unscientific and unprofessional persons generating data on the thylacine that were made by a scientist such as Gunn can be turned around and used against Gunn himself by other scientists 150 years later.

To support Gunn's claim for the existence of scientists and social structures that supported the sharing of scientific information on indigenous species in the early nineteenth century, a brief summary of the relevant points in the intellectual history of zoological science within Tasmania is required. The Hobart Mechanics Institute was founded in 1826 (Fitzpatrick, 1949) and in 1828 Institute members commenced planning a museum to include objects of natural history (Inkster and Todd, 1988). The Van Diemen's Land Scientific Society was then established in 1829 (Hoare, 1969; Moyal, 1986). Matching this early intellectual interest in Australian biodiversity was a growing community interest in Australian animals.

Probably from the time of their first arrival in Tasmania, the European colonists attempted to tame and keep as pets representatives of the local wildlife. Initially, people living at the edge of colonial expansion were advantaged in pet-keeping activities, but the obvious commercial possibilities associated with indigenous pet-ownership soon remedied this problem, and the first 'pet shop' in Hobart was opened at George Marsden's Livery Stables in 1831. Marsden charged admittance to view the live and dead animals and birds he had on display, but all were available for purchase to 'Connoisseurs in Natural Curiosities' for domestic use as companion animals, or as 'curios', either alive or dead, to be taken back to England by returning colonists and visitors (*Hobart Colonial Times*, 11 December 1832). The latter was the fate of the first live Tasmanian tiger on display in 1831 (*Hobart Town Courier*, 17 September 1831). Such was the interest and enthusiasm in keeping Australian animals as pets at this time in Tasmania, that in 1834 captive specimens of the mainland possum, the sugar glider *Petaurus breviceps*, 'easily bred and a delightful pet' (Troughton, 1941, p. 95), were deliberately released from captivity into the wild around Launceston for acclimatisation purposes, to 'improve' the existing Tasmanian fauna (Gunn, 1851). (An activity that, in terms of its stated objectives, proved remarkably successful as they have spread rapidly throughout the entire state.)

In 1835 land was formally set aside in Hobart for the establishment of a zoological garden (*Hobart Town Courier*, 3 April 1835), just seven years after the opening of London Zoo. Reflecting the wide-spread interest in indigenous fauna, the Tasmanian Society of Natural History was founded in 1837 (Hoare, 1969; Moyal, 1986). It is hardly surprising that in 1838 Gunn was prepared to argue that there were competent and knowledgeable naturalists, scientists and lay persons generating and sharing knowledge about Tasmania's flora and fauna who should be preferred, as primary sources of knowledge construction, over assumptions and analogies about the behaviour of species based on the anatomical dissection of preserved specimens sent back to Europe!

The first volume of the *Tasmanian Journal of Natural Science* was published in 1842. A private zoological collection was kept by the Governor of Tasmania, Eardley-Wilmot, from 1843 to 1846 (Guiler, 1985). The Royal Society of Tasmania was founded in 1844, and publication of the transactions of the Society began in 1849 (Hoare, 1969). The initial public display of animals that became the Launceston City Park Zoo commenced in 1850, and 1853 marked the opening of the first zoological gardens for public display in Hobart, containing both indigenous and exotic species, including a Tasmanian tiger (Propsting, 10/1/1854).

These activities that took place within the mid-nineteenth-century colonial community of Tasmania are indicative of a wide-spread public and professional interest in native flora and fauna. Given the small population of Tasmania at the time, such community awareness and enjoyment of Australian biodiversity have probably never been equalled; even when taking into consideration late twentieth- and early twenty-first-century environmental awareness in Australia, encouraged by both public and professional organisations, and taught within our education systems. On-going societies and the opening of museums with public access throughout the nineteenth century in Tasmania created an environment encouraging both amateur and professional scientists and naturalists to interact with one another, resulting in significant professional and public meetings, displays, exhibitions and publications (in which the thylacine was by no means under-represented). There was no rarity of scientists and naturalists in Tasmania in the nineteenth century, only an increasing rarity of thylacine specimens for them to study.

217

3 'INNOCENCE' DUE TO THE UNATTRACTIVE NATURE OF THE SPECIES

This argument for 'scientific innocence' suggests that scientists confronting the animal found it dull, boring and uninteresting. Three responses are presented against this argument.

The assessment of the characteristics of a species' behaviour from captive specimens is fraught with difficulty. Dick Rowe (26/9/1980) specifically distinguished between the thylacine being curious in the wild, but not being curious when restricted to a captive environment. Perceptions of 'dullness' in captivity are not so much related to the cognitive ability of the animal as to the restricted environment in which they are kept; the perceptions of keepers with whom they have not formed responsive relationships; and the cultural preconceptions of the zoo visitors who see them. A restricted zoo environment often encourages stereotyped behaviours of movement and restlessness, and given that the feeding process represents one of the few novel stimuli included in each day, a

218

Plate 9.5 Alert thylacine, Hobart Zoo, *ca* 1917. (Tasmanian Museum)

concentration upon food is often established that is motivated more by stimulus deprivation than gluttony. Such factors were not recognised by Ludwig Heck in his comments on the pair of thylacines that entered Berlin Zoo in 1902:

> Making allowance for the general marsupial dullness, the animals were trusting and would restlessly sniff along the bars if one stood in front of their cage. They were constantly greedy for food when they were not sleeping, and with their persistent stupidity they would try to bite through the iron bars . . . The clear, dark brown eyes stare vacantly at the observer. The expressiveness of the true carnivore is totally lacking. (Heck, 1912, translated by Grzimek, 1976, p. 86)

Admittedly, the analysis of an animal's expression is always going to be a personal, subjective one. Vacant staring, however, was not always typical of thylacines in zoos. I would suggest that most mammals, including, hopefully, the majority of members of my own species, should be able to read the message in the eyes of the thylacine in Plate 9.5 and recognise that it is not a casual glance from a gentle herbivore that may be ignored or approached at whim.

In contrast to the perception of thylacines in Berlin, Mary Roberts of Hobart Zoo, declared 'that these animals, in her opinion, were not of low intelligence' (P. C. Mitchell, 1910) and London Zoo captives were described as lively and extremely active specimens (Knight, 1855; W, 1855) that 'attracted much attention' (*Launceston Daily Telegraph*, 7 January 1885). As earlier considered, many thylacines in captivity, rather than being generalists, focussed their attention and responsiveness on a single, specific human. Consequently, while this could lead to a majority judgement of their nature as unresponsive and uninteresting, a small but select number of keepers in zoos, together with those keeping thylacines as pets in the Tasmanian community, discovered they were responsive, lively and interesting animals. A first argument against 'scientific innocence' due to the unattractive nature of the species is the readiness with which people approached the keeping of thylacines as pets.

Secondly, for a supposedly 'unattractive' species, it certainly generated a wealth of professional comment and illustration. As mentioned previously, recent scientific publication has tended to emphasise twentieth-century accounts (often at the expense of ignoring the work of nineteenth-century scientists). From the sources already used in this work and the attached bibliography (see also Paddle, November 1997), it is clear that there is a wealth of hitherto neglected nineteenth-century comment on the species, indicative of its having had a high profile amongst Tasmania's naturalists and scientists at that period. As Gould (1851) remarked 'so great is the interest which attaches to this singular species, that I have been induced to give a representation of the head, in addition to that of the entire animal'. Guiler and Godard (1998, p. 87) comment, with respect to nineteenth-century depiction of the species: 'The abundance of illustrations relating to the thylacine is confirmation of the very real fascination that naturalists, from the very beginning, have had for the animal'.

Thirdly, the numbers of thylacines purchased and placed on display in zoological gardens may also be used as an indication of the contemporary interest in the species. Fleay (1946w) and Yendall (1982) have suggested that thylacines were uninteresting, insignificant and uncared-for display specimens in Australian zoological gardens, attracting little professional or public interest. Guiler comments: 'It is noteworthy that the Australian zoos did not display them to any great extent, Beaumaris [Hobart] being the only one to have a long-term exhibit' (1986, p. 153). These opinions seem to be confirmed by the small number of specimens usually listed as having been exhibited in different Australian zoological gardens.

In reality, the published records of thylacine specimens in Australian zoos have significantly under-represented the number of specimens on display, thus lowering their real profile and attractiveness and their value to the zoological gardens in which they were displayed. Prior to my research, the Royal Zoological Society of New South Wales was only known to have ever had one thylacine on display (Guiler, 1985, 1986). I have now identified three specimens on display in Sydney between 1885 and 1923 (Paddle, 1993). Adelaide Zoo, variously recorded as having ten (Guiler, 1985), five (Guiler, 1986) or eight specimens on display (Rix, 1978), is now known to have exhibited seventeen thylacines between 1885 and 1903. To Guiler's detailed research on the number of thylacines in Hobart Zoo, estimated as fifteen plus (1985) and twenty-one plus (1986), I can add, an additional seven specimens. Amongst these twenty-eight, it is worth noting that

219

the thylacine, which died on 15 October 1922 was described as 'the pride of the collection' (*Hobart Mercury*, 16 October 1922). But Hobart was certainly not the only Australian zoo 'to have a long-term exhibit'. While Guiler has variously identified Melbourne Zoo as possessing either fifteen (1985) or only six (1986) thylacines on display, I have identified forty-four specimens held between 1864 and 1930. While Launceston is usually credited with only displaying two or three specimens (Claude, 1996; Guiler, 1985; Moeller, 1993) I have discovered that, Launceston City Park Zoo actually had the largest known collection of thylacines on display: some sixty-two specimens were held between 1885 and 1933.

The damning of the species as being dull, uninteresting and unvalued members of zoological garden collections, exciting no public or professional interest, was based on a maximum estimate, before I commenced my research, of only fifty known instances of thylacine display in Australian zoos. I have been able to increase the total by more than 200 per cent, identifying the separate display of 154 thylacines in Australian zoological gardens. In contrast to previous inferences of the low profile and unattractive nature of the species, the numbers now known to have been on display in Australia suggest that thylacines were significant, much-sought-after display specimens in zoological gardens, of high status to professional curators and the visiting public alike.

4 'INNOCENCE' RESULTING FROM AN IGNORANCE OF LIKELY EXTINCTION

The fourth argument in support of 'scientific innocence' suggests that, by the time scientists became aware of the thylacine's plight, it was too late to save the species anyway. Granted there were significant numbers of naturalists and scientists, associated support groups and captive thylacines to generate interest in the species, the central question is, to what extent were these individuals aware of the thylacine's impending extinction?

There are claims in the literature that the extinction of the thylacine apparently took the scientific community unawares. These reflect either a general lack of knowledge of the species, or specifically, a restricted knowledge of the species that did not entertain population dynamics. Grzimek (1976, p. 84) comments: 'It took a relatively long time for the Tasmanians to realise how near to extinction the largest marsupial carnivores really were'. Additionally, Vechtmann (1980) and Sayles (1980, p. 37), 'It was a long time before the Tasmanians recognised the plight of the thylacine', present similar points of view. The question of scientific awareness of the impending extinction of the thylacine needs to be considered in two parts. First, with the concept of extinction only receiving general support from a recognition of the work of Cuvier in the late eighteenth century, to what extent were Tasmanian scientists aware of the concept of extinction *per se*, and then, if they were aware of the concept, to what extent was it applied to the thylacine?

Conclusive evidence exists that nineteenth-century Tasmanian naturalists and scientists were well aware of the concept of extinction, principally because

220

one other large and significant Tasmanian vertebrate species preceded the thylacine into oblivion, namely the Tasmanian emu, *Dromaius diemenensis*.[7] The Tasmanian emu was not uncommon in the eastern half of Tasmania in 1804, was rare by 1830 (George Marsden, nevertheless, possessed a specimen for sale in 1832 – *Hobart Colonial Times*, 20 November 1832) and it became extinct during the early 1850s. The two emus donated by James Gibson of Circular Head, that formed part of the first recorded display of animals in the Launceston City Park in 1850, appear to be the last known representatives of the species (*Launceston Cornwall Chronicle*, 24 October 1850). Adam Amos (letter, 20/4/1826, cited in Barrett, 1944) penned an early warning from Oyster Bay: 'there is a kind of ostrich. We have only caught one. I believe they will soon be extinct'. In March 1832 Backhouse (1843, p. 30) declared 'The Emu is now extinct' from the midland region around Bothwell. In a letter to William Hooker (16/11/1836), Gunn outlined an unsuccessful attempt to get Lieutenant-Governor Arthur to respond to the plight of Tasmania's fauna, mentioning the emu's problem in particular:

> Many of our animals and Birds will become extinct . . . Emus are now extremely rare – and in a few years will be quite gone . . . a few pounds employed in collecting emus . . . would have been no great matter and their food, being grass alone, no expense would have been incurred beyond fencing in a piece of ground. (Gunn to Hooker, 16/11/1836)

221

Not without difficulty, Gunn sent specimens of the emu to Hooker who forwarded them to J. E. Gray at the British Museum: 'I have now packed an *immense chest* of sundries in the Natural History line . . . A pair of V.D. Land Emus, a bird now almost extinct, obliged my having it made 5 feet long' (Gunn, 31/3/1837). Unfortunately for science, Gray never responded. No acknowledgment of the specimens was made, Gunn's expenses in procuring them and in shipment were not met, nor was he acknowledged as a donor. In response to the plants he sent the Hookers, Gunn received letters of encouragement and enthusiasm, botanical books and journals and published acknowledgment of the utility of his botanical knowledge and specimens. From Gray, reflecting his 'blackballing' at the hands of the British zoological establishment at this time, Gunn received nothing but contemptuous silence. In a later letter to Hooker Gunn recounts how this lack of response had reduced his enthusiasm for zoology:

> Mr Gray of the British Museum has never written to me or sent me any Books. I regret this as it has dampened my ardour in many of those branches. The box which I now regret you gave to the British Museum was worth £50, and the specimens in it cost me nearly that amount! For some of the rarer birds I paid £2. each – and the two Emus were over here worth £10. Gould told me that he *bought* many of the skins from Gray and paid for them but even that amount never reached me. It is unfair, because with a large family I yet spend every spare shilling for the advancement of Science – & look for no return beyond that of

Books or similar things. The British Museum could afford to reward me liberally, & for their own sakes ought to have encouraged me as a Collector. (Gunn, 6/12/1843)

Gunn's loss to zoology was unfortunate; and while the loss of the emu species was tragic, and probably easily avoidable, given the potential for establishing the species in captivity, the eventual fate of the museum specimens was nothing short of criminal. Robert Hall, director of the Tasmanian Museum, summed the situation up in 1925: 'The true Tasmanian Emu has gone forever, excepting the few museum skins in Europe. Nothing beyond a dozen valuable eggs are left [in Tasmania]' (p. 225). In addition to Gunn's two specimens in the British Museum, there was also a skin in the Frankfurt Museum. These three specimens alone represented the entire preserved knowledge of the species. With respect to these specimens, Burns and Skemp (1961, p. 66) comment: 'Advice from both the British Museum and the Frankfurt Museum recently received reports that they are no longer in existence, probably being destroyed by bombing during the last war'. It is a great pity that the Tasmanian Museum's plea to the British Museum (Hall, 5/1/1908) for the return of just one of the emu skins to Tasmania was not acceded to. The eggs remain in the Tasmanian Museum, Hobart.

222

Matching the extinction of the Tasmanian emu was the attempted genocide of the native Tasmanian people. Both events occurred at the same time, aroused comment in the Tasmanian Society of Natural History and later the Royal Society, and had almost the same ending (Reynolds, 1995). Recognising the continuing destruction of the Tasmanian environment, and calling for more open, direct and public action on this by members of the Royal Society of Tasmania, the *Hobarton Guardian* commented: 'in a few years the changes introduced by civilisation will have obliterated many interesting, natural and physical records, [and] have placed many of the "feroe naturæ" on the list of those already, from various causes, utterly destroyed, the whole race forever gone, lost to the naturalist and the world' (23 September 1848). Unquestionably, by the middle of the nineteenth century Tasmanian naturalists and scientists, as well as the literate and educated public, were well aware of the potential extinction of indigenous species. Arguably, community awareness of, and close engagement with, Australia's fauna has never been greater. Yet the stark reality is, for all the knowledge and interest possessed by Tasmanian scientists and their society at the time, it was not good enough.

The second part of the question posed above now presents itself. Granted they had the knowledge of extinction, to what extent were Tasmania's scientists prepared to apply it to the thylacine? How aware were they of the impending extinction of the species? A detailed analysis of the historical records, rather than showing a lack of knowledge of the species and its impending extinction, tends to suggest exactly the opposite.

The year 1888, which marked the introduction of the government bounty scheme, was a crucial moment in the extinction of the species. Twenty-five different warnings by contemporary scientists and naturalists of the increasing scarcity and decline of the thylacine have been identified that predate this,[8] of which seven specifically refer to the likelihood of extinction. Twenty-five comments from eighteen different Tasmanian sources may not seem a particularly large number, but numerically it could have been much larger. In arriving at this total I have deliberately excluded all the comments on rarity and/or extinction by authors associated with published assumptions of thylacine primitiveness (for example, Cunningham, Flower and Owen) as these, and similar opinions of approaching extinction, may well have been motivated by the concept of marsupial inferiority, rather than a knowledge of the species' population dynamics. I have also excluded comments on rarity and extinction in general zoology texts that were published outside Australia, where the author gives no indication of any first-hand knowledge of the species. Finally, the list also excludes all those guides to Van Diemen's Land published outside Australia where the information presented was largely plagiarised and endlessly replicated from other source material. All those included in my reduced list of sources come from Tasmanian residents, with the exception of the two widely travelled Melbournians, Cambrian and Willoughby, or visiting professionals, such as Mueller, Gould, Broadbent and Krefft. Their position was summed up by Gould:

> When the comparatively small island of Tasmania becomes more densely
> populated, and its primitive forests are intersected with roads from the eastern
> to the western coast, the numbers of this singular animal will speedily diminish,
> extermination will have its full sway, and it will then like the wolf in England and
> Scotland, be recorded as an animal of the past . . . (J. Gould, 1851)

While the numbers of people acknowledging the decrease in the thylacine population and predicting the thylacine's extinction were to grow dramatically following the introduction of the bounty, these twenty-five comments represent the accepted point of view of the status of the thylacine in the eyes of the Tasmanian scientific community at the time. They express a viewpoint markedly different from that suggested by parliamentary representatives of the contemporary agricultural community. Considerable selectivity is again in operation in modern constructions of 'scientific innocence', when thylacine population estimates made in the nineteenth century by contemporary non-scientists are preferred over the population estimates made by contemporary scientists.

Tasmanian scientists were not ignorant of the thylacine destruction taking place around them, they were just very slow to react. They might have been politically naïve, but they were certainly not innocent of knowledge. The pathetic excuse – 'we did not know what was going on' – was as unreal in Tasmania in 1888, as it was in Germany over fifty years later. The reality is, scientists were well aware of the parlous state of the thylacine population and the species' impending extinction well before the government bounty scheme was introduced. Even in 1888, if the government was going to concern itself with the thylacine it should have been for preservation instead of persecution.

5 'INNOCENCE' THROUGH THE DENIAL OF CAPTIVE BREEDING

The final argument in the construction of 'scientific innocence' requires the rejection of all claims of successful captive breeding. Suggestions that the thylacine bred in captivity have been disputed and denied, or, more commonly, simply ignored by most contemporary authors. A primary reason for this is to preserve the precious argument that, as the thylacine would not breed in captivity, there was no way in which scientists could have saved the species from extinction anyway, even if they had been aware of the problem. When it comes to whitewashing the role scientists played in the extinction of the species, this is one of the more vital arguments. The point of view traditionally expressed in the literature is that 'these marsupials were never bred in captivity' (Collins, 1973, p. 137), and 'there is no record of any attempts being made to breed them in captivity, nor do we find any expressions of concern for their future' (Guiler, 1985, p. 66).

Occasionally the suggestion has been made in the recent literature that successful breeding of the thylacine did take place: in Hobart (E. A. Bell, 1965; *Burnie Advocate*, 25 August 1977); Launceston (Hodgkinson, 1988); Melbourne (Jenkins, 1977); and Washington (Bridges, 1974). More often than not, such suggestions have been vigorously denied in the scientific literature with little recourse to primary sources originating from these institutions.[9] No claim for successful captive breeding was ever made by Mary Roberts or Arthur or Alison Reid, the curators of Hobart Zoo; by William McGowan snr, curator of Launceston City Park Zoo; or by Frank Baker, superintendent, or William Blackburne, head keeper, of the National Zoological Park, Washington. Investigation of the archival records from these three zoos suggests that such breeding claims as were made originated from imprecise press reports, or from the impressions of zoo visitors observing the successful rearing to adulthood of specimens that had originally arrived with their mother as pouch or semi-independent young. Although a deliberate attempt at captive breeding was certainly made by Arthur Reid at Hobart (Lord, 1927), it was undertaken with little support from the Zoo's City Council administration and was unsuccessful. The National Zoological Park, Washington, also made determined efforts to purchase and house together unrelated adult pairs (Hamlet, 13/3/1985). Unfortunately, specimen death and poor introduction procedures prevented any success. However, the claim that thylacines were successfully bred at Melbourne Zoo is far less easily dismissed.

If the details of captive breeding at Melbourne were hidden away in the archival records of the zoo, it would not be possible to argue that scientists have deliberately ignored the data on captive breeding in order to 'save face'. But, at the time, the captive breeding of thylacines was recognised as a significant scientific event and was widely publicised. The Zoological Society proudly proclaimed its breeding success in print, and it was also mentioned by two different individuals in conference papers that were later published in full.

In October 1899 the Council of the Zoological Society wrote to the organisers of the forthcoming Melbourne meeting of the Australasian Association for the Advancement of Science, inviting members of the Association to visit the zoo. Prof. Baldwin Spencer accepted the offer on behalf of the Association, requesting that a representative from the Society present a paper to the Association, prior to the excursion, concerning the zoological specimens on display. The Council member, Frederick Godfrey, was elected to present the paper on the zoo's collection (Royal Melbourne Zoological Acclimatisation Society [RMZAS], 1903x, pp. 293, 295) on 12 January 1900 (Hall, 1901; *Melbourne Argus*, 13 January 1900). Godfrey mentioned the recent birth of the thylacine cubs as evidence for the contemporary progress of the Zoological Society, and noted the opportunity to view the newly independent young that afternoon when the Association members visited the zoo (RMZAS, 1903x, pp. 293, 297, 299). An updated and expanded version of this paper, with increased historical content, was published by the Zoological Society in February 1900. With reference to the marsupial wolf Godfrey noted that 'there are several specimens of this curious animal in the Gardens, including a she wolf with four young cubs which have until quite lately been in the pouch' (Godfrey, 1900, p. 20).

A draft Annual Report noting the significance of the birth of the cubs in 1899 was put before the Society's Council on 29 January 1900. It was then publicly read and accepted at the Annual General Meeting on 19 February 1900 (*Melbourne Argus*, 20 February 1900; RMZAS, 1903x, pp. 299, 304). When published, the Annual Report for the year 1899 prominently recorded the 'marsupial wolf' in 'a list of the more important births that have taken place during the past year' (RMZAS, 1900, p. 10).

W. H. D. Le Souëf succeeded his father as Director of Melbourne Zoo in May 1902. In February of that year he presented a paper to the Field Naturalists' Club of Victoria on Tasmanian fauna. The condensed version of this paper published in the *Victorian Naturalist* (1902) did not include reference to the thylacine or to its successful breeding. However, in the expanded version of the paper published some years later, Le Souëf referred to the actuality, while admitting also the rarity of thylacine breeding: 'They . . . do not as a rule breed in captivity' (1907, p. 176).

There is no evidence that, at the time, anyone questioned the probity of the claim. In fact, given the international demand for specimens, the proven ability of thylacines to breed in captivity was considered an important potential revenue source in the very first political moves made to establish a public zoo in Hobart on the Domain (*Hobart Daily Post*, 15 February 1910, 4 March 1910; Stimulus, 1910) – a political process that took thirteen years to reach fruition. The blanket denial of captive breeding only enters the scientific literature on the thylacine after the extinction of the species, suggesting that subjective, social factors associated with the extinction event have brought about this change in scientific knowledge construction.

The published record (Godfrey, 1900) suggests that the young born at Melbourne Zoo left the pouch permanently in January 1900. This is also identified as the significant date in the unpublished records. The details of thylacine specimen number C5600 in the National Museum of Victoria (Plate 9.6), a 'juvenile female' that on its death in the zoo was donated to the museum on

226

Plate 9.6 Museum specimen of young female thylacine, bred in Melbourne Zoo in 1899. (Author)

17 July 1901, describe it as '18 months' of age – backdating it to January 1900 – and, under the locality description, it is referred to as 'bred in zoo' (National Museum of Victoria, 1915). As this individual is the only known thylacine specimen with an accurately dated rather than estimated age, her basic measurements, as determined from the preserved, stretched skin, are of important comparative interest. While, as previously suggested, the average size of an adult female thylacine, measured along the body, was about 5 ft 8 in (172.8 cm), this juvenile female, 18 months after she permanently left the pouch, and therefore probably about 22 months of age, possessed a head and body length of only 72 cm and a tail length of 33 cm (totalling 105 cm). This is a long way short of currently accepted developmental estimates that suggest that, by the end of just their first twelve months, thylacines were probably around three-quarters grown, independent of their parents and even close to being sexually mature (Guiler, 1985; Joines, 1983; Nowak, 1991).

The definitive establishment of successful breeding at Melbourne Zoo requires an overview of the potential developmental history of thylacine young, plus the acquisition details of the adults concerned.

The vastly different developmental histories of the young in the two largest extant marsupi-carnivores, the spotted-tailed quoll or tiger cat (*Dasyurus maculatus*) – an agile, active predator with a maximum recorded head-to-tail

length of 130.9 cm and weight of 7 kg (Edgar, 1983), and the Tasmanian devil
(*Sarcophilus harrisii*) – a more stocky, slower scavenger with a maximum
recorded head-to-tail length of 91 cm and weight of 8 kg (Guiler, 1983), do not
allow of an easy extrapolation to the thylacine. The young of the spotted-tailed
quoll are born after a gestation period of only twenty-one days and by just seven
weeks of age their eyes are open, their fur reflects the full adult pattern, and, no
longer permanently attached to a teat, they venture outside the pouch for the
first time (Edgar, 1983; Fleay, 1940). After six to seven weeks of semi-independent existence, they leave the pouch permanently at three months of age,
returning only to suckle, with weaning achieved by five months of age (Tyndale-
Biscoe and Renfree, 1987). In contrast, Tasmanian devil young are born after
thirty-one days gestation, it is twelve weeks before their eyes are open and they
exhibit adult pelage, and it is thirteen to fifteen weeks before they detach themselves from the teat for the first time (Fleay, 1935, 1952; Guiler, 1983). A brief
period of semi-independence may follow, but while young devils have usually left
the pouch permanently by sixteen weeks of age, weaning may be protracted and
not achieved until eight months of age (Tyndale-Biscoe and Renfree, 1987). For
reasons possibly as simple as the shared common name for both species that
identifies the limits of their accepted recent distribution – or in recognition of
the need for an additional, developmental time factor, relative to the greater size
of the species – when developmental models have been hypothesised for Tasmanian tiger young, as minimum estimates, they have been based on the
lengthier developmental times known for the Tasmanian devil (for example,
Dixon, 1989; Guiler, 1985; Guiler and Godard, 1998) rather than those for the
similarly sized spotted-tailed quoll (unreservedly an Australian mainland as well
as Tasmanian resident).

227

Specific details known of the rearing of the second family of thylacines
acquired by Launceston City Park Zoo suggest a rapidity of development in
thylacine young more closely approximating the developmental times known
for the spotted-tailed quoll, rather than those of the Tasmanian devil.

The Public Reserves Committee report to the Launceston City Council on
10 July 1893 contained a list of recent additions to the zoological gardens which
did not include a thylacine (McGowan, 18/7/1893; *Launceston Examiner*, 15 July
1893). But shortly after 10 July, McGowan obtained a female thylacine from
Ringarooma 'carrying three unfurred young ones in her pouch' (*Launceston
Examiner*, 4 August 1893). From the analysis of preserved museum specimens it
is known that in thylacines a first growth of light hair commenced in the head
region of young as small as 7.5 cm in body length. It was uniformly 'very pale
brown in colour' (Boardman, 1945, p. 1) and not differentiated into stripes or
the typical adult pattern of distinctive light and dark patches on the head. By
the time thylacine young reached a head-to-body length of 28.8 cm, adult
patterns were present on the pelage, including the presence of well-developed
stripes (Boardman, 1945). Given the rapid behavioural development recorded
for these young in Launceston Zoo, it is suggested that the use of the word
'unfurred' in McGowan's description probably relates to the young arriving as

lightly furred but unbanded individuals, rather than their arrival in a completely immature, unfurred state. After less than four weeks in the zoo the female had readily adapted to captivity and allowed 'the very closest inspection' of her pouch and her now semi-independent young: 'the three little fellows occasionally venture forth from their cosy retreat and play about the cage, scuttling back to the protection of the maternal shelter at the slightest suspicion of danger' (*Launceston Examiner*, 4 August 1893). This record of rapid change and development, and of initial forays outside the pouch in, at the most, three and an half weeks of their arrival at Launceston Zoo as unbanded pouch young attached to a teat, is closer to the more rapid developmental history of young in the spotted-tailed quoll.

If we use the spotted-tailed quoll rather than the Tasmanian devil as the developmental model, adding in a factor for the increased size of the species, it might be suggested that thylacine young probably remained in the pouch attached to a teat for about eleven weeks before their first independent foray outside, and, this time subtracting a factor for the increased size of the species, that they then enjoyed one month of semi-independent life, sometimes in and sometimes out of the marsupium, with healthy specimens finally leaving the pouch permanently about four months after birth. These estimates are reflected in the known details for the family of thylacines sent to Washington Zoo in 1902. Three cubs, born in early April, were still attached to their teats ten weeks later on 20 June, but by early August appeared completely independent (Goding, 10/4/1902, 20/6/1902, 3/8/1902).[10]

A suggested eleven-week period of pouch dependency (close to three months) is also suggested in the literature. The earliest scientific publication that notes that thylacine young 'are carried in the pouch for about three months' is that of A. S. Le Souëf and Burrell (1926, p. 319). A. S. Le Souëf was Melbourne Zoo's veterinarian at the time of the birth of the thylacine cubs and later Assistant Director under his brother. A. S. Le Souëf's suggested three-month time-span in the pouch is consistently presented in the modern scientific literature. Amongst those who have accepted a three-month time-span in the pouch are some authors considered previously in note 9, who have denied the existence of captive breeding: Burton, Burton and Pearson (1987); Guiler (1985); Joines (1983); Moeller (1990); Nowak (1983); and Salvadori (1978). There is a certain piquancy about recent scientists accepting A. S. Le Souëf's figures for the length of time spent in the pouch – based on his observations as veterinarian at Melbourne Zoo – while at the same time denying the existence of the situation from which the data were obtained! It is this obvious dissonance that prompts the explanation of an unconscious motivation by present-day scientists to protect the 'integrity' of their profession, and the 'innocence' of their former colleagues, by suggesting that the preservation of the species through captive breeding was not a viable scenario.

Having considered details of the probable developmental history of thylacines, to establish that mating took place in the zoo it is necessary to turn to the arrival details of the two adult specimens concerned. The death of a thylacine on

228

13 June 1895 (National Museum of Victoria, 1897), temporarily terminated Melbourne Zoo's exhibition of thylacines after eleven years of continuous display. Finally, after a gap of more than three years, a thylacine of unknown sex arrived at the zoo in late December 1898, for which McGowan, at Launceston Zoo was paid £6.5.0 (RMZAS, 1903w, p. 66). If this specimen was the female, then the record of successful breeding in the zoo and the young becoming independent of the pouch in January 1900 is unassailable. The second specimen, also from Launceston Zoo, arrived the following year in late August/early September 1899, for which McGowan was paid £6.6.0 (RMZAS, 1903w, p. 68). If this specimen was the female, her record in captivity still suggests the necessity of accepting a successful mating, rearing and birth of thylacines in Melbourne Zoo.

The Launceston City Park Zoo was 'financed' by the Launceston City Council, but no money was provided in the annual budget for the purchase of stock until 1921 (City of Launceston, 1921). Before that time, McGowan developed the zoo to the point where it exhibited some 104 different species of animals, on the basis of donations from, and botanical or avian exchanges effected with, local animal collectors; and then the exchange or sale of these Tasmanian species to Australian mainland zoos for Australian or exotic specimens. By far the most important species for exchange and trading purposes was the thylacine. By December 1898 McGowan had already successfully reared two families of thylacines. It is highly unlikely that he would trade a female thylacine without checking her pouch first, either to keep the mother in captivity in Launceston, to rear the young and trade them all at a later stage of development, or merely to increase her immediate value to the zoo by selling her at an increased price as an adult-plus-young. With other marsupials, when a female carrying pouch young was offered for sale, the price of the specimen was increased by 50 per cent. The one shilling price difference for the two adults sold to Melbourne Zoo nine months apart suggests that, whichever specimen was the female, she was not carrying young in her pouch when she was sold.

The idea that she had already mated in the wild, but McGowan failed to notice she had immature young in her pouch on arrival at Launceston Zoo in July/August 1899 – and some time in Launceston needs to be posited while McGowan sought the highest bidder amongst the six zoological and scientific societies with which he is known to have traded thylacine specimens – coupled with the necessary suggestion that on arrival the curatorial staff at Melbourne Zoo also failed to notice the presence of young, which then stayed within the pouch for five or more months before quitting in January 1900, stretches the bounds of credibility.

The pair were introduced to each other in late August/early September. Taking a gestation period of about twenty-one days from the spotted-tailed quoll, the young were probably born in the last week of September. Eleven weeks later, during the second week of December, they emerged from their pouch for the first time, and four weeks later, in early January, before the Association's visit, the cubs had permanently quit the pouch. The apparent refusal of this female to tolerate a protracted period of semi-independence – as had been

229

recorded in captivity for separated families, consisting of just a lone mother and cubs – may well have been a reflection of the presence of the father and adult behaviours associated with the continued existence of the pair.

Written reports may well have been made of the reproductive behaviour of the two thylacines. A single 270-page volume (RMZAS, 1915) was set up in July 1896 to contain daily reports on the stock from individual keepers, the Director, Assistant Director and veterinarian. Because the volume needed to be accessible for written comment by a number of different people, it was centrally located and available for casual observation as well. Significant aspects of the behaviour of the animals were recorded, even down to the successful parturition of laboratory rats destined to be fed to the small carnivores and reptiles. Initially, the category of 'Births' (pp. 218–28) also included notes on the occurrence and incidence of reproductive and mating behaviours when observed.

However, during 1897 an unidentified individual in the zoo's administration decided that, given the necessity of open access, comments on the congress of lions and tigers and the like should not be on accidental display in the 'Births' section – presumably in case it corrupted the thoughts of the clerical staff. While the records of births were not considered offensive or corrupting, all entries of observed reproductive behaviour in this section were blotted out, scoured over or cross-hatched, and the notation added 'see page 242'. A new section was created in the rear of the volume restricted to the reproductive behaviour of the specimens. It is worth noting that, initially, these records were recognised as possessing some scientific value, as the early records were transferred and reproduced at the start of the new section. Records of the mating behaviour of the large carnivores figure prominently in the July 1896 to July 1897 records (underneath the attempted deletions). If mating behaviour of the thylacine pair was observed in 1899 it would have been recorded in the 'private' section of the record book, which continued in daily use until October 1909 (and again from June 1913 to January 1915). Regrettably, at some stage after the volume's initial closure in October 1909 some deeply disturbed individual, in order to prevent others from using the data for scientific or (almost inconceivably) sexually stimulating purposes, tore out and destroyed the 'offensive' section on reproductive behaviours (pp. 242–53) from the rear of the volume.

The specific histories of some of the individuals of this family are determinable. Seven months after leaving the pouch, one of the young cubs died, on 19 August 1900. One of the adults died on 24 May 1901 and another of the cubs (the specimen 'bred in zoo') on 17 July 1901. Unfortunately, the records of the deaths of the other members of this family are obscured by the arrival of additional juvenile and adult specimens from Tasmania. Given that adult carnivores frequently behave cavalierly towards young to which they are not related, it may have been fortunate for these captive-born cubs that the next four additions to the zoo's thylacine display were all juvenile specimens. An immature thylacine was added to the collection in November 1900, and three cubs of similar age and size to those bred in captivity were purchased in June 1901. Two adults, plus a specimen of unknown size, were added to the display between August and November 1901, and four additional thylacines entered the collection before the end of the year. It thus becomes impossible to ascribe the recorded deaths of thylacine specimens in the zoo after May 1901 specifically to the remaining locally bred cubs and one remaining parent.

One of the reasons for the rarity of captive breeding must lie in the conditions under which thylacines were typically displayed in zoos. They were usually housed *en masse* in traditional concrete or brick iron-barred cages, a situation which was tolerated with equanimity rather than excitement. Captive instances of severe acts of intraspecific aggression were rare (and reproductive behaviour even rarer). By chance, captive conditions of thylacines at Melbourne Zoo in 1899, of a single unrelated female and male pair housed together by themselves, encouraged a repertoire of pair bonding and reproductive behaviours that resulted in successful breeding. The ideal situation was relatively short-lived. As Melbourne Zoo purchased new thylacines they were merely added to the family group in the one enclosure.

All too frequently in the history of the zoological garden display of thylacines the purchase of a female and male pair may be traced to the capture of siblings taken together in the wild – not a situation likely to encourage breeding, given the social structure of the species. Apart from the breeding of thylacines at Melbourne Zoo in 1899, only seven other occasions have been identified where unrelated female and male thylacines are known to have been caged by themselves in captivity: in London, from 1850–53, 1856–57 and in 1910; at Melbourne from 1875–81, Adelaide in 1897, Washington from 1902–05 and Cologne from 1903–09. Given the highly unnatural constraints of traditional nineteenth-century and early twentieth-century cage design and display, successful breeding occurring only once from eight possible occasions does not appear remarkable.

The confirmation of successful breeding at Melbourne Zoo provides insights into the behaviour of the thylacine, as well as of the scientists who study them. For the thylacine, a much more rapid development in the pouch is suggested, as well a slower and more prolonged period of growth in the independent young. For modern-day scientists, no matter how unpalatable the data may be in destroying the 'innocence' of their forebears who did nothing practical to save the species from extinction, it is more than time that recognition be given to the successful breeding of thylacines in captivity at Melbourne Zoo – a recognition that, though supported by the archival records of Melbourne's Zoo and Museum, does nothing more than confirm the content of conference papers, scientific reports and publications at the time. The wonder is that these sources should ever have been successfully ignored or impeached in the first place.

Earlier, this chapter considered the victor rewriting history to conform to the dominant, successful viewpoint. Perhaps not only the victors but also the vanquished rewrite history when they can? In this case, the tendency of recent scientists to rewrite the history of the extinction of the thylacine away from one

231

of primary hominid responsibility into one combining marsupial inferiority with an hypothesised lack of knowledge possessed by earlier scientists, may be read as an example of the cultural construction of collective cowardice. If the rewriting of history by the victors is commonplace, perhaps collective cowardice may be commonplace amongst contemporary scientists confronting other impending extinction scenarios?

Notes

232

1   'Fiercely' is just one way to describe the competing colonial and metropolitan interests in obtaining the body of the last 'full-blood' male Tasmanian Aborigine, William Lanne. Before the Tasmanian Royal Society and Museum could obtain his body, William's head had already been removed by Dr Lodewyk Crowther at the Colonial Hospital, with the intention of sending it to London. Despite widespread public and scientific disgust at this mutilation, Lanne's head was never returned (Robson, 1985).

2   Such was the museum's misplaced confidence in its ability to obtain additional thylacines when desired, that not only did it fail to preserve the last thylacine in 1936, the year before, in the process of 'rationalising' the collection, it had destroyed five mounted specimens – one adult female and four juveniles – seen to be deteriorating and deemed unworthy of further preservation (annotations to the specimen catalogue, Tasmanian Museum, 1913, p. 120).

3   For the record, apart from Mary Roberts and the two Reids, the following individuals are identified in the above sources as being paid for full-time or part-time employment, or for providing casual or commission-based services to the zoo: Anders, Brett, Burbridge, Chatterton, Cooper, Cross, Dalgleish, Fleming, Fuller, Giblin, Harvey, King, Loone, Manson, Newman, Rathbone, Ritchie, Scanlon, Walter and Whiting. At best, if Darby ever had anything to do with the zoo, other than being just a casual visitor, he may have been one of the army of 'susso' workers, sent to work for the dole as gardeners, maintenance persons or pathway cleaners at the zoo.

4   This competitive scenario has an added dimension with the discovery in Victoria of one of the oldest known placental fossils, *Ausktriboshenos nyktos*, dated at 115 million years (Rich, Vickers-Rich, Constantine, Flannery, Kool and van Klaveren, 1997). While this fossil merely adds strength to the demonstrable success of the marsupial position in Australia, placental chauvinsim is in much greater trouble than this. 115 million years ago, two different mammalian types inhabited Australia. One of them, despite its success elsewhere, died out rapidly. The other, with a proud 120 million-year history behind it, remains to the present day. The ingestive process has barely begun, but the demonstrable early success in Australia of the egg-laying monotremes over their placental cousins is going to take a lot of swallowing by at least one placental species.

5 Lions tended to escape this labelling, with their prestige as the 'king of beasts', a product of medieval bestiaries, coupled with their presence as a symbol of British power and might (Ritvo, 1987, p. 26).

6 Huxley's commitment to progress led him to suggest a fourth mammalian reproductive type, the 'hypotheria' (1880, pp. 658–61), as the earliest mammals, evolving directly from amphibians. Necessarily sidestepping the known fossil reptiles, since including them would have implied, at best, a non-progressive, 'waiting stage', or, at worst, a regressive stage in mammalian development.

7 Because of its localised distribution and size, the rapid extinction in the early nineteenth century of the smaller, dwarf emu of King Island, *D. ater*, passed by without attracting significant professional notice (Leach, 1925).

8 The following list of pre-1888 inferences and comments from Tasmanian sources on the rarity of the thylacine, arranged chronologically (sources specifically referring to extinction are marked with an asterisk [*]); Jeffreys (1820), *Hobart Town Gazette* (2 August 1823), Henderson (1832), Gunn (1850w), (1850x), Meredith (1852), J. Gould (1851)*, Gunn (1852w)*, Cambrian (1855y)*, *Hobart Mercury*, (20 May 1858 republished in *Melbourne Argus*, 26 May 1858); Lloyd (1862), Mueller (16/6/1865)*, Hull (1871), Krefft (1871)*, Broadbent (20/2/1879, 15/3/1879, 15/10/1879), Meredith (1880), *Launceston Daily Telegraph* (24 August 1883), W. Crowther (1883)*, A. A. C. Le Souëf (26/11/1883), *Launceston Daily Telegraph* (26 September 1884), *Launceston Examiner* (17 April 1886), Willoughby (1886), and Sinurbe (1887)*.

9 Those who have actively disputed claims for breeding include: Burton, Burton and Pearson (1987); Collins (1973); Crandall (1964); Frauca (1957); Guiler (1985, 1986); Joines (1983); Moeller (1990); National Parks and Wildlife Service, Tasmania (*ca* 1982); Nowak (1983); Pink (1962, 1969); Salvadori (1978); Sayles (1980); Serventy (1966); and Serventy and Raymond (1973).

10 Don Coglan and Mike Archer of the Australian Museum, Sydney, have commenced the resurrection of extinct DNA from juvenile thylacines preserved in alcohol. From the replication and determination of select thylacine genes, to the cloning of different individuals to establish a small breeding population – creating an extant species – involves many giant steps along the way. But for the last stage in this process, choosing a species to carry a developing thylacine embryo, the overt similarities in developmental timing between tiger cat and thylacine young might initially suggest a female tiger cat, or spotted-tailed quoll (*Dasyurus maculatus*) as the most suitable surrogate parent.

233

Conclusion

234   The *Beagle* visited the Falkland Islands twice, in March 1833 and March 1834. While Darwin failed to see the marsupial wolf during the *Beagle*'s ten days in Hobart in February 1836, he was fortunate enough to collect four specimens of the anomalous Antarctic wolf or Warrah (*Dusicyon australis*) in the Falkland Islands: 'I entertain no doubt that the *Canis Antarcticus* is peculiar to this archipelago' (Darwin, 1839, p. 9). With different pelt patterns on different islands, the Antarctic wolf's origins are subject to much debate, including the possibility of its persistence as a Pleistocene remnant of the pre-glacial forests of the Falklands, and other sub-Antarctic and Antarctic islands (Berta, 1987). Darwin predicted:

> The number of these animals during the last fifty years must have been greatly reduced; already they are entirely banished from . . . half of . . . East Falkland . . . and it cannot, I think, be doubted, that as these islands are now becoming colonized, before the paper is decayed on which this animal has been figured, it will be ranked among those species which have perished from the face of the earth. (Darwin, 1839, p. 10)

Before Darwin's warning was even published, the Falklands' colonial government instituted a bounty scheme against the wolf which, combined with the fur trade, decimated the species. It may come as no surprise to discover that:

> As the numbers and threat of the Warrah as a predator decreased, tales of its destructive powers seem to have increased. The old vampire superstition surfaced again, and shepherds made unlikely claims of high sheep killings.

They insisted absurdly . . . that the Warrah killed only to suck blood from sheep; falling back on flesh only in times of need. The bounty was raised and hunting intensified once again. (Day, 1981, p. 167)

Three Antarctic wolves were displayed in London Zoo, in 1845, 1868 and 1870 (Allen, 1942; Garrod, 1878; Sclater, 1868w,x, 1870). The last known living representative of the species was killed at Shallow Bay, in the Hill Cove Canyon, in 1876 (Allen, 1942). The species' demise was far quicker than Darwin had imagined.

One small ray of hope may be found in the history of the extinction of the thylacine and similar species: the first casualty is always truth. The frequency with which those individuals hell-bent on environmental destruction for personal economic gain have been forced to lie blatantly in public about the behaviour and population numbers of their scapegoated species suggests their fear of the truth and the possibility that scientists may still be able to capture popular opinion. That this is rarely achieved before it is too late is part of the problem and bears on the intellectual, historical and moral education of scientists. That the serious presentation of the most ridiculous and unlikely suggestions – such as vampirism and increasing carnivore population numbers after decades of persecution – appear to be so easily accepted into popular opinion, rather than being laughed out of court, bears on the intellectual, historical and moral education of all people within a society.

The behaviour of those motivated towards the destruction of indigenous carnivores appears fairly restricted and stereotypical. The question as to whether the behaviour of local scientists in these situations is equally stereotypical is an important one to answer. A detailed analysis of contemporary responses by scientists to other instances of indigenous carnivore extinction is urgently required. From a psychological perspective, there is nothing harder to change than problem behaviour which is uniquely constrained by the different situations in which the problem is located. Conversely, if the contemporary behaviour of scientists confronted by rural mythology and impending extinction is as equally stereotypical as that shown by the environmental destroyers themselves, then the chance, through education and training, of effectively changing the collective conscience of scientists to help prevent further extinctions from taking place increases dramatically. As a first step, however, scientists must be sure of their data.

This book has attempted to distil some of the more significant changes in scientific perceptions of the thylacine made over the last two hundred years. It differs from previous attempts to discuss the behaviour of the species through its consideration of the social construction of scientific knowledge, exploring alongside the incorporation of accurate observations into scientific perceptions the incorporation of ideas from popular mythology as well. Its methodology has at least been unusual for science, covering as sources texts, extensive unpublished manuscripts and oral records of the species.

I have been forced to turn away from the perception of the thylacine as a primitive, solitary, asocial species – as commonly depicted in the twentieth-century literature – turning instead to a resuscitation of nineteenth-century perspectives on the social behaviour of the animal; and demonstrating conformity between these perspectives and scientific knowledge of environmental parameters, both in the past and in the present. Why these perspectives on the species' behaviour disappeared from the mainstream scientific literature was traced to the operation of two main factors. First, the colonial scientific organisations and structures developing in Australia at the end of the nineteenth century needed to demonstrate how advanced they were by wholeheartedly embracing contemporary European intellectual culture, at the cost of recognising and developing an Australian intellectual culture associated with indigenous entities. Secondly, for behavioural scientists in particular, this meant the adoption of the popular paradigm of progressive evolution, reflected in placental chauvinism with its prevailing assumption of the inferiority and second-rate – metatherian – status of marsupials in general, and the marsupial wolf in particular.

Inevitably, a major focus has been on extinction, and one of the most significant findings from my research lies in the illustration of how the narrative of extinction affects later scientific constructions and the conceptualisation of the extinction event itself. There is a marked alteration in the way information on the species has been handled prior to and after extinction. While myths about the thylacine, such as it being a marine predator, were effectively dealt with by contemporary scientists in a colonial context, and have stood the scientific test of time since extinction, other constructions of the thylacine have not been so fortunate. Whereas only two scientists published a belief in vampirism while the species was still extant, this idea has blossomed in scientific constructions of the species since extinction. The incorporation into modern scientific constructions of the muteness of thylacines, as suggested by the rural lobby opposed to protecting the species, also indicates how easily ideas, not from the scientific community but rather from those economically opposed to it, may override the acceptance of contemporary scientific opinion to become incorporated into scientific constructions of the animal after extinction has taken place.

Distinct from myths originating outside the scientific community are changes in the acceptance of ideas originating from within the scientific community. The dingo as the cause of the mainland extinction of the thylacine has recently become dramatically popular in the scientific mind, and it is amazing how the alternative explanation, which also originated in the nineteenth century – that humans were responsible for the mainland extinction – has yet to become widely accepted. So great has been the desire of twentieth-century scientists to rewrite the history of the extinction to absolve our own species and expose supposed inadequacies in the thylacine itself, that even contemporary accounts of successful captive breeding in the species have been suppressed in the literature.

236

Considerable time has been spent in a consideration of the relationship between thylacines and sheep, concluding that the thylacine was never more than an occasional, opportunistic predator on them. Only six published first-hand accounts of sheep predation are known before the 1880s. As a result, contemporary naturalists and scientists in the 1880s either held to an 'occasional predation' position, or believed that, while they may have been a problem in the past, thylacines were so reduced in numbers as to no longer present a problem to sheep farmers in the present. Nevertheless, these perspectives were completely ignored by the rural lobby in parliament and a bounty scheme was introduced against the species, overturning not just the contemporary scientific knowledge of the predatory behaviour of the thylacine, but also the general scientific recognition that already, prior to the bounty, the thylacine appeared headed for extinction. What is disturbing is that twentieth-century scientific constructions of the thylacines' predatory behaviour have ignored the position of nineteenth-century scientists on the unimportance of sheep predation, in favour of the picture painted by the species' destroyers in the rural lobby. Our past scientists deserve better treatment from their intellectual inheritors.

Sadly, there is nothing unique about the history of the destruction of the thylacine recorded here: predictable and disturbing parallels in the treatment of indigenous carnivores may be found almost at will in the invasive history of human colonisation over the past millennium. European scientists and their colonial and post-colonial counterparts have proved pathetically ineffective in countering such destruction. It is questionable whether, granted a knowledge of the extinction of the thylacine, Australian zoologists and comparative psychologists are any better equipped to deal with conservative political beliefs and aggressive economic rationalism today. The complete inability of early twentieth-century scientists to save the thylacine from extinction by achieving even a modicum of protection in time is disillusioning. When powerful economic pressure groups are arraigned against scientists and their knowledge claims it is difficult to see how scientists will ever win the battle.[1]

In many respects the insensitivity of treatment handed out to the last thylacines in the wild and the scientists who tried to protect them was reflected in the treatment handed out to the last thylacine in captivity and the last curators of Hobart Zoo. The professional knowledge, experience and concerns of both scientists and curators were undermined by administrative inaction or deliberate obstruction by individuals and committees unconcerned with the welfare of native animals and opposed to the preservation of Tasmanian bio-diversity. The treatment of the last curator of the zoo, with her intimate knowledge of the thylacine, as a second-class citizen because of her sex, parallels the castigation of scientists attempting to protect the species as second-class thinkers, espousing unrealistic ideas unattached to the economic realities of life. Both scientists as a whole, and the female half of the human species, deserve better treatment from their peers.

To learn from past mistakes and to prevent their repetition in the future, a recognition is required that Australia's scientists have failed to protect species and conserve environments in the past, and that if they continue to behave in the same old, ivory-tower way in the present, they will continue to fail in the future. To save other species from similar fates and to save the professional and public dignity of science, it is necessary to destroy the false, optimistic belief that knowledge alone will win an economic argument about the environment. Conservative interpretations of scientific methodology and professional practice need to be replaced with a redefinition of the self-image and construction of the scientist as a radical political animal. One of the most obvious steps to achieve this is through educating scientists in the history and philosophy of their chosen discipline so that they become aware of the social construction of scientific knowledge, and prepared to explore the moral as well as the cost-benefit consequences of conducting research. Scientists must come to accept the moral imperative of acting and speaking publicly on the implications of scientific data, rather than speaking softly, or just to the converted, or not even speaking out loud at all.

Konrad Lorenz once suggested: 'If you wish to really study an animal, you must first love it' (Fox, 1987, p. ix). He was not referring to a blind, subjective and unquestioning emotional response to a species, but rather to the development of an empathy and understanding for a species that, through such affection and knowledge, constructs questions about the behaviour of that species and provides a context for understanding the answers to them. Part of that love is naturally expressed in a commitment to the fair treatment, care and continued existence of that species.

Past naturalists and scientists, and the societies in which they lived, failed the thylacine. The loss of any species is regrettable, primarily because it represents a loss of knowledge. A 3.9 billion-year history of life lies behind each species alive today. The contingent events of history have shaped each species differently, representing a different path of evolution; a different way of responding to the same environments in which we now interact. We need to respect and explore such difference to gain vital knowledge in understanding and determining our own behaviour in different environments. Rather than encouraging an increasing sameness of expression and adaptation, we need to learn to love and appreciate the knowledge that lies behind the concept of difference: different histories, different ways of adaptation, and different ways of knowing. Operating on this level, the loss of the thylacine represents a significant loss of knowledge. But beyond this loss of knowledge lies another level.

Arguably, the extinction of the marsupial wolf is the most significant loss we have, as a species, experienced to date. No other deliberately organised, human-based extinction has so narrowed our own experience of the world and quality of life.

Over the last 100 000 years the placental wolf has accompanied our own species on this planet, wherever we have wandered. With a social group structure based around a recognition of individuals and the different roles they play within the family group, they have easily integrated with our own species' social structures and they have been domesticated and bred as dogs to meet the needs and inclinations of their companions and owners. For many humans, past and present, the domesticated placental wolf has provided an offer of perception and friendship, qualitatively different from that enjoyed with members of our own species.

The marsupial wolf, with a similar social group structure based around a recognition of individuals and the different roles they played within the family group, possessed enormous potential for similar integration. For a handful of humans in the past, the captive marsupial wolf provided an experience of perception and friendship, qualitatively different from that enjoyed with our own species, but a relationship which apparently met similar expectations to those associated with the domesticated placental wolf. The quality of this inter-species relationship can be gauged through the positive comments made about these specimens, alongside the quantitative argument that points out that keeping marsupial wolves in captivity – not for sale or profit, but for the relation-ship they established with their human owners – was not a transient, passing, social fad, but something practised continuously for over one hundred years, by perceptive individuals with a love and interest in the species. The privilege of living with a marsupial wolf is no longer possible. This is not just a loss for the present generation, but for generations ahead of us as well. We have all lost the potential of another species, seemingly so well adapted to integration and domestication with our own social structures and society. For this reason, I weep for my own species, as much as for the thylacine.

For the thylacine has not been the only species under consideration in this book. To the behavioural scientist, it is just as important to be able to love our own species as well. Possibly above all, the narrative pursued in this book demonstrates the need for us to know and love ourselves better. Not with an uncritical love, accepting of faults and divorced from any developmental or comparative contexts, but with a species-specific love seen to operate in the interspecific context in which we had our genesis; a love that desires to change our faults and make of ourselves a better species with which to co-habit. Only through a heterogeneous education that supports unfettered self-knowledge and the encouragement of constant questioning will the groundwork for effective change in human behaviour be achieved.

There is no longer room in the world for the effects of a narcissistic intel-lectual orientation that focusses only on the human; postulating for ourselves a uniqueness of mind, body and geophysical ownership. For we are all brothers to dragons and companions to owls, and whenever we lose sight of this in our personal, political and spiritual philosophies, we descend to the depths of self-deceit and contempt for others, and we destroy the world as well as ourselves.

239

1 To those who naïvely declare that such a thing as the thylacine's extinction could
never happen again today, I beg them to read the history of the dusky seaside
sparrow which became extinct when the last known specimen died in Disney
World, Florida, on 17 June 1987. Despite the existence of the appropriate
knowledge for saving the sparrow, supposedly the toughest protective legislation
in the world, and government organisations and committees theoretically
designed to achieve the sparrow's conservation, scientists were unable to cope
with developers, the hunting lobby and opposing government agencies all with
competing designs on the sparrow's remaining habitat (Walters, 1992).

Bibliography

Abbreviations

AC	Author's Collection	
AoT	Archives Office of Tasmania	
HCC	Hobart City Council papers	
LCC	Launceston City Council Archives	
LSD	Lands and Surveys Department papers	
MtL	Mitchell Library, New South Wales	
PRO	Public Records Office, Melbourne	
PWL	Parks, Wildlife and Heritage Library, Hobart	
QVM	Queen Victoria Museum	
RCG	Ronald Campbell Gunn papers	
StH	St Helens Local History Room	
TAB	Tasmanian Animals and Birds' Protection Board papers	
ThP	Thylacine papers	
TMG	Tasmanian Museum and Art Gallery	
TsL	Tasmaniana Library	
VDL	Van Diemen's Land Company papers	

MANUSCRIPTS AND INTERVIEWS

Archer, M. (letters, 28/5/1996, 17/5/1998) AC.
Bailey, C. (letters, 29/9/1996, 20/11/1996) AC.
Baldcock, A. (interview, 1/5/1953) ThP, QVM.
Bart [of Tenalga]. (interview, 6/8/1952) ThP, QVM.
Bayley, E. J. (letter, 22/8/1981) ThP, AoT.
Beechey, T. R. (interviews, 2–5/1/1989) StH.

Bell, D. A. (letter, 1/9/1981) ThP, AoT.

Bell, E. A. (1975). *Thylacine*, ThP, AoT.

Blackburne, W. H. (letter, 7/10/1905) Smithsonian Archives.

Blackwell, W. (interview, 27/11/1951) ThP, QVM.

Braddon, E. N. C. (letters, 29/2/1888; 28/5/1888) LSD, AoT.

Brain, W. A. (numerous letters, 1935–1936) HCC, AoT.

Branagan, J. G. (interview, 18/6/1992) AC.

Broadbent, K. (letters, 20/2/1879, 15/3/1879, 15/10/1879) Ramsey Papers, MtL.

Bryden, W. (letter, 28/8/1959) ThP, TMG.

Bunce, J. M. (1991). *Van Diemen's Land Company. Letters, Dispatches, Minutes, Reports, 1829–1847*, State Reference Library, Burnie.

Bunce, J. M. (1994). *Woolnorth. Selected documents, 1826–1845*, State Reference Library, Burnie.

Butler, E. J. (letter, 6/8/1934) TAB, AoT.

Carter, P. (interview, 22/10/1992) AC.

Carter, P. (interview, 22/10/1992) ThP, QVM.

Cartledge, I. (letter, 30/9/1970; questionnaire, 17/10/1970) ThP, QVM.

Cattley, H. (1863). *The Van Diemen's Land Company*, (handwritten manuscript) State Reference Library, Launceston.

Chaplin, A. (letter, 31/8/1954) ThP, TMG.

Chaplin, A. (letter, 6/11/1956) TAB, AoT.

Chapple, P. (letter, 3/6/1999) AC.

City of Launceston. (1898–1938) *Mayor's Valedictory Addresses and Annual Departmental Reports*, LCC.

Clucas, E. (1978). *Tracing the Decline of the Thylacine*, TsL.

Collins, G. M. (letter, 21/8/1981) ThP, AoT.

Collins, H. (interview, 6/8/1952) ThP, QVM.

Cooper, J. (questionnaire, 14/10/1970) ThP, QVM.

Cooper-Maitland, S. (*ca* 1968). *List of Specimens donated to Royal Society of Tasmania Museum 1849–1886*. Compiled from the Papers and Proceedings of the Royal Society of Tasmania, ThP, TMG.

Cotton, W. J. (interview, 1980) PWL.

Court. (letters, 1831–1835) VDL, AoT.

Cowburn, J. J. (letter, 30/8/1981) ThP, AoT.

Crawford, K. M. (letter, 2/9/1981) ThP, AoT.

Crawford, L. D. (1952w). *Tasmanian Tiger*, ThP, QVM.

Crawford, L. D. (1952x). *Interview notes, November 1952*, ThP, QVM.

Curr, E. (letters, 1827–1839) VDL, AoT.

Ditmars, R. L. (daily reports, 29/12/1902; 22/5/1903; 31/12/1917; 21/10/1918) New York Zoological Park.

Doherty, K. (letter, 16/12/1972) ThP, QVM.

Doherty, K. (letter, 2/9/1981) ThP, AoT.

Dransfield, M. (letter, 25/8/1981) ThP, AoT.

Dyer, S. J. (letters, 31/3/1843; 30/4/1843; 30/6/1843) VDL, AoT.

Edwards, T. (letters, 17/7/1928; 2/8/1928) TAB, AoT.

Ferrar, W. L. (letter, 24/8/1981) ThP, AoT.

Fleay, D. (letter, 9/6/1934) W. L. Crowther Library, Hobart.

Gardner, J. (letter, June 1972) ThP, QVM.

Gibson, J. A. (letters, May–September 1843) VDL, AoT.

Gossage, E. L. (letter, 24/8/1981) ThP, AoT.

Goding, F. W. (letters, 10/4/1902; 20/6/1902; 3/8/1902) Smithsonian Archives.

Gould, D. (interview, 23/6/1994) AC.

Graham, G. (letter, 1/9/1981) ThP, AoT.

Green, L. R. (*ca* 1975). *Tasmania's Unique 'Tigers'*, ThP, QVM.

Griffith, J. (1971). *Extinct? Or the rarest mammal in the world?* ThP, QVM.

Griffiths, K. E. (interview, 21/5/1980) PWL.

Guiler, E. R. (1964). *The Thylacine Investigation, 1963–1964*, TAB, AoT.

Gunn, R. C. (letters, 16/11/1836; 31/3/1837; 6/12/1843) RCG, QVM.

Gunn, R. C. (1838x). *Notes of Dissection Performed at Hobart Town, March 23rd 1838, by Dr. Bedford*, RCG, MtL.

Gunn, R. C. (1850x). *Zoology*, RCG, MtL.

Gunner, E. (1928). *North Eastern District Files re Tasmanian Marsupial Wolf*, TAB, AoT.

Hall, R. (letters, 5/1/1908, 14/12/1910) ThP, TMG.

Hamlet, B. (1985). *Sure, We Had a Thylacine, [13/3/1985]*, Smithsonian Archives.

Hardacre, J. (questionnaire, 1970) ThP, QVM.

Harrison and Harrison [of Mawbanna]. (interview, 9/10/1952) ThP, QVM.

Harrower, G. D. (letter, 26/4/1928) TAB, AoT.

Holmes, E. (letter, 26/8/1981) ThP, AoT.

House, O. F. (letter, 2/9/1981) ThP, AoT.

Hutchinson, J. H. (letters, 16/12/1833; 4/8/1834) VDL, AoT.

Lane, R. (interview, 21/11/1996) AC.

Larkins, K. (letter, 2/9/1981) ThP, AoT.

Lawrence, R. W. (1829–1831). *Diary, Journal and Notebooks, 1829–1833*, ThP, QVM.

Le Fevre, J. (letter, 9/6/1938) ThP, AoT.

Le Fevre, P. (ABC television interview, 1976) StH.

Le Souëf, A. A. C. (26/11/1883). Director's report. *Minute Book, 1880–1884*, PRO.

Le Souëf, A. S. (1945). *Vanishing Wild Life*, PRO.

Le Souëf, J. C. (*ca* 1928). *History of the Zoo, 1861–1927* (four handwritten notebooks) PRO.

Le Souëf, J. C. (1934). *3LO Children's Hour Broadcasts (1932–1934): Drafts of Stories and Articles*, PRO.

Le Souëf, J. C. (1970). *Draft Manuscript of Children's Book*, PRO.

Ling, W. J. (questionnaire, 1970) ThP, QVM.

Lord, C. E. (letters, 27/3/1928; 3/7/1928; 23/7/1928; 21/8/1928) TAB, AoT.

Lord, J. E. C. (letters, 1928–1936) TAB, AoT.

Lorkin, R. (letter, 20/8/1981) ThP, AoT.

McGaw, A. K. (letter, 29/5/1903) VDL, AoT.

McGowan, W. (letter, 18/7/1893) LCC.

McGowan, W. (letters, 21/6/1909; 31/8/1911) Museum of Victoria.

McGowan, W. (letter, 10/7/1912) State Records Office, Adelaide.

243

Marthick, R. (letter, 10/2/1937) TAB, AoT.

Martin, R. (interview, 11/1/1953) ThP, QVM.

Mitchell, S. (letter, 25/8/1981) ThP, AoT.

Mollison, B. C. (interviews, 25/11/1951; 14/3/1952) ThP, QVM.

Mooney, N. (interview, 12/10/1990) AC.

Morgan, W. S. (letter, 9/4/1928) TAB, AoT.

Mueller, F. J. H. von. (letter, 16/6/1865) RCG, MtL.

Murray, H. V. (1957). *I Chased a Tasmanian Tiger*, TAB, AoT.

Museum Trustees. (1928–1937). *Museum Trustees Minute Books*, ThP, TMG.

Museum Trustees. (1945). *Cash Book. Museum and Gardens. 13/1/1908–30/6/1945*, ThP, TMG.

National Museum of Victoria (1897). *Donations Book*, 6/1/1871–9/3/1897.

National Museum of Victoria (1915). *Specimens. Acquisitions*, 5/7/1899–29/2/1915.

Nibbs, L. R. (letter, 9/4/1928) TAB, AoT.

O'Shea, L. V. (letter, 25/8/1981) ThP, AoT.

Owen, R. (letter, 4/1/1837) RCG, MtL.

Paddle, R. N. (November 1997). *Changing Scientific Perceptions of the Thylacine, or Tasmanian tiger* (Thylacinus cynocephalus). PhD thesis, University of Melbourne.

Patman, W. (letter, 27/11/1949) TAB, AoT.

Pearse, R. (1976). *Thylacines in Tasmania*. Paper presented at the Annual General Meeting of the Australian Mammal Society, Launceston, May, 1976, PWL.

Propsting, R. (letter, 10/1/1854) RCG, MtL.

Reid, A. M. (interviews, 27/2/1992; 25/6/1992; 24/6/1996) AC.

Reid, A. M. (letters, 10/6/1992; 16/12/1993) AC.

Reid, A. R. (letters, 16/5/1935, 23/8/1935) HCC, AoT.

Reserves Committee. (1924–1937). *Minute Books*, HCC, AoT.

Roberts, M. G. (letters, 19/11/1909; 25/2/1910; 14/7/1910; 16/11/1917) Wellington City Council Archives.

Robinson, G. A. (diary entries, 1830–1834) Robinson archives, MtL.

Rowe, A. R. (interview, 25/11/1951) ThP, QVM.

Rowe, A. R. (interview, 26/9/1980) PWL.

Royal Melbourne Zoological and Acclimatisation Society. (1903w). *Zoological and Acclimatisation Society Ledger, 1898–1903*, PRO.

Royal Melbourne Zoological and Acclimatisation Society. (1903x). *Eighth Minute Book of the Zoological and Acclimatisation Society, 1893–1903*, PRO.

Royal Melbourne Zoological and Acclimatisation Society. (1910). *Ninth Minute Book of the Zoological and Acclimatisation Society, 1903–1910*, PRO.

Royal Melbourne Zoological and Acclimatisation Society. (1915). *Presentations, Births and Deaths at the Zoological Gardens, 1896–1915*, Private collection.

Sawford, B. (questionnaire, 1970) ThP, QVM.

Schayer, A. (letters, 1836–1842) VDL, AoT.

Scott, H. H. (1903). *Newspaper Cuttings Book, 1898–1903*, ThP, QVM.

Scott, H. H. (letters, 1901–1904). *Letter Book, 28/4/1899–25/5/1906*, ThP, QVM.

Scott, T. (surveyor) (1823). *Sketch of a tyger trap*. MtL.

Scott, T. (bushman) (interview, 8/5/1953) ThP, QVM.

Semmens, I. (interview, 24/10/1992) AC.

Sharland, M. S. R. (letter, 17/12/1972) ThP, QVM.

Simmons, O. (letter, 31/8/1981) ThP, AoT.

Simpson, E. (letter, 14/11/1999) AC.

Slebin, J. W. (letter, 22/4/1937) TAB, AoT.

Smith, M. (letter, 16/9/1972) ThP, QVM.

Stanfield, V. (letter, 31/8/1981) ThP, AoT.

Stevenson, G. (letters, September/October 1941; 21/10/1941; 13/11/1941) ThP, QVM.

Stevenson, L. F. (interview, 1/12/1972) ThP, QVM.

Tasmanian Animals and Birds' Protection Board. (1934–1948). *Minute Books*, TAB, AoT.

Tasmanian Animals and Birds' Protection Board. (1936). *Statement re Protection of Thylacine*, TAB, AoT.

Tasmanian Museum. (1913). *Register of Specimens, 1892–1913*, ThP, TMG.

Tasmanian Museum. (1914). *List of Animals Purchased, 1914*, ThP, TMG.

Tubb, J. E. (letter, 27/8/1981) ThP, AoT.

Turner, M. (interview, 25/10/1992) AC.

Unsigned. (questionnaire, 1970) ThP, QVM.

Wainwright, H. (questionnaire, 10/12/1970) ThP, QVM.

Wainwright, H. (interview, 1/10/1972) ThP, QVM.

Walker, B. W. (letter, 20/8/1981) ThP, AoT.

Welsh, J. (interview, 18/1/1952) ThP, QVM.

West [of Burnie]. (interview, *ca* 1953) ThP, QVM.

Williams, O. (letter, 10/9/1981) ThP, AoT.

Willoughby, K. T. (interview, 4/10/1970) ThP, QVM.

Willoughby, K. T. (questionnaire, 15/10/1970) ThP, QVM.

Wilson, A. (*ca* 1922w). *Notes on the Tiger*, TsL.

Wilson, A. (*ca* 1922x). *Hyenas or Tasmanian Tigers*, TsL.

Wilson, A. (*ca* 1922y). *The Tasmanian Tiger or Hyena*, TsL.

Wilson, A. (*ca* 1922z). *Untitled Typed Manuscript*, TsL.

PUBLICATIONS

Aflalo, F. G. (1896). *A Sketch of the Natural History of Australia, with Some Notes on Sport*. London: Macmillan.

Agricultural Gazette. (1897). 'Tasmanian tigers'. June 1897, 178–9.

Allen, G. M. (1942). *Extinct and Vanishing Mammals of the Western Hemisphere with the Marine Species of all the Oceans*. New York: American Committee for International Wild Life Protection.

Allport, M. (1868). 'Remarks on Mr. Krefft's "Notes on the Fauna of Tasmania"'. *Papers and Proceedings of the Royal Society of Tasmania*, 33–6.

Amos, A. (1826). Letter to William Pringle, 20/4/1826. Reproduced in C. Barrett, *Isle of Mountains. Roaming through Tasmania* (1944). Melbourne: Cassell.

Amos, C. (1963). *Family History of Adam Amos of 'Glen Gala' and William Lyne of 'Apsley'*. Hobart: A. G. Amos and P. Benson Walker.

Anderson, R. (1967). *On the Sheep's Back*. Adelaide: Rigby.

Andrews, A. P. (*ca* 1975). '*Thylacine* Thylacinus cynocephalus (*Tasmanian tiger, marsupial wolf*)'. *Education Leaflet No. 8*. Hobart: Tasmanian Museum and Art Gallery.

Andrews, A.P. (1985). '*Thylacine*. Thylacinus cynocephalus'. Hobart: Tasmanian Museum and Art Gallery.

Angas, G. F. (1862). *Australia; A Popular Account of its Physical Features, Inhabitants, Natural History, and Productions, . . .* London: Society for the Promotion of Christian Knowledge.

Anon. (1970). 'The rare thylacine'. *Nature Walkabout*, 6, (2), 5–7.

Anon. (1974). 'Lost world of Mt. Brockman'. *Northern Territory Newsletter*, April 1974, 1–8.

Archer, M. (1974). 'New information about the Quaternary distribution of the thylacine (Marsupialia, Thylacinidae) in Australia'. *Journal of the Royal Society of Western Australia*, 57, (2), 43–50.

Archer, M. (1993). 'The Murgon monster'. *Australian Natural History*, 24, (4), 60–1.

Archer, M. and Baynes, A. (1972). 'Prehistoric mammal faunas from two small caves in the extreme south-west of Western Australia'. *Journal of the Royal Society of Western Australia*, 55, 80–9.

Archer, M., Flannery, T. F., Ritchie, A. and Molnar, R. E. (1985). 'First Mesozoic mammal from Australia – an early Cretaceous monotreme'. *Nature*, 318, 363–6.

Atkinson, J. (1826). *An Account of the State of Agriculture & Grazing in New South Wales . . .* London: J. Cross.

Australian Museum. (1964). *The Thylacine, or 'Tasmanian Wolf'*. Leaflet No. 49. Sydney: V. C. N. Blight.

Australian National Parks and Wildlife Service. (1978). '*Thylacine*, Thylacinus cynocephalus'. Rare and Endangered Species Leaflet. Mammals No. 9. Canberra: Australian Government Publishing Service.

B, A. N. I. (1946). 'Puzzle of the Tasmanian tiger'. *Country Life*, 23/8/1946, 355.

Backhouse, J. (1843). *A Narrative of a Visit to the Australian Colonies*. London: Hamilton, Adams.

Baird, M. (1858). *A Cyclopædia of the Natural Sciences*. London: Richard Griffin.

Baird R.F. (1991). 'The Quaternary avifauna of Australia'. In P. Vickers-Rich, J. M. Monaghan, R. F. Baird and T. H. Rich (eds), *Vertebrate Palaeontology of Australia*. Melbourne: Monash University Publications Committee.

Barrett, C. (1943). *An Australian Animal Book*. Melbourne: Oxford University Press.

Barrett, C. (1944). *Isle of Mountains. Roaming through Tasmania*. Melbourne: Cassell.

Barrett, J. (1925). 'General introduction'. In J. Barrett (ed.), *Save Australia. A Plea For the Right Use of Our Flora and Fauna*. Melbourne: Macmillan.

Basalla, G. (1967). 'The spread of western science'. *Science*, 156, (3775), 611–22.

Batchelor, C. (1991). 'Anyhow, I enjoy the scrub'. In S. Cubit (ed.), *What's the Land For? People's Experiences of Tasmania's Central Plateau Region. Volume 2*. Launceston: Central Plateau Oral History Project.

Baudement, G. (1849). 'Thylacine'. In M. C. D'Orbigny (ed.), *Dictionnaire Universel D'Histoire Naturelle, Vol. 12*. Paris: Renard, Martinet et Cie.

246

Baulch, W. (1961). 'Ronald Campbell Gunn., F.R.S. F.L.S. A biographical note'.
    In T. E. Burns and J. R. Skemp (eds), *Van Diemen's Land Correspondents. Letters
    from R. C. Gunn, R. W. Lawrence, Jorgen Jorgenson, Sir John Franklin and others to
    Sir William J. Hooker. 1827–1849.* Launceston: Queen Victoria Museum.
Beatty, B. (1952). *Unique to Australia.* Sydney: Ure Smith.
Beddard, F. E. (1891). 'On the pouch and brain of the male thylacine'. *Proceedings of
    the Zoological Society of London*, 138–45.
Beddard, F. E. (1902). *Mammalia.* London: Macmillan.
Beddard, F. E. (1903). 'Exhibition and remarks upon sections of the ovary in
    Thylacinus'. *Proceedings of the Zoological Society of London*, 116.
Beddard, F. E. (1907). 'On the azygous veins in the mammalia'. *Proceedings of the
    Zoological Society of London*, 181–223.
Beddard, F. E. (1908). 'On the anatomy of *Antechionomys* and some other marsupials,
    with special reference to the intestinal tract and mesenteries of these and other
    mammals'. *Proceedings of the Zoological Society of London*, 561–605.
Bell, E. A. (1965). *An Historic Centenary. Roberts, Stewart & Co. Ltd. 1865–1965.* Hobart:
    Mercury Press.
Beresford, Q. (1985). 'Tragedy of the Tasmanian tiger'. *The Islander*, 12–14.
Beresford, Q. and Bailey, G. (1981). *Search for the Tasmanian Tiger.* Hobart: Blubber
    Head Press.
Berta, A. (1987). 'Origin, diversification, and zoogeography of the South American
    Canidae'. In B. D. Patterson and R. M. Timms (eds), *Studies in Neotropical
    Mammalogy: Essays in honour of Philip Hershkovitz.* Chicago: Field Museum of
    Natural History.
Betts, T. (1830). *An account of the Colony of Van Diemen's Land . . .* Calcutta: Baptist
    Mission Press.
Bischoff, J. (1832). *Sketch of the History of Van Diemen's Land, Illustrated by a Map of the
    Island, and an Account of the Van Diemen's Land Company.* London: John
    Richardson.
Boardman, W. (1945). 'Some points in the external morphology of the pouch young
    of the marsupial, *Thylacinus cynocephalus* Harris'. *Proceedings of the Linnean Society
    of New South Wales*, 70, 1–8.
Bowden, K. M. (1964). *Captain James Kelly of Hobart Town.* Melbourne: Melbourne
    University Press.
Bozman, E. F. (1958). 'Tasmania'. *Everyman's Encyclopædia. Volume 12* (4th edition).
    London: J. M. Dent.
Brazenor, C. W. (1950). *The Mammals of Victoria.* Melbourne: Brown, Prior Anderson.
Breckwoldt, R. (1988). *A Very Elegant Animal: the Dingo.* Sydney: Angus and
    Robertson.
Breton, W. H. (1835). *Excursions in New South Wales, Western Australia, and Van Diemen's
    Land, During the Years 1830, 1831, 1832 and 1833.* London: Richard Bentley.
Breton, W. H. (1846). 'Excursion to the Western Range, Tasmania'. *Tasmanian Journal*,
    2, 121–41.
Breton, W. H. (1847). 'Exhibit of a large specimen of Thylacinus Harrisii'. *Tasmanian
    Journal*, 3, 125–6.

Bridges, W. (1974). *Gathering of Animals. An Unconventional History of the New York Zoological Society.* New York: Harper and Row.

Brogden, S. (1948). *Tasmanian Journey.* Melbourne: Pioneer Tours.

Brown, P. L. (1952). *Clyde Company Papers, volume 2, 1836–1840.* London: Oxford University Press.

Brown, P. L. (1971). *Clyde Company Papers, volume 7, 1859–1873.* London: Oxford University Press.

Brown, R. (1973). 'Has the thylacine really vanished?' *Animals,* 15, (9), September 1973, 416–19.

Brown, R. (1983). 'Has the last thylacine gone to ground or . . . is there hope for our tiger?' *Tasmanian Mail,* 16 August 1983, 8.

Buchman, O. L. K. and Guiler, E. R. (1977). 'Behaviour and ecology of the Tasmanian devil, *Sarcophilus harrisii*'. In B. Stonehouse and D. Gilmore (eds), *The Biology of Marsupials.* London: Macmillan.

Bunce, D. (1857). *Australasiatic Reminiscences of Twenty-three Years' Wanderings in Tasmania and the Australias . . .* Melbourne: J.T. Hendy.

Burns, T. E. and Skemp, J. R. (eds) (1961). *Van Diemen's Land Correspondents. Letters from R.C. Gunn, R.W. Lawrence, Jorgen Jorgenson, Sir John Franklin and others to Sir William J. Hooker. 1827–1849.* Launceston: Queen Victoria Museum.

Burrell, H. (1921).'The Tasmanian tiger or wolf'. *Australian Museum Magazine,* 1, (3), 62.

Burton, J. A., Burton, V. G. and Pearson, B. (eds) (1987). *Collins Guide to the Rare Mammals of the World.* London: Collins.

Button, H. (1909). *Flotsam and Jetsam floating fragments of life in England and Tasmania . . .* Launceston: A. W. Birchall & Sons.

Calaby, J. H. (1971). 'Man, fauna and climate in Aboriginal Australia'. In D. G. Mulvaney and J. Golson (eds), *Aboriginal Man and Environment in Australia.* Canberra: Australian National University Press.

Cambrian. (1855w). 'Notes on the natural history of Australasia, letter first'. *Melbourne Monthly Magazine,* 1, (2), June, 95–101.

Cambrian. (1855x). 'Notes on the natural history of Australasia, letter second'. *Melbourne Monthly Magazine,* 1, (3), July, 164–9.

Cambrian. (1855y). 'Notes on the natural history of Australasia, letter third'. *Melbourne Monthly Magazine,* 1, (6), October, 360–2.

Carpenter, W. B. (1848). *Zoology: A Systematic Account of the General Structure, Habits, Instincts, and Uses of the Principal Families of the Animal Kingdom. Volume I.* London: Wm. S. Orr and Co.

Carter, P. (1991). 'We was always 'venturers'. In S. Cubit (ed.), *What's the Land For? People's Experiences of Tasmania's Central Plateau Region.* Volume 3. Launceston: Central Plateau Oral History Project.

City of Launceston. (1921). 'Mayor's Valedictory Address and Annual Departmental Reports'. Launceston: *Examiner.*

Claude, C. (1996). '*Der Beutelwolf.* Thylacinus cynocephalus *Harris,* 1808'. *Leben und Sterben einer Tierart.* Zurich: Zoologisches Museum der Universität Zürich.

Clutterbuck, L. (1989). 'Tiga. An animated film'. Melbourne: Wilderness Society.

Colbron-Pearse, D. (1968). 'Tame tiger'. *Walkabout,* December 1968, 8.

248

Collins, L.R. (1973). *Monotremes and Marsupials. A Reference Guide for Zoological Institutions*. Washington: Smithsonian Institution Press.

Commissioners of the Victorian Intercolonial Exhibition. (1875). *Victorian Intercolonial Exhibition 1875. Preparatory to the Philadelphia Exhibition 1876. Opened 2nd September, 1875. Official catalogue of exhibits*. Melbourne: McCarron, Bird and Co.

Constant Reader. (1885). 'Reward for capturing tigers'. *Tasmanian Mail*, 8 August 1885, 18.

Cornish, C. J. (1917). *Mammals of Other Lands*. New York: University Society Inc.

Cotton, J. (1979). *Touch the Morning. Tasmanian Native Legends*. Hobart: O.B.M.

Crandall, L. S. (1964). *The Management of Wild Mammals in Captivity*. Chicago: Chicago University Press.

Crisp, E. (1855). 'On some points relating to the anatomy of the Tasmanian wolf (*Thylacinus*) and of the Cape hunting dog'. *Proceedings of the Zoological Society of London*, 188–91.

Crowther, W. L. (1883). 'Letter to Dr P. L. Sclater, Secretary, Zoological Society of London, 23/2/1883'. *Proceedings of the Zoological Society of London*, 252.

Cunningham, D. J. (1882). 'Report on some points in the anatomy of the thylacine (*Thylacinus cynocephalus*), cuscus (*Phalangista maculata*), and phascogale (*Phascogale calura*) . . .' In C. W. Thompson and J. Murray (eds), *Report on the Scientific Results of the Voyage of the H.M.S. Challenger During the Years 1873–1876. Zoology*. Part XVI, 5 (ii), 1–192.

Cuppy, W. (1950). *How to Attract the Wombat*. London: Dennis Dobson.

Curr, E. (1824). *An Account of the Colony of Van Diemen's Land, Principally Designed for the Use of Emigrants*. London: George Cowie and Co.

Curr, E. (1835). 'Letter to Dr James Ross'. *Hobart Town Courier*, 27/3/1835, 4.

Cuvier, G. (1827). *The Animal Kingdom. . . . Volume III. The Class Mammalia*. London: Geo. B. Whittaker.

Cuvier, G. (1840). *Cuvier's Animal Kingdom . . .* London: W. M. S. Orr and Co.

Czechura, G. V. (1984). 'Modern sightings of the thylacine – what do they tell us?' *Skeptic*, 4, (4), 1–3, 6.

Dando, W. P. (1913). *More Wild Animals and the Camera*. London: Jarrold & Sons.

Darwin, C. R. (1839). *The Zoology of the Voyage of H.M.S. Beagle. During the years 1832–1836. Part II. Mammalia*. London: Smith Elder.

David, T. W. E. (1923). 'Trip to National Park'. *Hobart Mercury*, 24 January 1923, 10.

Davies, J. L. (ed.) (1965). *Atlas of Tasmania*. Hobart: Mercury Press.

Day, D. (1981). *The Doomsday Book of Animals. A Unique Natural History of Three Hundred Vanished Species*. London: Ebury Press.

de Beer, G. (1963). *Charles Darwin. Evolution by Natural Selection*. London: Thomas Nelson.

Desmarest, A. G. (1822). *Encyclopédie Méthodique: Mammalogie, ou Description des espèces des Mammifères*. Paris: Agasse.

Desmond, A. (1982). *Archetypes and Ancestors. Palaeontology in Victorian London, 1850–1875*. London: Blond and Briggs.

Dixon, J. (1822). *Narrative of a Voyage to New South Wales, and Van Dieman's Land . . .* Edinburgh: John Anderson, Jun., and Longman, Hurst, Rees, Orme and Brown.

Dixon, J. M. (1989). 'Thylacinidae'. In D. W. Walton and B. J. Richardson (eds), *Fauna of Australia. Mammalia. Volume 1B*. Canberra: Australian Government Printing Service.

Dodson, J., Fullagar, R., Furby, J., Jones, R. and Prosser, I. (1993). 'Humans and megafauna in a late Pleistocene environment from Cuddie Springs, north western New South Wales'. *Archaeology in Oceania*, 28, 94–9.

Doherty, K. (1977). 'When we caught a tiger'. In *N.W. Tasmania Short Stories and Articles*. Boat Harbour: Tasmanian Fellowship of Australian Writers, North West Branch.

East Coaster. (1929). Letter to the editor. *Hobart Mercury*, 6 June 1929, 11.

Eaves, R. (1989). 'On the tiger's trail!' *Launceston Examiner*, 3 February 1989.

Edgar, R. (1983). 'Spotted-tailed quoll. *Dasyurus maculatus*'. In R. Strahan (ed.), *The Australian Museum Complete Book of Australian Mammals*. Sydney: Angus and Robertson.

Edwards, A. B. and Edwards, E. M. (1941). ' "That curious rock" – and the Van Diemen's Land Company'. *Walkabout*, 1 January 1941, 34–6.

Eldredge, N. (1991). *The Miner's Canary. Unravelling the mysteries of extinction*. Princeton, NJ: Princeton University Press.

Evans, G. W. (1822). *A Geographical, Historical, and Topographical Description of Van Diemen's Land . . .* London: John Souter.

Evans, K. and Jones, M. D. (1996). *The Beaumaris Zoo Site Conservation Plan*. Hobart: Hobart City Council.

Farmer. (1887). Letter to the editor. *Tasmanian News*, 6 September 1887, 4.

Figuier, L. (1870). *Mammalia*. London: Chapman and Hall.

Finch, M. E. and Freedman, L. (1982). 'An odontometric study of the species of *Thylacoleo* (Thylacoleonidae, Marsupialia)'. In M. Archer (ed.), *Carnivorous Marsupials, Vol. II*. Sydney: Royal Zoological Society of New South Wales.

Fitzpatrick, K. (1949). *Sir John Franklin in Tasmania, 1837–1843*. Melbourne: Melbourne University Press.

Flannery, T. F. (1994). *The Future Eaters. An ecological history of the Australasian lands and people*. Sydney: Reed Books.

Fleay, D. (1935). 'Notes on the breeding of Tasmanian devils'. *Victorian Naturalist*, 52, 100–5.

Fleay, D. (1940). 'Breeding of the tiger-cat'. *Victorian Naturalist*, 56, 159–63.

Fleay, D. (1946w). 'On the trail of the marsupial wolf. (Tracking and trapping in the wild and picturesque west of Tasmania)'. *Victorian Naturalist*, 63, 129–35.

Fleay, D. (1946x). 'On the trail of the marsupial wolf. Part II'. *Victorian Naturalist*, 63, 154–9.

Fleay, D. (1952). 'The Tasmanian or marsupial devil – its habits and family life'. *Australian Museum Magazine*, 10, 275–80.

Fleay, D. (1956). *Talking of Animals*. Brisbane: Jacaranda.

Fleay, D. (1979). 'The poor Tassie tiger'. *Brisbane Courier-Mail*, 10 April 1979, 5.

Fleming, A. L. (1939). 'The thylacine. Reports on two expeditions in search of the thylacine. (Tasmanian tiger, or Tasmanian wolf.)'. *Journal of the Society for the Preservation of the Fauna of the Empire*, 38, 20–5.

Fleming, D. (1964). 'Science in Australia, Canada, and the United States: Some comparative remarks'. *Proceedings of the Tenth International Conference of the History of Science, 1962,* 1, 179–96.

Flower, W. H. (1867). 'On the development and succession of the teeth in the Marsupialia'. *Philosophical Transactions of the Royal Society of London,* 157, 631–41.

Flower, W. H. (1883). 'Mammalia'. In *The Encyclopædia Britannica. A Dictionary of Arts, Sciences and General Literature,* (9th edition), *Vol. 15,* 347–446. Edinburgh: Adam and Charles Black.

Flynn, T. T. (1914). 'The mammalian fauna of Tasmania'. In *Tasmanian Handbook. British Association for the Advancement of Science.* Hobart: John Vail.

Fox, M. W. (1987). 'Foreword'. In H. Frank (ed.), *Man and Wolf. Advances, issues and problems in captive wolf research.* Dordrecht: Dr W. Junk.

Frauca, H. (1957). 'How to catch a Tasmanian tiger'. *Australian Outdoors,* September, 22–5, 77–9.

Frauca, H. (1963). *Encounters with Australian Animals.* Melbourne: Heinemann.

Frauca, H. (1965). *The Book of Australian Wild Life. A panoramic view of Australian animals from insects to the high mammals.* Melbourne: Heinemann.

Fullagar, R. L. K., Price, D. M. and Head, L. M. (1996). 'Early human occupation of northern Australia: archaeology and thermoluminescence dating of Jinium rock-shelter, Northern Territory'. *Antiquity,* 70, 751–3.

Garrod, M. A. (1878). 'Notes on the visceral anatomy of *Lycaon pictus* and *Nyctereutes procyonides'. Proceedings of the Zoological Society of London,* 373–7.

Geoffroy Saint-Hilaire, E. (1810). 'Déscription de deux espèces du Dasyurus. (D. Cynocephalus et ursinus)'. *Annales du Muséum Histoire Naturel,* 15, 301–6.

Godfrey, F. R. (1900). 'The history and progress of the Zoological and Acclimatisation Society of Victoria'. In Royal Melbourne Zoological and Acclimatisation Society, *The Thirty-sixth Annual Report,* Melbourne: Rae Bros., Photo-Process House.

Godthelp, H., Archer, M., Ciselli, R., Hand, S. J. and Glikeson, C. F. (1992). 'Earliest known Australian Tertiary mammal fauna'. *Nature,* 356, 514–16.

Godwin, T. (1823). *Godwin's Emigrant's Guide to Van Diemen's Land, More Properly Called Tasmania . . .* London: Sherwood, Jones and Co.

Goodrick, J. (1977). *Life in Old Van Diemen's Land.* Adelaide: Rigby.

Gossage, C. (1966). 'Tiger may become a "bunyip"'. *Launceston Examiner,* 20 August 1966.

Gotch, A. F. (1979). *Mammals, Their Latin Names Explained.* Poole, Dorset: Blandford.

Gould, J. (1851). *The Mammals of Australia. Volume 1, Part iii.* London: Taylor and Francis.

Gould, S. J. (1989). *Wonderful Life. The Burgess Shale and the nature of history.* New York: W. W. Norton.

Gould, S. J. (1991). *Bully for Brontosaurus.* London: Hutchinson Radius.

Gould, S. J. (1996). *Life's Grandeur. The spread of excellence from Plato to Darwin.* London: Jonathan Cape.

Grant, J. E. (1831). 'Notice of the Van Diemen's Land tiger'. *Gleanings in Science,* 3, 175–7.

251

Grant, J. E. (1846). '*Thylacinus harrisii*'. *Tasmanian Journal of Natural Science, Agriculture, Statistics, &c*, 2, 311.

Gray, J. (1884). 'Native tigers; Petition'. *Tasmania. Journals and Papers of Parliament*, 3, (165), 1–3.

Gray, J. (1885). 'Eagles and native tigers; Petition for power to levy rate for destruction of, from Municipality of Spring Bay'. *Tasmania. Journals and Papers of Parliament*, 6, (100), 1–3.

Gray, J. E. (1841). 'Appendix C. Contributions towards the geographical distribution of the Mammalia of Australia, with notes on some recently discovered species, by J. E. Gray, F. R. S. &c. &c., in a letter addressed to the author.' In G. Grey *Journals of Two Expeditions of Discovery in North–West and Western Australia . . . Volume II*. London: T. and W. Boone.

Gray, J. E. (1849). 'On *Cypræa umbilicata* and *C. eximia* of Sowerby'. *Proceedings of the Zoological Society of London*, 125.

Green, R. H. (1973). *The Mammals of Tasmania*. Launceston: R. H. Green.

Gregory, W. K. (1921). 'Australian mammals and why they should be protected'. *Australian Museum Magazine*, 1, (3), 65–74.

Gribble, C. (1978). 'In the wild'. *People*, 4 May 1978, 23.

Griffith, J. (1972). 'The search for the Tasmanian tiger. Do the continuing reports of sightings mean that this primitive carnivore is not extinct?' *Natural History*, 81, 70–9.

Grzimek, B. (1967). *Four-Legged Australians. Adventures with animals and men in Australia*. London: Collins.

Grzimek, B. (1976). 'Subfamily Thylacinidae'. In B. Grzimek, *Grzimek's Animal Life Encyclopædia. Vol. 10*. New York: Van Nostrand Reinhold.

Guiler, E. R. (1958w). 'Observations on a population of small marsupials in Tasmania'. *Journal of Mammalogy*, 39, 44–58.

Guiler, E. R. (1958x). 'The thylacine'. *Australian Museum Magazine*, 12, 352–4.

Guiler, E. R. (1960). *Marsupials of Tasmania*. Hobart: Tasmanian Museum and Art Gallery.

Guiler, E. R. (1961w). 'The former distribution and decline of the thylacine'. *Australian Journal of Science*, 23, (7), 207–10.

Guiler, E. R. (1961x). 'Breeding season of the thylacine'. *Journal of Mammalogy*, 42, 396–7.

Guiler, E. R. (1970). 'Observations on the Tasmanian devil, *Sarcophilus harrisii* (Marsupialia: Dasyuridae). II. Reproduction, breeding, and growth of pouch young'. *Australian Journal of Zoology*, 18, 63–70.

Guiler, E. R. (1973w). 'Some thoughts on man and the mammals in the Central Plateau'. In M. R. Banks (ed.), *The Lake Country of Tasmania*. Hobart: Royal Society of Tasmania.

Guiler, E. R. (1973x). 'The importance of the north west to Tasmanian fauna'. *Proceedings of Seminar, Arthur to Pieman River Area*. Smithton: Municipality of Circular Head.

Guiler, E. R. (1978). 'Observations on the Tasmanian devil, *Sarcophilus harrisii* (Dasyuridae: Marsupialia) at Granville Harbour, 1966–75'. *Papers and Proceedings of the Royal Society of Tasmania*, 112, 161–88.

Guiler, E. R. (1980), 'The thylacine – an unsolved case'. *Australian Natural History*, 20 (2), 69–70.

Guiler, E. R. (1983). 'Tasmanian devil'. In R. Strahan (ed.), *The Australian Museum Complete Book of Australian Mammals*. Sydney: Angus and Robertson.

Guiler, E. R. (1985). *Thylacine: The Tragedy of the Tasmanian Tiger*. Melbourne: Oxford University Press.

Guiler, E. R. (1986). 'The Beaumaris Zoo in Hobart'. *Papers and Proceedings of the Tasmanian Historical Research Association*, 33, (4), 121–71.

Guiler, E. R. (1991). *The Tasmanian Tiger in Pictures*. Hobart: St David's Park Publishing.

Guiler, E. R. and Godard, P. (1998). *Tasmanian Tiger. A Lesson to be Learnt*. Perth: Abrolhos Publishing.

Guiler, E. R. and Heddle, R. W. L. (1970). 'Testicular and body temperatures in the Tasmanian devil and three other species of marsupial'. *Comparative Biochemistry and Physiology*, 33, 881–91.

Gulson, L. (1991). *City Park Launceston. National Estate Objection Assessment Report*. Hobart: Register of the National Estate.

Gunn, R. C. (1838w). 'Notices accompanying a collection of quadrupeds and fish from Van Diemen's Land. (Addressed to Sir W. J. Hooker, and by him transmitted to the British Museum.) With notes and descriptions of the new species, by J. E. Gray, F. R. S., &c'. *Annals and Magazine of Natural History: Zoology, Botany and Geology*, 1, 101–11.

Gunn, R. C. (1849). 'On the habitat of *Cypræa umbilicata*, Sowerby'. In a letter to J. E. Gray, Esq. *Proceedings of the Zoological Society of London*, 124–5.

Gunn, R. C. (1850w). 'Letter to D. W. Mitchell, Esq., Secretary Zoological Society, 29/12/1849'. *Proceedings of the Zoological Society of London*, 90–1.

Gunn, R. C. (1850y). 'Letter to J. Gould, 12/11/1850'. Reproduced in J. Gould, *The Mammals of Australia. Volume 1, Part iii*, (1851). London: Taylor and Francis.

Gunn, R. C. (1851). 'On the introduction and naturalization of *Petaurus sciureus* in Tasmania'. *Papers and Proceedings of the Royal Society of Van Diemen's Land*, 1, 253–5.

Gunn, R. C. (1852w). 'Zoology'. In J. West *The History of Tasmania* . . . Launceston: Henry Dowling.

Gunn, R. C. (1852x). 'A list of the mammals indigenous to Tasmania'. *Papers and Proceedings of the Royal Society of Van Diemen's Land*, 2, 77–90.

Gunn, R. C. (1852y). 'Thylacinus cynocephalus – introduced alive into England'. *Papers and Proceedings of the Royal Society of Van Diemen's Land*, 2, 184.

Gunn, R. C. (1852z). 'Notes on natural history'. *Papers and Proceedings of the Royal Society of Van Diemen's Land*, 2, 156–157.

Gunn, R. C. (1863). 'Letter. Jan. 19. 1863'. *Proceedings of the Zoological Society of London*, 103–4.

Gunther, A. C. L. G. (1900). 'The unpublished correspondence of William Swainson with contemporary naturalists (1806–1840)'. *Proceedings of the Linnean Society of London*, (1899/1900), 14–61.

Hall, R. (1910). 'The Tasmanian Museum. The Dasyures'. *Hobart Mercury*, 3 November 1910, 3.

253

Hall, R. (1925). 'The protection of fauna in Tasmania'. In J. Barrett (ed.),
  Save Australia. A Plea for the Right Use of Our Flora and Fauna. Melbourne:
  Macmillan.
Hall, T. S. (1901). *Report of the Eighth Meeting of the Australasian Association for the
  Advancement of Science held at Melbourne, Victoria, 1900.* Melbourne: Australasian
  Association for the Advancement of Science.
Harman, I. (1949). 'Tasmania's wolf and devil'. *Zoo Life,* 4, (3), 87.
Harper, F. (1945). *Extinct and Vanishing Mammals of the Old World.* New York: American
  Committee for International Wild Life Protection.
Harris, G. P. (1808). 'Description of two new species of *Didelphis* from Van Diemen's
  Land'. *Transactions of the Linnean Society of London,* 9, 174–8.
Harrison, W. (1884). 'Chief Inspector of Sheep: Report for 1883, (30/6/1884)'.
  Tasmania. Journals and Printed Papers of the Parliament of Tasmania, 3, Report #110.
Haswell, W. A. (1914). 'The animal life of Australia'. In G. H. Knibbs (ed.), *Federal
  Handbook: Prepared in connection with the eighty-fourth meeting of the British
  Association for the Advancement of Science, held in Australia, August, 1914.*
  Melbourne: Albert J. Mullett.
Haswell, W. A. (1926). 'Tasmanian wolf'. In A. W. Jose and H. J. Carter (eds),
  The Australian Encyclopædia. Volume II. Sydney: Angus and Robertson.
Healy, T. and Cropper, P. (1994). *Out of the Shadows. Mystery Animals of Australia.*
  Chippendale, New South Wales: Ironbark Pan Macmillan.
Heazlewood, I. C. (1992). *Old Sheep for New Pastures. A Story of British Sheep in the Hands
  of Tasmanian Colonial Shepherds.* Launceston: Australian Society of Breeders of
  British Sheep, Tasmanian Branch.
Henderson, J. (1832). *Observations on the Colonies of New South Wales and Van Diemen's
  Land.* Calcutta: Baptist Mission Press.
Heuvelmans, B. (1958). *On the Track of Unknown Animals,* (first English edition).
  London: Rupert Hart-Davis.
Hickman, V. V. (1955). 'The Tasmanian tiger'. *Etruscan,* 5, (2), 8–11.
Hier, H. (1976). 'The thylacine'. *Thylacinus: Journal of the Australasian Society of
  Zookeepers,* 1, 29–31.
Hoare, M. E. (1969). ' "All things queer and opposite": Scientific societies in Tasmania
  in the 1840's'. *Isis,* 60, 198–209.
Hodgkinson, D. (1982). 'Fred often saw tiger eyes in firelight'. *Northern Scene,* 9 June
  1982, 28.
Hodgkinson, D. (1988). 'Killing the tiger threat'. *Launceston Examiner,* 14 January
  1988, 11.
Home, R. W. (1988). 'Introduction'. In R. W. Home (ed.), *Australian Science in the
  Making.* Cambridge: Cambridge University Press.
Horton, H. (1996). 'The pursuit of natural history and attitudes to the land'.
  Proceedings of the Royal Society of Queensland, 106, (1), 31–5.
Howitt, W. (1855). *Land, Labour and Gold or Two Years in Victoria. With Visits to Sydney
  and Van Diemen's Land.* London: Longman, Brown, Green and Longmans.
Hull, H. M. (1859). *The Experience of Forty Years in Tasmania.* London: Orger and
  Meryon.

254

Hull, H. M. (1871). *Practical Hints to Emigrants Intending to Proceed to Tasmania . . .*
  Hobart: W. Fletcher.
Huxley, T. H. (1880). 'On the application of the laws of evolution to the arrangement
  of the Vertebrata, and more particularly the Mammalia'. *Proceedings of the
  Zoological Society of London*, 649–62.
Inkster, I. (1985). 'Scientific enterprise and the colonial "model": Observations on
  Australian experience in historical context'. *Social Studies of Science*, 15, 677–704.
Inkster, I. and Todd, J. (1988). 'Support for the scientific enterprise, 1850–1900'. In
  R. W. Home (ed.), *Australian Science in the Making*. Cambridge: Cambridge
  University Press.
Ireland, A. (1865). *Geography and History of Oceania, Comprising a Detailed Account of the
  Australian Colonies . . .* (2nd edition). Hobart: W. Fletcher.
Jeffreys, C. H. (1820). *Van Dieman's Land. Geographical and descriptive delineations of the
  island of Van Dieman's Land*. London: J. M. Richardson.
Jenkins, C. F. H. (1977). *The Noah's Ark Syndrome. (One Hundred Years of Acclimatization
  and Zoo Development in Australia.)*. Perth: Zoological Gardens Board.
Jenkinson, M. (1980). *Beasts Beyond the Fire*. New York: E. P. Dutton.
Jenks, S. M. and Ginsburg, B. E. (1987). 'Socio-sexual dynamics in a captive wolf pack'.
  In H. Frank (ed.), *Man and Wolf. Advances, Issues and Problems in Captive Wolf
  Research*. Dordrecht: Dr W. Junk.
Joines, S. (1983). 'The mysterious tiger from Tasmania'. *Zoonooz*, September, 4–11.
Jones, R. (1970). 'Tasmanian Aborigines and dogs'. *Mankind*, 7, 256–71.
Jordan, A. M. (1987). *The Tiger Man. The Thylacine – Yesterday, Today and Tomorrow.
  My Study and Findings*. Perth: Wordswork Express.
Keast, A. (1966). *Australia and the Pacific Islands. A Natural History*. New York: Random
  House.
Kentish Times. (1984). 'Tassie tiger's tale'. 7, (4), 4–5.
Kershaw, J. A. (1912). 'The Tasmanian devil in Victoria'. *Victorian Naturalist*, 29 (5),
  75–6.
Kleiman, D. G. (1977). 'Monogamy in mammals'. *Quarterly Review of Biology*, 52,
  39–69.
Knight, C. (1855). 'Marsupiata'. *The English Cyclopaedia. A new dictionary of universal
  knowledge, Vol. III*. London: Bradbury and Evans.
Knopwood, R. (1838). *The Diary of the Reverend Robert Knopwood, 1803–1838. First
  Chaplain of Van Diemen's Land*, (ed. by M. Nicholls, 1977). Hobart: Tasmanian
  Historical Research Association.
Kolar, K. (1965). *Continent of Curiosities*. London: Souvenir Press.
Kolig, E. (1973). 'Aboriginal man's best foe?' *Mankind*, 9, 122–4.
Krefft, G. (1868w). 'Description of a new species of thylacine (*Thylacinus breviceps*)'.
  Annals and Magazine of Natural History, 2, (4), 296–7.
Krefft, G. (1868x). 'Notes on the fauna of Tasmania'. *Papers and Proceedings of the Royal
  Society of Tasmania*, 91–6.
Krefft, G. (1871). *Mammals of Australia*. Sydney: Thomas Richards.
La Billardière, J. J. H. de (1799). *Relation du Voyage à la Recherche de la Pérouse*. Paris:
  H. J. Jansen.

255

Labillardière, J. J. H. de (1800). *Voyage in Search of La Pérouse, 1791–1794* . . . Tr. from the French. London: J. Stockdale.

Lang, W. H. (*ca* 1910). *Romance of Empire: Australia.* London: T. C. and E. C. Jack.

Le Souëf, A. S. (1926). 'Notes on the habits of certain families of the order Marsupialia'. *Proceedings of the Zoological Society of London,* 935–7.

Le Souëf, A. S. and Burrell, H. (1926). *The Wild Animals of Australasia* . . . London: George Harrap.

Le Souëf, W. H. D. (1902). 'A visit to the Furneaux Group of Islands'. *Victorian Naturalist,* 18, 181–8.

Le Souëf, W. H. D. (1907). *Wild Life in Australia.* Melbourne: Whitcombe and Tombs.

Leach, J. A. (1925). 'Australian birds: their place in nature'. In J. Barrett (ed.), *Save Australia. A Plea for the Right Use of Our Flora and Fauna.* Melbourne: Macmillan.

Leach, J. A. (1929). *Australian Nature Studies. A Book of Reference for Those Interested in Nature-Study.* Melbourne: Macmillan.

Leahy, C. (1960). 'Thylacine or Tasmanian tiger'. *Parklan News,* 1, (12), 1, 17–18.

Lewis, D. J. (1977). 'More striped designs in Arnhem Land rock paintings'. *Archaeology and Physical Anthropology in Oceania,* 12, (2), 98–111.

Lloyd, G. T. (1862). *Thirty-three Years in Tasmania and Victoria* . . . London: Houlston and Wright.

Loone, A. W. (1928). *Tasmania's North-East: A Comprehensive History of North-eastern Tasmania and Its People.* Launceston: *Examiner* and *Launceston Weekly Courier.*

Lord, C. E. (1919). 'Notes on the mammals of Tasmania'. *Papers and Proceedings of the Royal Society of Tasmania,* 53, 16–52.

Lord, C. E. (1927). 'Existing Tasmanian marsupials'. *Papers and Proceedings of the Royal Society of Tasmania,* 61, 17–24.

Lord, C. E. and Scott, H. H. (1924). *A Synopsis of the Vertebrate Animals of Tasmania.* Hobart: Oldham, Beddome and Meredith.

Lord, J. E. C. (1936). *Animals and Birds, Tasmania: Close Seasons, Prohibitions, Restrictions.* Hobart: W. E. Shimmins.

Lorenz, K. (1952). *King Solomon's Ring.* London: Methuen.

Lorenz, K. (1954). *Man Meets Dog.* London: Methuen.

Lost. (1908). Letter to the editor. *Hobart Mercury,* 22 July 1908, 3.

Lucas, A. H. S. and Le Souëf, W. H. D. (1909). *The Animals of Australia. Mammals, Reptiles and Amphibians.* Melbourne: Whitcombe and Tombs.

Lycett, J. (1824). *Views in Australia or New South Wales, and Van Diemen's Land* . . . London: J. Souter.

Lydekker, R. (1894). *A Hand-book to the Marsupialia and Monotremata.* London: W. H. Allen.

Lydekker, R. (1895). *The Royal Natural History. Volume III.* London: Frederick Warne.

Lydekker, R. (1915). *Wildlife of the World. A Descriptive Survey of the Geographical Distribution of Animals, Volume III.* London: Frederick Warne.

Macdonald, D. W. and Moehlman, P. D. (1982). 'Cooperation, altruism and restraint in the reproduction of carnivores'. In P. P. G. Bateson and P. Klopfer (eds), *Perspectives in Ethology, vol. 5.* New York: Plenum Press.

256

McGrath, A. (1995). 'Tasmania: 1'. In A. McGrath (ed.), *Contested Ground. Australian Aborigines under the British Crown*. Sydney: Allen and Unwin.

McKay, A. (1962). *Journals of the Land Commissioners for Van Diemen's Land 1826–28*. Hobart: University of Tasmania.

Macleay, W. (1886). 'Zoology of Australia'. *Papers and Proceedings of the Royal Society of Tasmania for 1885*, 26, 285–308.

MacLeod, R. (1988). 'From imperial to national science'. In R. MacLeod (ed.), *The Commonwealth of Science: ANZAAS and the scientific enterprise in Australasia, 1888–1988*. Melbourne: Oxford University Press.

McMichael, D. F. (1968). 'Conserving our wildlife'. *Hemisphere*, March, 2–8.

McMichael, D. F. (1982). 'What species, what risk?' In R. H. Groves and W. D. L. Ride (eds), *Species at Risk: Research in Australia*. Canberra: Australian Academy of Science.

McOrist, S., Kitchener, A. C. and Obendorf, D. L. (1993). 'Skin lesions in two preserved thylacines, *Thylacinus cynocephalus*'. *Australian Mammalogy*, 16, (1), 81–2.

Macphail, E. M. (1982). *Brain and Intelligence in Vertebrates*. Oxford: Clarendon Press.

M, L. G. (1916). 'Too Stupid to Tame. Tasmanian wolf at the Zoological Gardens'. Unidentified New York newspaper/magazine article from November 1916. Thylacine file, Taronga Park Zoo, Sydney.

Marshall, A. J. (1966). 'The world of Hopkins Sidthorpe'. In A. J. Marshall (ed.), *The Great Extermination. A Guide to Anglo-Australian cupidity, wickedness & waste*. Melbourne: Heinemann.

Marshall, L. G. (1984). 'Monotremes and marsupials'. In S. Anderson and J. Knox Jones, Jr (eds), *Orders and Families of Recent Mammals of the World*. New York: John Wiley.

Marshall Graves, J. A., Hope, R. M. and Cooper, D. W. (eds) (1990). *Mammals from Pouches and Eggs: Genetics, Breeding and Evolution of Marsupials and Monotremes*. Canberra: CSIRO.

Martin Duncan, P. (1884). 'Marsupialia'. In P. Martin Duncan (ed.), *Cassell's Natural History, Vol. III*. London: Cassell and Co.

Martin, P. S. (1967). 'Prehistoric overkill'. In P. S. Martin and H. E. Wright (eds), *Pleistocene Extinctions; the Search for a Cause*. New Haven: Yale University Press.

Martin, P. S. (1973). 'The discovery of America'. *Science*, 179, 969–74.

Martin, R. M. (1839). *History of Austral-Asia: Comprising New South Wales, Van Diemen's Island . . .* London: Whittaker.

Melville, H. (1833). *Van Diemen's Land; Comprehending a Variety of Statistical and other Information . . .* London: Smith, Elder and Co.

Menkhorst, P. W. (ed.) (1995). *Mammals of Victoria. Distribution, Ecology and Conservation*. Melbourne: Oxford University Press.

Mercer, P. (1963). *The Story of Burnie. 1823 to 1910*. Ridgley: Peter Mercer.

Meredith, L. A. (1852). *My Home in Tasmania, During a Residence of Nine Years*. London: John Murray.

Meredith, L. A. (1880). *Tasmanian Friends and Foes. Feathered, Furred, and Finned. A family chronicle of country life, natural history and veritable adventure*. London: Marcus Ward.

257

Merrilees, D. (1968). 'Man the destroyer: late Quaternary changes in the Australian marsupial fauna'. *Journal of the Royal Society of Western Australia*, 51, 1–24.

Meston, A. L. (1958). 'The Van Diemen's Land Company, 1825–1842'. *Records of the Queen Victoria Museum*, 9, 1–62.

Miles, D. (1991). 'I still go back to my mountains'. In S. Cubit (ed.), *What's the Land For? People's Experiences of Tasmania's Central Plateau Region*. Volume 5. Launceston: Central Plateau Oral History Project.

Milligan, J. (1853). 'Remarks upon the habits of the wombat, the hyaena, and certain reptiles'. *Papers and Proceedings of the Royal Society of Van Diemen's Land*, 2, 310.

Milligan, J. (1859). 'Vocabulary of dialects of the Aboriginal tribes of Tasmania'. *Papers and Proceedings of the Royal Society of Tasmania*, 3, 239–74.

Mitchell, P. C. (1910). 'Exhibition of photographs of a thylacine (*Thylacinus cynocephalus*) and three cubs'. *Proceedings of the Zoological Society of London*, 385.

Mivart, St. G. J. (1869). 'Difficulties of the theory of natural selection. I'. *The Month: A Magazine and Review*, 11, 35–53.

Mivart, St. G. J. (1871). *On the Genesis of Species*, (2nd edition). London: Macmillan.

Moeller, H. F. (1968). 'Zur Frage der Parallelerscheinungen bei Metatheria und Eutheria. Vergleichende Untersuchungen an Beutelwolf und Wolf'. *Zeitschrift für wissenschaftliche Zoologie*, 177, (3/4), 283–392.

Moeller, H. F. (1970). 'Vergleichende untersuchungen zum evolutionsgrad der gehirne großer Raubbeutler (Thylacinus, Sarcophilus und Dasyurus). I. Hirngewicht. II. Hirnform und Furchenbild'. *Zeitschrift für Zoologische Systematik und Evolutionsforschung*, 8, 69–80.

Moeller, H. F. (1990). 'Tasmanian wolf'. In B. Grzimek (ed.), *Grzimek's Encyclopaedia of Mammals. Volume 1*. New York: McGraw-Hill.

Moeller, H. F. (1993). 'Beutelwölfe *Thylacinus cynocephalus* (Harris, 1808) in Zoologischen Gärten und Museen'. *Zeitschrift des Kölner Zoo*, 36, (2), 67–71.

Moeller, H. F. (1997). *Der Beutelwolf*, Thylacinus cynocephalus. Magdeburg, Germany: Westarp Wissenschaften.

Molnar, R. E. (1991). 'Fossil reptiles in Australia'. In P. Vickers-Rich, J. M. Monaghan, R. F. Baird and T. H. Rich (eds), *Vertebrate Palaeontology of Australia*. Melbourne: Monash University Publications Committee.

Money. (1884). 'A sheepowners' mutual protection union'. *Hobart Mercury*, 3 November 1884, 3.

Montgomery, A. E. (1912). 'Currickbilly conundrum'. *Sydney Sun*, 24 November 1912, 13.

Morgan, S. (1992). *Land Settlement in Early Tasmania. Creating an Antipodean England*. Cambridge: Cambridge University Press.

Morrison, P. C. (1950). 'Tasmanian Fauna Board'. *Wild Life, Australian Nature Magazine*, 12, (2), 55.

Morrison, P. C. (1957). 'Wildlife: tiger'. *Melbourne Argus (Weekender Section)*, 12 January 1957, 3.

Mother. (1908). Letter to the editor. *Hobart Mercury*, 23 July 1908, 3.

Moyal, A. (1986). *'A Bright and Savage Land': Scientists in Colonial Australia*. Sydney: William Collins.

258

Mudie, R. (1829). *The Picture of Australia: Exhibiting New Holland, Van Dieman's Land, and all the settlements . . .* London: Whittaker, Treacher.

Muirhead, J. and Wroe, S. (1998). 'A new genus and species, *Badjcinus turnbulli* (Thylacinidae: Marsupialia), from the late Oligocene of Riversleigh, northern Australia, and an investigation of thylacinid phylogeny'. *Journal of Vertebrate Paleontology*, 18, 612–26.

Mullan, B. and Marvin, G. (1987). *Zoo Culture.* London: Weidenfeld & Nicolson.

Murray, P. (1991). 'The Pleistocene megafauna of Australia'. In P. Vickers-Rich, J. M. Monaghan, R. F. Baird and T. H. Rich (eds), *Vertebrate Palaeontology of Australia.* Melbourne: Monash University Publications Committee.

National Parks and Wildlife Service, Tasmania. (*ca* 1982). *Wildlife Notes: Tasmanian Tiger.* Hobart: National Parks and Wildlife Service.

Naylor, G. F. K. (1935). A Preliminary Investigation of the Behaviour of Certain Australian Marsupials. MA thesis, University of Sydney.

Nowak, R. M. (1983). 'Marsupialia; Family Thylacinidae'. In *Walker's Mammals of the World. Volume 1*, (4th edition). Baltimore: Johns Hopkins University Press.

Nowak, R. M. (1991). 'Marsupialia; Family Thylacinidae'. In *Walker's Mammals of the World. Volume 1*, (5th edition). Baltimore: Johns Hopkins University Press.

Nyman, L. (1976). *The Lyne Family History.* Hobart: Mercury-Walch.

Nyman, L. (1990). *The East Coasters.* Launceston: Regal.

Ogilby, W. (1841). 'Notice of certain Australian quadrupeds, belonging to the order Rodentia'. *Transactions of the Linnean Society of London*, 18, 121–32.

O'Lochlainn, C. (1946). *Irish Street Ballads.* Dublin: Sign of the Three Candles.

Olsen, P. D. (1983). 'Water-rat. *Hydromis chrysogaster*'. In R. Strahan (ed.), *The Australian Museum Complete Book of Australian Mammals.* Sydney: Angus and Robertson.

Osborn, A. R. (1917). *Almost Human. Reminiscences from the Melbourne Zoo, told by A. A. W. Wilkie.* Melbourne: Whitcombe and Tombs.

Oscar. (1882). 'The native tiger'. *Hobart Mercury*, 19 September 1882, 3.

O'Shea, L. V. (1981). 'The day a tiger growled at me'. *Tasmanian Mail*, 13 October 1981, 21.

Owen, R. (1834). 'On the generation of the marsupial animals, with a description of the impregnated uterus of the kangaroo'. *Philosophical Transactions of the Royal Society of London*, (part 2), 333–64.

Owen, R. (1837). 'Abstract (read 24/4/1834): On the generation of the marsupial animals, with a description of the impregnated uterus of the kangaroo'. *Abstracts of the Papers Printed in the Philosophical Transactions of the Royal Society of London, Volume 3 (1830–1837).* London: Royal Society.

Owen, R. (1838w). 'On the osteology of the Marsupialia, Part I'. *Proceedings of the Zoological Society of London*, 117.

Owen, R. (1838x). 'On the osteology of the Marsupialia, Part II'. *Proceedings of the Zoological Society of London*, 120–47.

Owen, R. (1839w). 'Abstract of a paper presented to the Zoological Society, 22/1/1839: Outlines of a classification of the marsupialia (conclusion)'. *Proceedings of the Zoological Society of London*, 5–19.

259

Owen, R. (1839x). 'Outlines of a classification of the Marsupialia'. *Philosophical Transactions of the Royal Society of London*, 315–33.

Owen, R. (1840). *Odontography; or a Treatise on the Comparative Anatomy of the Teeth; their Physiological Relations, Mode of Development, and Microscopic Structure, in the Vertebrate Animals.* London: Hippolyte Baillière.

Owen, R. (1841). 'Marsupialia'. In R. B. Todd (ed.), *The Cyclopædia of Anatomy and Physiology. Volume 3.* London: Marchant, Singer and Smith.

Owen, R. (1842). 'Account of a *Thylacinus*, the great dog-headed opossum, one of the rarest and largest of the Marsupiate family of animals'. *Report of the Eleventh Meeting of the British Association for the Advancement of Science; held at Plymouth in July 1841.* London: John Murray.

Owen, R. (1843). 'On the rudimental marsupial bones in the *Thylacinus*'. *Proceedings of the Zoological Society of London*, 148–9.

Owen, R. (1846). 'On the rudimental marsupial bones in the Thylacinus'. *Tasmanian Journal*, 2, 447–9.

Owen, R. (1847). 'Letter to R. C. Gunn, 1847'. *Tasmanian Journal*, 3, [2 September].

Owen, R. (1877). *Researches on the Fossil Remains of Extinct Mammals of Australia, Volume 1.* London: J. Erxleben.

Oxley, J. (1810). 'Account of the settlement of Port Dalrymple, 1810'. In F. Watson (ed.), *Historical Records of Australia, Series III. Despatches and papers relating to the settlement of the States. Volume 1.* (1921). Sydney: Library Committee of the Commonwealth Parliament.

Oyama, S. (1985). *The Ontogeny of Information. Developmental Systems and Evolution.* Cambridge: Cambridge University Press.

Packard, A. S. (1889). *Zoology for High Schools and Colleges*, (7th edition). New York: Henry Holt and Co.

Paddle, R. N. (1989). 'The pedantic monopoly on primate monogamy'. *Australian Primatology*, 4, (3), 2–3.

Paddle, R. N. (1991). 'Deconstructing primate monogamy: increasing its incidence, and requiring global models for its phylogenetic origins'. *Primatology Today*, 255–8.

Paddle, R. N. (1992). 'Last resting place of a thylacine'. *Nature*, 360, (6401), 215.

Paddle, R. N. (1993). 'Thylacines associated with the Royal Zoological Society of New South Wales'. *Australian Zoologist*, 29, (1/2), 97–101.

Paddle, R. N. (1996). 'Mueller's magpies and marsupial wolves'. *Victorian Naturalist*, 113, (4), 215–18.

Papini, M. R. (1986). Psicologia comparada de los marsupiales. *Revista Latinoamerica de Psicologia*, 18, 215–46.

Park, A. (1986). 'Tasmanian tiger. Extinct or merely elusive?' *Australian Geographic*, 1, (3), 66–83.

Parker, H. W. (1833). *The Rise, Progress and Present Status of Van Dieman's Land; with Advice to Emigrants*, (1st edition). London: J. Cross and Simpkin and Marshall.

Parker, T. J. and Haswell, W. A. (1897). *Text-book of Zoology. Vol. II*, (1st edition). London: Macmillan.

Parker, T. J. and Haswell, W. A. (1910). *Text-book of Zoology. Vol. II*, (2nd edition). London: Macmillan.

Parker, T. J. and Haswell, W. A. (1921). *Text-book of Zoology. Vol. II*, (3rd edition). London: Macmillan.

Pascual, R., Archer, M., Ortiz Jaureguizar, E., Prado, J. L., Godthelp, H. and Hand, S. J. (1992). 'First discovery of monotremes in South America'. *Nature*, 356, 704–6.

Paterson, W. (1805). 'An animal of a truly singular and nouvel description'. *Sydney Gazette and New South Wales Advertiser*, 21 April 1805, 3.

Pink, K. G. (1962). 'Our first "tiger" for 30 years'. *Burnie Advocate*, 1 December 1962, 17, 19.

Pink, K. G. (1969). 'The "tiger" – one of the world's rarest animals'. *Burnie Advocate*, 1969, Thylacine file, State Reference Library, Burnie.

Pink, K. G. (1982). *The West Coast Story. (A history of Western Tasmania and Its Mining Fields)*. Burnie: Advocate Printers.

Pink, K. G. (1990). *And Wealth for Toil. A History of the North-West and Western Tasmania. 1825–1900*. Burnie: Advocate Marketing Services.

Pizzey, G. (1966). *Animals and Birds in Australia*. Melbourne: Cassell.

Pizzey, G. (1968). 'Victim of the dingo'. *Melbourne Herald, Weekend Magazine*, 25 May 1968, 29.

Plomley, N. J. B. (1966). *Friendly Mission. The Tasmanian Journals and Papers of George Augustus Robinson, 1829–1834*. Hobart: Tasmanian Historical Research Association.

Pocock, R. I. (1926). 'The external characters of *Thylacinus, Sarcophilus*, and some related marsupials'. *Proceedings of the Zoological Society of London*, 1037–84.

Powell, J. M. (1988). 'Protracted reconciliation: society and the environment'. In R. MacLeod (ed.), *The Commonwealth of Science: ANZAAS and the scientific enterprise in Australia, 1888–1958*. Melbourne: Oxford University Press.

Prinsep, E. (1833). *The Journal of a Voyage from Calcutta to Van Diemen's Land: Comprising a Description of that Colony during a Six Months Residence*. London: Smith and Elder.

Raethel, H. (1992). 'Bemerkenswertes über die beutelwölfe des Berliner Zoologischen Gartens'. *Bongo. Beiträge zur Tiergärtnerei und Jahresberichte aus dem Zoo Berlin*, 20, 61–4.

Reader. (1971). 'Reader remembers the "hyenas"'. *Burnie Advocate*, 8 May 1971.

Regan, C. T. (1936). *Natural History*. London: Ward, Lock & Co.

Reid, A. R. (1907). 'Tasmanian quail and game propagation'. *Tasmanian Naturalist*, 1, (2), 3–8.

Rembrantse, D. (1682). 'A short relation out of the journal of Captain Abel Jansen Tasman, upon the discovery of the *South Terra incognita*; not long since published in the Low Dutch'. *Philosophical Collections of the Royal Society of London*, (6), 179–86.

Renshaw, G. (1905). *More Natural History Essays*. London: Sherratt and Hughes.

Renshaw, G. (1938). 'The thylacine'. *Journal of the Society for the Preservation of the Fauna of the Empire*, 35, 47–9.

Reynolds, H. (1995). *Fate of a Free People*. Melbourne: Penguin.

Reynolds, J. (1956). 'Premiers and political leaders'. In F. C. Green (ed.), *A Century of Responsible Government in Tasmania, 1856–1956*. Hobart: Tasmanian

Government.

Rich, T. H. (1991). 'The history of mammals in *Terra Australis*'. In P. Vickers-Rich, J. M. Monaghan, R. F. Baird and T. H. Rich (eds), *Vertebrate Palaeontology of Australia*. Melbourne: Monash University Publications Committee.

Rich, T. H., Vickers-Rich, P., Constantine, A., Flannery, T. F., Kool, L. and van Klaveren, N. (1997). 'A tribosphenic mammal from the Mesozoic of Australia'. *Science*, 278, (21 November 1997), 1438–42.

Ride, W. D. L. (1964). 'A review of Australian fossil marsupials'. *Journal of the Royal Society of Western Australia*, 47, (4), 97–131.

Ride, W. D. L. (1970). *A Guide to the Native Mammals of Australia*. Melbourne: Oxford University Press.

Ritvo, H. (1987). *The Animal Estate. The English and Other Creatures in the Victorian Age*. Cambridge, Ma: Harvard University Press.

Rix, C. E. (1978). *The Royal Zoological Society of South Australia. 1878–1978*. Adelaide: Royal Zoological Society of South Australia.

Robinson, A. C. and Young, M. C. (1983). *The Toolache Wallaby* (Macropus greyi Waterhouse). Adelaide: Department of Environment and Planning, South Australia.

Robson, I. F. (1928). *Minutes of Conference of Representatives of State Committees to Discuss the Administration of the Proclamation Prohibiting the Exportation of Birds and Other Native Fauna Unless with the Consent of the Minister for Trade and Customs*. Melbourne: Victorian State Advisory Committee.

Robson, L. L. (1985). *A Short History of Tasmania*. Melbourne: Oxford University Press.

Rœdelberger, F. A. and Groschoff, V. I. (1967). *Wildlife of the South Seas*. London: Constable.

Rolls, E. C. (1969). *They All Ran Wild. The Story of Pests on the Land in Australia*. Sydney: Angus and Robertson.

Ross, J. (1829). *The Hobart Town Almanack for the Year 1829*. Hobart: J. Ross.

Ross, J. (1830). *The Hobart Town Almanack for the Year 1830*. With Embellishments. Hobart: James Ross.

Ross, J. (1831). *The Van Diemen's Land Anniversary and Hobart-Town Almanack for the Year 1831*. With Embellishments. Hobart: James Ross.

Roth, H. L. (1891). *Crozet's Voyage to Tasmania, New Zealand, etc . . . 1771–1772*. London: Truslove and Shirley.

Rounsevell, D. E. (1983). 'Thylacine. *Thylacinus cynocephalus*'. In R. Strahan (ed.), *The Australian Museum Complete Book of Australian Mammals*. Sydney: Angus and Robertson.

Rowcroft, C. (1843). *Tales of the Colonies; or, the Adventures of an Emigrant, Edited by a Late Colonial Magistrate*. London: Saunders and Otley.

Royal Melbourne Zoological and Acclimatisation Society. (1900). *The Thirty-sixth Annual Report . . .* Melbourne: Rae Bros., Photo-Process House.

Sadlier, R. (1970). *Animals of Australia and New Zealand*. London: Hamlyn.

Salvadori, F. B. (1978). *Ces Animaux qui Disparaissent*. Paris: Bordas.

Saward, A. (1990). 'Experiences with the Tasmanian tiger'. *Circular Head Local History Journal*, 3, (2), 24–5.

262

Sayles, J. (1980). 'Stalking the Tasmanian tiger'. *Animal Kingdom*, December 1979/
    January 1980, 35–40.
Scholes, K. (1990). *Tiga. A Teacher's Guide*. Melbourne: Wilderness Society.
Science News. (1966). 'Rare tiger trace seen'. 90, (20), 20 August 1966, 118.
Sclater, P. L. (1868w). 'Additions to the menagerie, June to October'. *Proceedings of the
    Zoological Society of London*, 526–30.
Sclater, P. L. (1868x). 'List of additions to the Society's menagerie'. *Proceedings of the
    Zoological Society of London*, 638–54.
Sclater, P. L. (1870). 'Additions to the menagerie, October to November'. *Proceedings of
    the Zoological Society of London*, 796–7.
Sclater, P. L. (1901). 'Report on the additions to the Society's menagerie in March
    1901'. *Proceedings of the Zoological Society of London*, 324.
Sclater, P. L. (1904). 'Rare living animals in London'. *Knowledge and Scientific News*,
    (April 1904), 59–60. [Republished as 'The thylacine. (*Thylacinus cynocephalus*)',
    Scientific American Supplement, (1488), 9 July 1904, 23845.
Scott, T. (1829). 'From the descriptive itinerary of Van Diemen's Land, now preparing
    for publication. – Excursion of his Excellency to the west'. *Hobart Town Courier*,
    28 February 1829, 2–3.
Scott, T. (1830). 'Excursion to the westward, of his excellency Lieutenant Governor
    Arthur in January 1829'. In J. Ross (ed.), *The Hobart Town Almanack for the Year
    1830*. Hobart: James Ross.
Serventy, V. (1966). *Australia. A Continent in Danger*. London: Andre Deutsch.
Serventy, V. and Raymond, R. (1973). 'The Tasmanian tiger'. *Australia's Wildlife
    Heritage*, 2, 628–31.
Seth-Smith, D. (1926). Footnote. *Proceedings of the Zoological Society of London*, 937.
Sharland, M. S. R. (1924). 'Preserving our native animals and birds. Phenomenal
    success of the Beaumaris Zoo'. *Hobart Mercury (Anniversary Supplement)*, 5 July
    1924, 43–5.
Sharland, M. S. R. (1937). 'Tasmanian tiger. Marsupial's stand'. *Sydney Morning Herald*,
    20 February 1937, 13.
Sharland, M. S. R. (1939). 'In search of the thylacine'. *Proceedings of the Royal Society of
    New South Wales, 1938–1939*, 20–38.
Sharland, M. S. R. (1941). 'Tasmania's rare "tiger"'. *Bulletin of the New York Zoological
    Society*, 44, (3), 84–8.
Sharland, M. S. R. (1956). 'Farm hands paid to snare tiger'. *Hobart Mercury*, 9 June 1956.
Sharland, M. S. R. (1957w). 'Tiger secrets need probing'. *Hobart Mercury*, 12 January
    1957, 17.
Sharland, M. S. R. (1957x). 'In search of the vanished "tiger"'. *People*, 3 April 1957, 25–6.
Sharland, M. S. R. (1960). 'Hunting the thylacine'. *Hemisphere*, May, 7–11.
Sharland, M. S. R. (1962). *Tasmanian Wild Life. A Popular Account of the Furred Land
    Animals, Snakes and Introduced Mammals of Tasmania*. Carlton: Melbourne
    University Press.
Sharland, M. S. R. (1966). *Tasmania*. Sydney: Nelson Doubleday.
Sharland, M. S. R. (1971w). 'Bream Creek's tiger toll was heavy'. *Hobart Mercury*,
    13 February 1971, 6.

Sharland, M. S. R. (1971x). *A Pocketful of Nature. An Anthology of Notes and Articles by the Natural History Columnist of* The Mercury, Hobart, 'Peregrine' (Michael Sharland) *during the Past Half-Century*. Hobart: *Mercury*.

Sharland, M. S. R. (1971y). 'Gone, but not forgotten'. *Hobart Mercury*, 6 March 1971.

Sharland, M. S. R. (1980). 'Has the thylacine "evolved" into the tiger cat?' *Hobart Mercury*, 29 March 1980, 6.

Shaw, S. (1972). 'Man who caught that tiger'. *Burnie Advocate (Weekender)*, 15 April 1972.

Silver, S. W. (1874). *Handbook for Australia and New Zealand* . . . (2nd edition). London: S. W. Silver.

Simpson, S. (1980). 'Chasing our elusive tiger with a computer printout'. *Hobart Mercury*, 28 January 1980, 5.

Sinurbe. (1887). 'A city stroll'. *Launceston Examiner (Supplement)*, 19 November 1887, 2.

Small, M. (1985). 'Tasmanian tiger mystery. Is this creature extinct, or has it learnt from harsh experience to avoid successfully its worst enemy – man?' *This Australia*, 5, (1), 6–11.

Smith, G. W. (1909w). *A Naturalist in Tasmania*. Oxford: Clarendon.

Smith, G. W. (1909x). 'Notes on Tasmanian crustacea'. *Tasmanian Naturalist*, 2, (1), 5–9.

Smith, J. (1862). 'Tasmanian tigers'. *Launceston Examiner*, 22 November 1862, 2.

Smith, M. (1982). 'Review of the thylacine (Marsupialia, Thylacinidae)'. In M. Archer (ed.), *Carnivorous Marsupials, vol. I*. Sydney: Royal Zoological Society of New South Wales.

Smith, P. A. (1968). *Tiger Country*. Adelaide: Rigby.

Smith, S. J. (1981). *The Tasmanian Tiger – 1980. A report on an investigation of the current status of thylacine Thylacinus cynocephalus, funded by the World Wildlife Fund Australia*. Hobart: National Parks and Wildlife Service, Tasmania.

Special Correspondent. (1884). 'The traveller, through Tasmania, No. 60'. *Tasmanian Mail*, 25 October 1884, 27–8.

Spencer, W. B. (1927). *The Arunta: A Study of Stone Age People by Sir Baldwin Spencer and the late F. J. Gillen*. London: Macmillan.

Sprent, J. F. A. (1970). '*Baylisascaris tasmaniensis* sp. nov. in marsupial carnivores: heirloom or souvenir?' *Parasitology*, 61, 75–86.

Sprent, J. F. A. (1971w). 'Speciation and development in the genus *Lagochilascaris*'. *Parasitology*, 62, 71–112.

Sprent, J. F. A. (1971x). 'A new genus and species of ascaridoid nematode from the marsupial wolf (*Thylacinus cynocephalus*)'. *Parasitology*, 63, 37–43.

Sprent, J. F. A. (1972). '*Cotylascaris thylacini*: a synonym of *Ascaridia columba*'. *Parasitology*, 64, 331–2.

Spurling, S. (1943). 'The Tasmanian tiger or marsupial wolf, *Thylacinus cynocephalus*'. *Journal of the Bengal Natural History Society*, 18, (2), 55–9.

Stancombe, G. H. (1968). *Highway in Van Diemen's Land*. Glendessary, Western Junction: G. H. Stancombe.

Steers, B. (1991). 'There was not a hunter has ever lived who has chucked a good possum away'. In S. Cubit (ed.), *What's the Land For? People's Experiences of*

Tasmania's Central Plateau Region. Volume 6. Launceston: Central Plateau Oral History Project.

Sterland, W. J. (1892). *The Hand-Book of Natural History. Mammalia. For Teachers and Pupil Teachers in Schools, Colleges, &c.* London: Jarrold & Sons.

Stevenson, R. (1928). Letter to the editor: the Tasmanian wolf. *Hobart Mercury,* 20 July 1928, 3.

Stewart, H. W. (1919). 'Aboriginalities'. *Bulletin,* 11 November 1919, 20.

Stimulus. (1910). 'Zoological gardens'. *Hobart Daily Post,* 15 February 1910, 7.

Stivens, D. (1973). 'The thylacine mystery'. *Animal Kingdom,* July, (76), 18–23.

Strahan, R. (1981). *A Dictionary of Australian Mammal Names.* Sydney: Angus and Robertson.

Strahan, R. (ed.) (1983). *The Australian Museum Complete Book of Australian Mammals.* Sydney: Angus and Robertson.

Strahan, R. (1991). *Beauty and the Beasts. A History of Taronga Zoo, Western Plains Zoo and their Antecedents.* Sydney: Surrey Beatty.

Strzelecki, P. E. de. (1845). *Physical Description of New South Wales and Van Diemen's Land. Accompanied by a Geological Map, Sections, and Diagrams, and Figures of the Organic Remains.* London: Longman, Brown, Green and Longmans.

Swainson, W. (1834). 'The dog-faced opossum'. In H. Murray, W. Wallace, R. Jameson, W. J. Hooker and W. Swainson (eds), *An Encyclopædia of Geography* . . . London: Longman, Rees, Orme, Brown, Green and Longman.

Swainson, W. (1846). 'The dog-faced opossum'. In H. Murray, W. Wallace, R. Jameson, W. J. Hooker and W. Swainson (eds) *An Encyclopædia of* . . . (revised edition). London: Longmans, Green and Co.

Tabart, T. A. (1885). 'Chief Inspector of Sheep: Report for 1884, (30/6/1885)'. *Tasmania. Journals and Printed Papers of the Parliament of Tasmania,* 6, Report #89.

Tabart, T. A. (1886). 'Chief Inspector of Sheep: Report for 1885, (30/6/1886)'. *Tasmania. Journals and Printed Papers of the Parliament of Tasmania,* 8, Report #44.

Tabart, T. A. (1887). 'Inspector of Sheep: Report for year ending 30th June, 1887'. *Tasmania. Journals and Printed Papers of the Parliament of Tasmania,* 12, Report #85.

Tabart, T. A. (1888). 'Chief Inspector of Sheep: Report for the year ending June 30, 1888'. *Tasmania. Journals and Printed Papers of the Parliament of Tasmania,* 15, Report #101.

Tabart, T. A. (1889). 'Chief Inspector of Sheep. Report for the year ending June 30, 1889'. *Tasmania. Journals and Printed Papers of the Parliament of Tasmania,* 17, Report #59.

Tabart, T. A. (1890). 'Chief Inspector of Stock: Report for 1889, (30/6/1890)'. *Tasmania. Journals and Printed Papers of the Parliament of Tasmania,* 21, Report #92.

Tabart, T. A. (1891). 'Chief Inspector of Stock: Report for 1890, (30/6/1891)'. *Tasmania. Journals and Printed Papers of the Parliament of Tasmania,* 23, Report #98.

Tasman. (1884). 'The ferae of Tasmania'. *Tasmanian,* 10 May 1884, 28.

Tasmanian Government Tourist Bureau. (1934). *Tasmania. The Wonderland.* Hobart: Walter E. Shimmins, Government Printer.

265

Tasmanian Parliament. (1886). *Tasmania. Journals and Printed Papers of the Parliament of Tasmania, 1886. Vol. 8.* Hobart: William Thomas Strutt.

Tasmanian Parliament. (1887). *Tasmania. Journals and Printed Papers of the Parliament of Tasmania, 1887. Vol. 10.* Hobart: William Thomas Strutt.

Tasmanian Parliament. (1888). *Tasmania. Journals and Printed Papers of the Parliament of Tasmania, 1887. Vol. 15.* Hobart: William Thomas Strutt.

Temminck, C. J. (1824). *Monographies de Mammalogie, ou description de Quelques Genres de Mammifères, dont les espèces ont été observées dans les différens musées de l'Europe, Tome premier* (1st edn). Paris: G. Dufour et Ed. D'Ocagne.

Temminck, C. J. (1827). *Monographies de Mammalogie, ou description de Quelques Genres de Mammifères, dont les espèces ont été observées dans les différens musées de l'Europe, Tome premier* (2nd edn). Paris: G. Dufour et Ed. D'Ocagne.

Thomas, N. (1977). Letter to the editor. *Hobart Mercury,* 2 March 1977, 4.

Thomas, O. (1888). *Catalogue of the Marsupialia and Monotremata in the Collection of the British Museum (Natural History).* London: British Museum.

Thomas, R. W. (1996). 'Tasmanian tiger photograph'. *South Australian Naturalist,* 71 (1/2), 18–19.

Thorne, J. C. (1970). Letter to the editor. *Sydney Sun-Herald,* 13 September 1970, 146.

Triggs, B. (1984). *Mammal Tracks and Signs. A Fieldguide for South-eastern Australia.* Melbourne: Oxford University Press.

Trot, T. (1889). 'Echoes of the streets'. *Launceston Examiner,* 11 July 1889, 4.

Troughton, E. (1941). *Furred Animals of Australia,* (1st edition). Sydney: Angus and Robertson.

Troughton, E. (1944). *Furred Animals of Australia,* (2nd edition). Sydney: Angus and Robertson.

Troughton, E. (1965). *Furred Animals of Australia,* (8th edition). Sydney: Angus and Robertson.

Tunbridge, D. (1991). *The Story of the Flinders Ranges Mammals.* Sydney: Kangaroo Press.

Turner, A. (1991). 'I think anybody that was reared in the Lake Country, they never get out of that love'. In S. Cubit (ed.), *What's the Land For? People's Experiences of Tasmania's Central Plateau Region. Volume 2.* Launceston: Central Plateau Oral History Project.

Tyndale-Biscoe, C. H. (1973). *Life of Marsupials.* London: Edward Arnold.

Tyndale-Biscoe, C. H. and Renfree, M. B. (1987). *Reproductive Physiology of Marsupials.* Cambridge: Cambridge University Press.

VandeBerg, J. L. (1990). 'The grey short-tailed opossum (*Monodelphis domesticus*) as a model didelphid species for genetic research'. In J. A. Marshall Graves, R. M. Hope and D. W. Cooper (eds), *Mammals from Pouches and Eggs: Genetics, Breeding and Evolution of Marsupials and Monotremes.* Canberra: CSIRO.

Vaughan, H. M. (1914). *An Australian Wander-Year.* London: Martin Secker.

Vechtmann, N. (1980). 'Hoe "uitgestorven" is Tasmaanse buidelwolf?' *Het Vrije Volk,* 13 June 1980.

Veitch, A. (1979). '$1 million . . . if you catch a tiger'. *Australasian Post,* 17 May 1979, 10–11.

Vogt, C. and Specht, F. (1887). *The Natural History of Animals (Class Mammalia – Animals which Suckle their Young) in Word and Pictures, Vol II.* London: Blackie and Son.

Von Stieglitz, K. R. (1966). *Sketches in Early Van Diemen's Land by Thomas Scott. Assistant Surveyor General of Van Diemen's Land, 1821–1836.* Hobart: Fullers Bookshop.

W, A. (1855). 'The tiger-wolf'. *Excelsior: Helps to Progress in Religion, Science and Literature*, 3, 246–9.

W, J. (1857). 'Mammalia'. In *The Encyclopædia Britannica, or Dictionary of Arts, Sciences, and General Literature, Vol. XIV*, (8th edition). Edinburgh: Adam and Charles Black.

Wall, L. E. (1955). 'History of the Tasmanian Field Naturalists' Club'. *Tasmanian Naturalist*, 11, (3), 33–6.

Walters, A. (1991). 'I never asked for a job in my life – I never'. In S. Cubit (ed.), *What's the Land For? People's Experiences of Tasmania's Central Plateau Region. Volume 2.* Launceston: Central Plateau Oral History Project.

Walters, M. J. (1992). *A Shadow and a Song – The Struggle to Save an Endangered Species.* Post Mills, Vermont: Chelsea Green Publishing Company.

Waterhouse, G. R. (1841). 'Marsupialia or pouched animals'. In *The Naturalist's Library, Vol. 25, (Mammalia, Vol. 11)*, 123–8. Edinburgh: Jardine.

Waterhouse, G. R. (1846). *A Natural History of the Mammalia. Volume 1 . . .* London: Hippolyte Baillière.

Watts, D. (1993). *Tasmanian Mammals. A Field Guide*, (2nd edition). Hobart: Tasmanian Conservation Trust.

Wayback. (1911). 'An unwelcome visitor'. *Launceston Weekly Courier*, 7 December 1911, 34.

Wedge, J. H. (1962). *The Diaries of John Helder Wedge, 1824–1835.* Hobart: Royal Society of Tasmania.

Weindorfer, G. and Francis, G. (1920). 'Wild life in Tasmania, I'. *Victorian Naturalist*, 36, 157–60.

Wendt, H. (1956). *Out of Noah's Ark. The Story of Man's Discovery of the Animal Kingdom.* London: Weidenfeld and Nicolson.

Wentworth, W. C. (1819). *Statistical, Historical, and Political Description of the Colony of New South Wales, and its Dependent Settlements in Van Diemen's Land . . .* London: G. and W. B. Whittaker.

West, J. (1852). *The History of Tasmania . . .* Launceston: Henry Dowling.

White, D. V. (1917). *Geoffrey Watkins Smith.* Oxford: Printed for private circulation.

White, P. and Flannery, T. (1992). 'The impact of people on the Pacific world. In J. Dodson (ed.), *The Naïve Lands. Prehistory and Environmental Change in Australia and the Southwest Pacific.* Melbourne: Longman Cheshire.

Whitley, G. P. (1966). 'T. J. Lempriere, an early Tasmanian naturalist'. *Australian Zoologist*, 13, 350–5.

Whitley, G. P. (1970). *Early History of Australian Zoology.* Sydney: Royal Zoological Society of New South Wales.

Whitley, G. P. (1973). 'I remember the thylacine'. *Koolewong*, 2, 10–11.

Widowson, H. (1829). *Present State of Van Diemen's Land . . .* London: S. Robinson, W. Joy, J. Cross and J. Birdsall.

Willis, P. M. A. (1997). 'Review of fossil crocodilians from Australasia'. *Australian Zoologist*, 30, (3), 287–98.

Willoughby, D. P. (1966). 'The vanished quagga'. *Natural History*, 75, (2), 60–3.

Willoughby, H. (1886). *Australian Pictures. Drawn with pen & pencil.* London: The Religious Tract Society.

Wilson, E. O. (1992). *The Diversity of Life.* London: Allen Lane.

Wood Jones, F. (1923). *The Mammals of South Australia. Part 1. The Monotremes and the Carnivorous Marsupials. (The Ornithodelphia and the didactylous Didelphia.).* Adelaide: British Science Guild (South Australia Branch).

Woodburne, M.O. and Case, J. A. (1996). 'Dispersal, vicariance, and the Late Cretaceous to Early Tertiary land mammal biogeography from South America to Australia'. *Journal of Mammalian Evolution*, 3, (2), 121–61.

Woods, A. (1977). 'Tiger, tiger in the night'. *Sydney Morning Herald*, 25 August 1977, 7.

Wright, E. P. (1892). *Cassell's Concise Natural History, Being a Complete Series of Descriptions of Animal Life.* London: Cassell.

Wright, R. R. (1888). 'Marsupialia'. In J. S. Kingsley (ed.), *The Riverside Natural History. Volume V. Mammals.* London: Kegan Paul, Trench & Company.

Wynne, C. D. L. and McLean, I. G. (1999). 'The comparative physiology of marsupials'. *Australian Journal of Psychology*, 51, (2), 111–16.

Yendall, D. (1982). 'Search for the thylacine'. *Wildlife (International)*, May, 182–3.

Zahavi, A. and Zahavi, A. (1997). *The Handicap Principle. A missing piece of Darwin's puzzle.* New York: Oxford University Press.

Ziswiler, V. (1967). *Extinct and Vanishing Animals. A Biology of Extinction and Survival.* New York: Longmans.

Index

270

272

273

LaVergne, TN USA
18 February 2010
73466LV00003B/1/P